Focus on Computer Graphics

Tutorials and Perspectives in Computer Graphics

Edited by W.T. Hewitt, R. Gnatz, and D.A. Duce

B. Falcidieno I. Herman
C. Pienovi (Eds.)

Computer Graphics and Mathematics

With 159 Figures

Springer-Verlag
Berlin Heidelberg New York
London Paris Tokyo
Hong Kong Barcelona
Budapest

Focus on Computer Graphics

Edited by W. T. Hewitt, R. Gnatz, and D. A. Duce
for EUROGRAPHICS –
The European Association for Computer Graphics
P. O. Box 16, CH-1288 Aire-la-Ville, Switzerland

Volume Editors

Bianca Falcidieno
Caterina Pienovi
Institute for Applied Mathematics
CNR, Via L. B. Alberti 4
I-16132 Genoa, Italy

Ivan Herman
Centre for Mathematics and Computer Sciences
Kruislaan 413
1098 SJ Amsterdam, The Netherlands

ISBN-13:978-3-642-77588-8 e-ISBN-13:978-3-642-77586-4
DOI: 10.1007/978-3-642-77586-4

Softcover reprint of the hardcover 1st edition 1992

Typesetting: Camera ready by authors
45/3140 – 5 4 3 2 1 0 – Printed on acid-free paper

Preface

Most fields in computer science hardly make any use of the machinery offered by classical and modern mathematics. They might use some combinatorics, graph theory and, of course, formal logic; however, it is very rare for the results of the formidable development of mathematics of the past centuries to be extensively used in computer science.

One notable exception to this rule is computer graphics. Since its very existence as a separate field within computer science, computer graphics, by its very nature, had to make extensive use of non-trivial mathematics, for example, projective geometry, solid modelling and approximation theory. This interplay of mathematics and computer science makes research in this area especially exciting and gives it a special flavour. However, it is also the source of a number of problems; university students find it difficult to assimilate the necessary mathematics when confronted by computer graphics. Active researchers in the field also have problems in gaining a view of all the necessary (mathematical) tools. From these two major problems it has been recognized that the possibilities offered by an interdisciplinary approach are still not fully utilized and further steps should be taken to induce cooperation between mathematicians and computer scientists active in computer graphics.

In October 1991, in the beautiful North-Italian village of Santa Margherita Ligure, near Genoa, a group of mathematicians and computer scientists gathered at a Workshop promoted by the European Association of Computer Graphics to explore ways of extending the cooperation between mathematics and computer graphics. This book gives a selection of the best and most relevant contributions to this workshop. The extreme diversity of the topics made the workshop intellectually extremely exciting and challenging; this diversity is also reflected in the papers published in this volume.

Some of the application areas of mathematics in computer graphics have their tradition already, although the workshop organizers tried to stimulate new and not yet explored application areas as well. Computational geometry (duly represented, e.g., by the paper of M. Mukherjee et al.) has already found its way as an accepted area within computer graphics as well as discrete mathematics. Also, surface modelling has always made extensive use of approximation theory in the past, but the papers of H. Prautzsch, H.-P. Seidel and of X. Zhou give a vivid example of the amount of work still to be done in this area as well as the necessity to use sometimes intricate mathematics to overcome the problems arising. Application of algebraic geometry might lead to very exiting and still largely unexplored results; the papers of M. Alonso et al., P. Cellini et al. are of particular interest in this respect. Additionally, the paper of H. Du and R. Goldmann, or the one of D. Richardson present mathematical techniques (previously unused in computer graphics) which might have their application in surface modelling.

Topology is one of the most abstract fields of mathematics both in the form of point-set topology and of combinatorial topology. It is one of the remarkable developments of the past years to realize that this branch of mathematics can have direct usage in computer graphics and modelling. The papers by Y. Luo and G. Lukács and the survey of T. Takala as well as the resulting discussion were undeniably among the highlights of the workshop.

There are a number of papers which are quite diverse in their applications as well as in their use of different mathematical tools. The use of Markov processes has already found its way into fractal theory; the paper of M. Coevoet shows some new results. Probability theory is also the main tool in the paper of X. Wu on optimal colour quantizing. Geometric constraint solving, which is still a largely unexplored area, was attacked by different means in the papers of N. Kin et al. as well as R. Veltkamp and F. Arbab. The former is particularly interesting in that it looks at the use of neural computing for computer graphics; we should remember that neural computing is also one of the fields in computer sciences which makes extensive use of mathematics.

The paper by C. Hoffmann showed how fruitful an interdisciplinary approach to computer graphics might be; some of the tools he used were created by mathematicians long before anybody could dream about computers. It was remarkable to realize that these forgotten mathematical tools were very well adapted to the problems at hand nowadays. In the same view, different techniques from geometry, e.g., differential geometry, computational geometry, and fractal geometry, are used as basic tools for natural surface generation and understanding in the paper of B. Falcidieno et al. Application of quite non-trivial differential geometry by J. Lemordant leads to new tools in image warping; a possible application not only in computer graphics but also in image processing! Finally, some well known theorems in classical mathematical analysis have led to new digitalization techniques in the paper of C. Wüthrich.

All papers cited above have shown how fruitful it may be to use higher-level mathematical tools in computer graphics. The paper of C. Gunn is, in a sense, just the opposite: it shows how useful it may be for mathematicians to have visualizing tools to look at the sometimes fairly abstract constructions they use.

No all possible applications of mathematics for computer graphics were addressed at the workshop. Hence there are notable fields missing in this book, such as projective geometry, use of quaternions and linear programming. However, as a first attempt for such a meeting the workshop was certainly very successful and we hope that the reader will be able to catch some of the unique flavour we could share at this event.

We would like to thank all authors for their contributions, and the workshop participants in general for their active participation in the discussions. The work of the international programme committee is very much appreciated; their work made it possible to ensure a high quality for the event as a whole. We would also like to express our gratitude to the sponsor of the meeting, the National Research Council of Italy, and to its organizer, the Institute for Applied Mathematics of Genoa. Our thanks are due to Marinella Pescaglia for running the Workshop secretariat so effectively, and to all assistants and students of the I.A.M. Computer Graphics group, who have considerably lightened our workload with their usual cheerfulness and energy. Without their collaboration the organization of the workshop would not have been possible.

Amsterdam and Genoa, January 1992

Bianca Falcidieno
Ivan Herman
Caterina Pienovi

First Eurographics Workshop on Mathematics and Computer Graphics

Co-Chairpersons

B. Falcidieno	Institute for Applied Mathematics, Genoa, Italy
I. Herman	Centre for Mathematics and Computer Sciences, Amsterdam, The Netherlands

Programme Committee

F. Arbab	CWI, Amsterdam, The Netherlands
W. Böhm	TU Braunschweig, Braunschweig, FRG
K. Brodlie	Leeds University, Leeds, UK
V. Capasso	SASIAM, Valenzano, Italy
M. Cohen	University of Utah, Salt Lake City, USA
F. Dévai	University of Ulster, Londonderry, UK
T. Hewitt	Manchester University, Manchester, UK
Ch. Hoffmann	Purdue University, West Lafayette, USA
G. Krammer	MTA SzTAKI, Budapest, Hungary
S. Mudur	National Center for Software Technology, Bombay, India
M. Overmars	Rijksuniversiteit Utrecht, Utrecht, The Netherlands
D. Rogers	U.S. Naval Academy, Annapolis, USA
T. Sederberg	Brigham Young University, Provo, USA
H.-P. Seidel	University of Waterloo, Waterloo, Canada
T. Takala	The George Washington University, Washington, USA
A. Vinacua	University of Barcelona, Barcelona, Spain

Table of Contents

1 Integer Approximation to the Intersection of Three Planes with Planar Constraints

Maharaj Mukherjee
George Nagy
Shashank Mehta

ABSTRACT
The intersection point of three planes specified with rational coordinates is approximated by a point with integer-valued coordinates so that the *integer point* is constrained within the pyramid specified by three half-spaces of the planes. An $O(N \log N)$ algorithm guarantees that the obtained integer point is closest to the apex of the pyramid. An $O(\log N)$ algorithm, not yet proven to guarantee the closest point, yields promising results.

1.1 Introduction

We consider the following problem:

> given three oriented planes in R^3 with rational coefficients, intersecting at a point, and a specified region constrained by these three planes, find a point with integer-valued coordinates that is inside the constrained region and is the closest *Manhattan-distance* approximation to the intersection of the three planes.

We are investigating the problem of vertex approximation to find a consistent *finite-precision* representation of a geometric object. The adoption of computational-geometry algorithms for massive data manipulation in *geographic information systems* and in *solid modeling* has resulted in additional emphasis on *robustness* in their implementation using finite-precision arithmetic.

The above problem arises in approximating a polyhedron with *rational–valued* vertices by a polyhedron with *integer-valued* vertices, such that the rational–polyhedron encloses (alternatively, is enclosed by) the integer–polyhedron (Figure 1.1). Rational coefficients arise as a result of exact Boolean operations on geometric entities with integer coefficients. The numerators and denominators of the rational coefficients grow with each successive operation.

> A vertex created by four or more planes can be approximated by dividing the pyramid adjacent to it, systematically, into three-sided pyramids. Pyramids with f faces can be divided into at most $f-2$ three-sided pyramids. We may find the approximation in all such three-sided pyramids and then choose the best approximation, in the sense of Manhattan-distance, from all contender points.

Topological consistency in finite-precision approximations is maintained, in the methods below, through non-redundant judgements. *Incidence* is resolved by either maintaining a tolerance region, or by arbitrarily merging features within certain region. Milenkovic uses *hidden variable* methods for consistent ordering of lines and planes [13]. Segal and Sequin tested some *minimum feature separation* to distinguish between incident and non-incident features [15]. Hoffmann, Hopcroft, and Karasick have built a solid modeler using

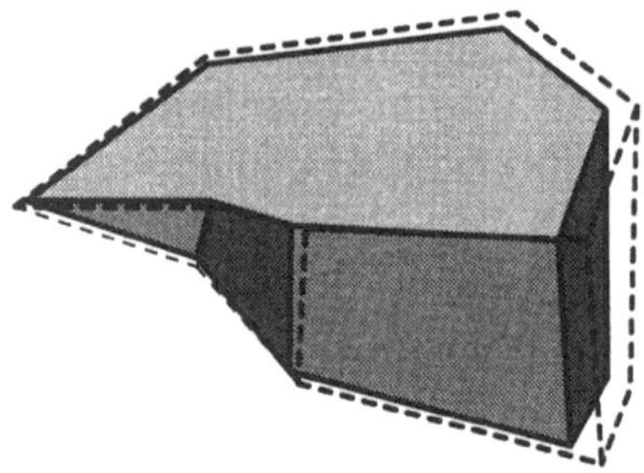

FIGURE 1.1. A rational polyhedron with its constrained integer approximating polyhedron. The integer polyhedron (shaded) is constrained to remain inside the rational-polyhedron (dotted outlined).

minimum feature separation and one-way judgement for the topology [4]. Guibas, Salesin, and Stolfi defined *epsilon-geometry* to reason about geometric relationships with a fuzzy region maintained for "unknown" [3].

There are other methods where numerical inaccuracy is handled at a much lower level. These methods resort to traditional round-off analysis techniques [5, 17], which include interval-arithmetic [14], infinite or finite-precision rational arithmetic [10], and statistical error estimation [16]. Greene and Yao use rational arithmetic for a line-arrangement problem [2]. Karasick, Lieber, and Nackman use both rational and interval arithmetic to determine the sign of a determinant [9]. Dobkin and Silver apply statistical error estimation to the iterated-pentagon problem [1]. These methods often come very close to number–theoretic analysis.

The problem we address in this paper is reminiscent of *simultaneous Diophantine approximation.* A polynomial time algorithm was developed by Lenstra, Lenstra, and Lovász for *polynomial factorization* [8]. Lenstra used the *basis reduction method* to find a short vector in the lattice using Gram-Schmidt orthogonalization [7]. Even though a theoretically fast solution to our problem might be possible using the basis-reduction-method, programming complexity and additional overhead would make it impractical [1]. Because of the associated constraints our problem is also similar to the integer-programming problem, which is known to be NP-complete [6]. Since we are restricting our problem to 3-D with only *three* planar constraints, it does not encompass the general theory of integer programming, and the solution is simpler.

We first describe a 2-D version of the problem. This problem with two straight lines can be solved in $O(\log N)$ time, where N is the largest representable integer [11, 12]. (In a typical computer N can be as large as 2^{32}.) We present an $O(N \log N)$ solution for 3-D which is provably guaranteed to obtain the closest integer point. In 3-D, we believe that there exists a solution of $O(\log N)$ arithmetic complexity. We present an $O(\log N)$ extension of the 2-D solution for 3-D. We show how to obtain the optimal solution point for the special case of a *right pyramid*, and transform all other cases to this one. Though we have not proved that this algorithm guarantees the closest solution in all cases, we have not been able to find an example where it does not. Thus, we believe that it is a practical solution to the 3-D problem stated above.

[1] Prof. Lovász, personal communication to the first author.

1.2 2-D Version of the Problem

In 2-D the problem can be stated as follows:

> given two intersecting straight lines with rational slopes and intercepts, and a sector specified by these two lines, find the integer point inside the given sector, nearest to the point of intersection of the two lines.

The required point is found in $O(\log N)$ time, where N is the largest representable integer [11, 12]. For the sake of completeness, we recapitulate the solution here.

Two lines $y = ax + b$ and $y = cx + d$ are specified by rational numbers a, b, c, d, with $a \neq c$. The coordinates of the intersection point (x_0, y_0) are given as rational numbers by

$$x_0 = (b-d)/(c-a) \tag{1.1}$$

$$y_0 = (ad-bc)/(c-a) \tag{1.2}$$

If the angle of the given sector is obtuse, then the required point can be obtained by checking the four points with $x = \{\lfloor x_0 \rfloor, \lceil x_0 \rceil\}$ and $y = \{\lfloor y_0 \rfloor, \lceil y_0 \rceil\}$. We consider only the case where the angle of the sector is acute.

Without loss of generality, consider the sector:

$$\{(x,y) : ax + b \leq y \leq cx + d\} \tag{1.3}$$

with $c \geq 0$ and $a < c$. Any other acute-angle sector can be mapped to this one by rotations of $\pi/2$ and reflections about the x or y axis. These transformations do not change the distance of the integer points inside the sector from the intersection-point.
Case 1. $a \leq 0$: In this case the required point lies on a horizontal line with either $y = \lceil y_0 \rceil$, or $y = \lfloor y_0 \rfloor$. Its x-coordinate is given by

$$\min \left\{ \left\lceil \frac{\lfloor y_0 \rfloor - b}{a} \right\rceil, \left\lceil \frac{\lceil y_0 \rceil - d}{c} \right\rceil \right\} \tag{1.4}$$

If both terms inside the *min* function are equal, then the required point is the one nearer to (x_0, y_0).

> Example 1. Consider lines $y = -3/4x + 7/8$ and $y = 1/2x + 1/4$ (Figure 1.2a). We have $(x_0, y_0) = (1/2, 1/2)$, $\lfloor y_0 \rfloor = 0$, and $\lceil y_0 \rceil = 1$. The corresponding x-coordinates are 7/6 and 3/2. The x-coordinate of the required point is obtained as min ($\lceil 7/6 \rceil$, $\lceil 3/2 \rceil$) $= 2$. Both contender points $(2,0)$ and $(2,1)$ are at the same distance from the intersection point, so we may choose either.

Case 2. $c \geq \lceil a \rceil > a > 0$: Subtracting $\lceil a \rceil$ from the slopes transforms the problem to Case 1. The required point can be obtained as in Case 1 and then transformed back to the original domain. If two points qualify for the minimum x-coordinates in the transformed problem (i.e., in Case 1), then we transform back the point with the *smaller* y-coordinate. (This is not necessarily the solution point of Case 1. For example, consider $y = 19/11x + 81/110$ and $y = 42/19x + 47/95$.)

> Example 2. Consider lines $y = 5/4x + 7/8$ and $y = 5/2x + 1/4$. Subtracting $\lceil 5/4 \rceil = 2$ from the slopes gives the lines $y = -3/4x + 7/8$ and $y = 1/2x + 1/4$. These are the same lines as in Example 1 (see Figure 1.2a). Transforming back the solution point (2, 0) of Example 1, obtain the required solution point $(2, 4)$.

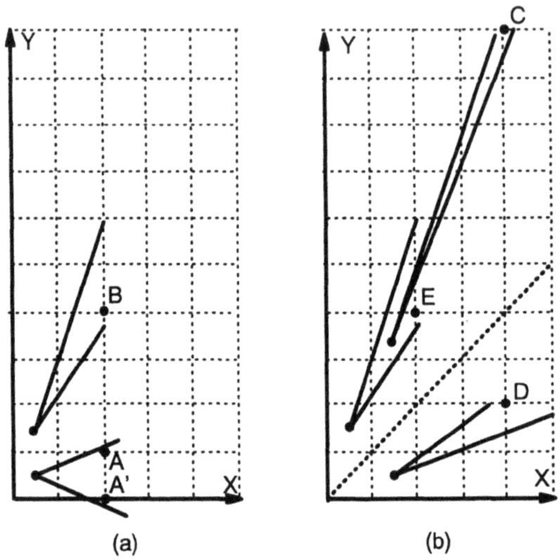

FIGURE 1.2. A and A' are both solutions of Example 1 (Case 1). Point B is the solution point of Example 2 (Case 2). Point C, the solution point after the first reduction of the original problem in Example 3, is mapped by Step 1 into point D, and then by Step 2 to point E (which is the same as point B). (The original sector in Example 3 is too narrow to display at the same scale.)

Case 3. $\lceil a \rceil > c > a > 0$: This is the remaining, and also the most difficult, case. Reduce this case to Case 2 as follows.

Algorithm 1.

Input: lines $y = ax + b$ and $y = cx + d$ satisfying $\lceil a \rceil > c > a > 0$.

while $\lceil a \rceil > c > a > 0$ **do**

1. $c \leftarrow c - \lfloor a \rfloor$, $a \leftarrow a - \lfloor a \rfloor$
2. $d \leftarrow -b/a$, $b \leftarrow -d/c$

   swap (a, c)

   $a \leftarrow 1/a$, $c \leftarrow 1/c$

This algorithm increases the angle of the sector in each iteration and eventually yields Case 2 with $c \geq \lceil a \rceil > a$. We then find the solution point as described in Case 2, and back-transform the solution point to the original domain.

Example 3. Consider lines $47/14x + 1/4$ and $41/12x + 1/24$ (Figure 1.2b).
At Step 1 of the first iteration we subtract $\lfloor 41/12 \rfloor = 3$ from each slope. After Step 2 we obtain the new set of lines $12/5x - 1/10$ and $14/5x - 7/10$.
In the second iteration we subtract $\lfloor 12/5 \rfloor = 2$ from both slopes. After the second step we obtain the new set of lines $5/4x + 7/8$ and $5/2x + 1/4$. These lines satisfy the condition for Case 2.
As in Example 2, we find the solution for the reduced problem as $(2, 4)$. Transforming this solution back to the original domain, we obtain the solution as $(10, 34)$.

Thus the complete algorithm in 2-D is

Algorithm 2.

Input: lines: $y = ax + b,\ y = cx + d$, *specified sector.*

1. Transform *specified sector* to the one in Eqn. (1.3).

2. Apply Algorithm 1 to transform the sector to Case 2.

3. Transform Case 2 to Case 1 by subtracting $\lceil a \rceil$ from both the slopes.

4. Obtain the solution using Eqn. (1.4).

5. Transform the solution point back to the *specified sector.*

In 2-D, the optimal solution point is unique in Case 2 and Case 3. The x-coordinate of the optimal point is the *minimum* x for which there exists an integer within the interval $[ax+b, cx+d]$, and the y-coordinate is the *minimum* integer within that interval. Alternatively, the y-coordinate of the optimal solution point is the minimum y for which there exists an integer within the interval $[y/c - d/c, y/a - b/a]$, and the x coordinate is the minimum integer in that interval. Thus, the solution point remains the same even if the x and y axes are switched.

The result is correct, because each step of the algorithm maps the solution point in each domain to the solution point in the next reduced domain. Step 1 does not change the x-coordinate of the required point (m, n), but only reduces its y-coordinate by $\lfloor a \rfloor m$. The *minimum* integer y at the *minimum* integer x maps to the corresponding point in the reduced pyramid. The second step reflects the plane about the line $y = x$. This step switches the x and the y-coordinates of the solution point.

The iterations of Step 1 and Step 2 in Algorithm 1 generate the *continued fraction expansion* of the slopes. For instance, in Example 3,

$$47/14 = 3 + \frac{1}{2 + \frac{1}{5/4}} \quad \text{and} \quad 41/12 = 3 + \frac{1}{2 + \frac{1}{5/2}}$$

Eventually, the corresponding terms will differ by an integer at some stage of their continued fraction expansions (because $a \neq c$). Thus they will satisfy the conditions for Case 2. Since both the slopes a and c are rational numbers, their continued fraction expansions will terminate after a finite number of steps. In the worst case, the number of terms in the continued fraction expansion is $O(\log N)$, where N is the largest representable integer [5]. We assume that integer division and floor function are constant-time operations. Hence, Algorithm 1 has $O(\log N)$ run-time in the worst case.

Step 2 of Algorithm 2 is Algorithm 1. Steps 1, 3, and 4 are constant time operations. Step 5 has the same complexity as Step 2. Thus, Algorithm 2 has $O(\log N)$ arithmetic complexity.

1.3 3-D Version of the Problem

Now we describe the problem in *3-D.* We first define an *open pyramid* to represent the constrained region. Then we define *integer shear transformations* for the pyramid. These transformations are used to open up the pyramid (as in Algorithm 1 in *2-D*) to facilitate obtaining an integer point inside the pyramid.

A point (x, y, z) is defined by a vector $\mathbf{p}$. When x, y, z are rational numbers, we call $\mathbf{p}$ a *rational point.* When x, y, z are integers, we call $\mathbf{p}$ an *integer point.* Note that all integer points are rational points, but not *vice versa.*

1.3.1 Representation of a Pyramid

An *open pyramid* is defined by a rational point $\mathbf{g}$, denoting the apex, and three linearly independent vectors $\mathbf{a} = [a_x, a_y, a_z]^T$, $\mathbf{b} = [b_x, b_y, b_z]^T$, and $\mathbf{c} = [c_x, c_y, c_z]^T$. The vectors $\mathbf{a}$, $\mathbf{b}$, and $\mathbf{c}$ denote the directions of the edges of the pyramid. The elements of these vectors are integers. This open pyramid is represented by $(\mathbf{M}, \mathbf{g})$, where $\mathbf{M} = [\mathbf{a}, \mathbf{b}, \mathbf{c}]$.

Since $\mathbf{a}$, $\mathbf{b}$, and $\mathbf{c}$ are linearly independent, they form a *basis* for the 3-*D* space. Any point $\mathbf{p}$ in the space has a unique representation with respect to $(\mathbf{M}, \mathbf{g})$, i.e., $\mathbf{p} = \mathbf{M}.\mathbf{e} + \mathbf{g}$. We say that $\mathbf{p}$ is *inside* the pyramid if all the elements of vector $\mathbf{e}$ are non-negative. The *Manhattan distance* between $\mathbf{p}$ and $\mathbf{g}$ is defined by

$$|\mathbf{p} - \mathbf{g}| = |p_x - g_x| + |p_y - g_y| + |p_z - g_z|.$$

Example 4. Consider a pyramid with

$$\mathbf{M} = \begin{bmatrix} 56 & 54 & 55 \\ 36 & 38 & 31 \\ 28 & 25 & 30 \end{bmatrix} \quad \text{and} \quad \mathbf{g} = \begin{bmatrix} 3/2 \\ 2/5 \\ 1/4 \end{bmatrix}.$$

An integer point $\mathbf{p}^T = [43, 27, 21]$ inside the pyramid can be represented by $\mathbf{e}^T = [3/56, 11/28, 11/35]$
The distance of $\mathbf{p}$ from $\mathbf{g}$ is

$$|43 - 3/56| + |27 - 11/28| + |21 - 11/35| = 90\frac{67}{280}$$

1.3.2 Transformation of the Pyramid

An *integer shear transformation* (IST) in the *3-D* space is defined by a 3×3 matrix:

$$\mathbf{T} = \begin{bmatrix} 1 & l & 0 \\ 0 & 1 & 0 \\ 0 & 0 & 1 \end{bmatrix}$$

All diagonal elements of an IST are 1. It has only one non-zero off-diagonal element. For the sake of conciseness we express this IST as $\mathbf{T}(l_{j,k})$, where j and k are the row and the column numbers, respectively, of the non-zero off-diagonal element with value l. This is a type of coordinate transformation which maps an integer point in the original domain to an integer point in the transformed domain. The inverse of an IST exists and maps the transformed point back to the original point. Thus, the inverse of an IST is also an IST. The IST has the following properties:

1. It maps the origin onto itself.
2. The magnitude of the determinant of an IST is 1.

3. Parallel planes remain parallel after an IST.
4. A point in a specified half-space w.r.t. a plane remains so after an IST. A point on a plane remains on the plane after an IST.

The product of several ISTs inherits properties 1-4 above.

It is worthwhile to consider what happens, under the transformation, to the *cross-sectional triangles* formed by the intersection of the pyramid with planes parallel to the coordinate planes. Since only the $\mathbf{M}$ matrix determines how wide a pyramid is, we consider a pyramid with its apex $\mathbf{g} = \mathbf{0}$. An IST of the form $\mathbf{T}(l_{3,1})$ translates the cross-sectional triangle of such a pyramid at the $x = 1$ plane by l in the z direction. The cross-sectional triangle at $x = k$ is translated by kl in the z direction.

If there is an integer point $\mathbf{p}(p_x, p_y, p_z)$ in the cross-sectional triangle at $z = p_z$, then after an IST by $\mathbf{T}(l_{3,1})$, point $\mathbf{p}$ is mapped to $\mathbf{p}'(p_x, p_y, p_z + l \cdot p_x)$. The cross-sectional triangle at the $z = p_z$ plane is mapped to the triangle formed by the intersection of the plane $z = p_z + lx$ with the pyramid. The "tilted" triangle formed by the intersection of the plane $z = p_z + l \cdot (p_x - x)$ with the pyramid is mapped to the cross-sectional triangle at the $z = p_z + l \cdot p_x$ plane, which now contains the transformed point $\mathbf{p}'$.

To summarize, after an IST $\mathbf{T}(l_{3,1})$, planes parallel to the y-z planes remain so, and planes parallel to $z = -lx$ become parallel to the x-y plane. Depending on whether l is negative or positive, the pyramid expands or contracts.

For a pyramid confined to the positive octant with the apex at the origin, the planes $x = 1$, $y = 1$, and $z = 1$ play a role in visualization of the ISTs. We call these three planes *one-planes*.

There is another type of transformation similar to an IST, which rotates the cross-sectional triangles at the one-planes by multiples of $\pi/2$, or reflects the same triangles about any coordinate axis. For example, a reflection of the cross-sectional triangle at the $x = 1$ plane about the y-axis is

$$\mathbf{T} = \begin{bmatrix} 1 & 0 & 0 \\ 0 & 1 & 0 \\ 0 & 0 & -1 \end{bmatrix}$$

Similarly, a rotation of the same triangle by $\pi/2$ about the origin is

$$\mathbf{T} = \begin{bmatrix} 1 & 0 & 0 \\ 0 & 0 & -1 \\ 0 & 1 & 0 \end{bmatrix}$$

These transformations also have the properties 1-4 of IST above.

To relate the *3-D* problem to the *2-D* problem, we describe Algorithm 1 in terms of ISTs. We express the slopes of the lines by a 2×2 $\mathbf{M}$ matrix. The series of transformations from Case 3 to Case 2 in Algorithm 1 can be represented as a matrix product of the form $\mathbf{T}_n \ldots \mathbf{T}_2\mathbf{T}_1\mathbf{M}$. The off-diagonal element of the $\mathbf{T}_i$ is the ith term in the continued fraction expression of the slopes.

Example 5. Consider Example 3. The slopes of the lines are $a = 47/14$ and $c = 41/12$. Write the $\mathbf{M}$ matrix as

$$\mathbf{M} = \begin{bmatrix} 47 & 41 \\ 14 & 12 \end{bmatrix}$$

After ISTs $\mathbf{T}(-3_{1,2})$ and $\mathbf{T}(-2_{2,1})$, we obtain the matrix

$$\mathbf{M}' = \mathbf{T}(-2_{2,1}).\mathbf{T}(-3_{1,2}).\mathbf{M} = \begin{bmatrix} 5 & 5 \\ 4 & 2 \end{bmatrix}$$

The slopes 5/4 and 5/2 now satisfy the condition for Case 2, i.e., $5/4 < 2 < 5/2$.

1.3.3 Spanning Points

To emphasize the difference between the *2-D* and the *3-D* problem we introduce the notion of *spanning points.*

A *spanning point* inside a pyramid $(\mathbf{M}, \mathbf{g})$ is defined as an integer point $\mathbf{q}$ such that there is no integer point $\mathbf{p}$ inside the pyramid with $|p_x - g_x| \leq |q_x - g_x|$, $|p_y - g_y| \leq |q_y - g_y|$, and $|p_z - g_z| \leq |q_z - g_z|$.

In *2-D*, if the sector is confined to one quadrant, then the closest integer point is the *only* spanning point inside the sector. On the other hand, in *3-D*, the number of spanning points inside a pyramid can be more than one. The closest integer point is one of the spanning points. Other (non-optimal) spanning points can be thought of as a set of local minima. When a pyramid is expanded using an IST with a negative off-diagonal element, spanning points are mapped to spanning points. It is not clear whether the same holds for an IST with a positive off-diagonal element.

1.4 An O(N log N) Solution to the 3-D Problem

We present the $O(N \log N)$ solution to the *3-D* Problem for a pyramid with the following characteristics. The pyramid is confined to the positive octant (i.e., all the elements of $\mathbf{M}$ are non-negative) and the apex $\mathbf{g}$ is in the unit-cube: $0 \leq g_x, g_y, g_z \leq 1$. Any other pyramid can be systematically divided, rotated, and translated to pyramids of this form.

Parallel planes $x + y + z = d$ with integer $d = 1, 2, \ldots$, are at constant Manhattan-distance from the origin. These planes span all the integer points in the octant. They cut the pyramid, forming a sequence of triangles. Since the apex of the pyramid $\mathbf{g}$ is inside the unit-cube adjacent to the origin, the solution point must be inside the triangle at *minimum* d that contains an integer point.

Simplify the problem by the following transformation, that transforms the $x+y+z = d$ plane to a plane parallel to the x-y plane.

$$\mathbf{T} = \mathbf{T}(1_{3,1}).\mathbf{T}(1_{3,2}) = \begin{bmatrix} 1 & 0 & 0 \\ 0 & 1 & 0 \\ 1 & 1 & 1 \end{bmatrix}$$

Now we must find the minimum integer z such that the cross-sectional triangle includes a point with integer-valued x and y coordinates. To check if a triangle on the x-y plane contains an integer point, we need the following lemma about a triangle on the x-y plane having an integer point within it.

Lemma 1. A triangle with rational point vertices on a coordinate plane includes an integer point within it, only if the *closest integer point* of at least one of the three sectors (formed by two edges) of the triangle is within the triangle.

Proof: Assume that there is an integer point inside the triangle. Draw a vertical (or horizontal) line from the median x (or y) vertex, to divide the triangle in two parts

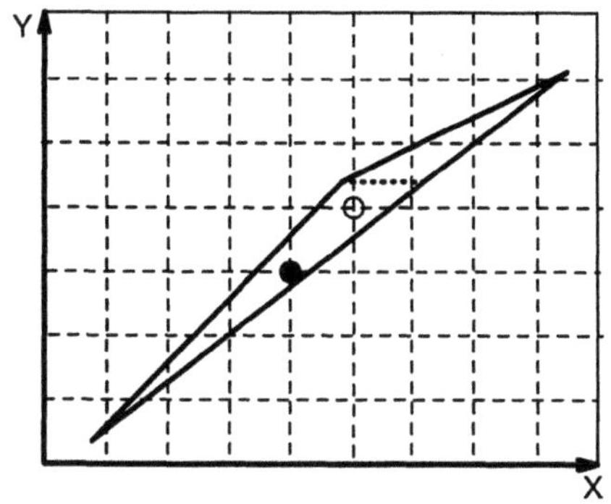

FIGURE 1.3. Proof of Lemma 1. A triangle with $x = k$ is divided by a horizontal line through its median y vertex. The presence of an integer point in one part of the triangle enforces the presence of the closest-integer point inside the corresponding sector.

(Figure 1.3). Now, if the integer point inside the triangle is left of (below) the dividing line, then either this point is closest to the vertex on the left (below), or there is an integer point further left (down).

If the integer point is right of (above) the dividing line, a similar argument applies to the vertex on the right (above).

The above lemma can be used to determine whether a triangle contains an integer point within it:

Algorithm 3.

```
Input: Triangle
for each vertex of the triangle
    for the sector adjacent to the vertex
        obtain the closest integer approximation using Algorithm 2.
        if the closest integer point is within the triangle report(yes)
```

The $O(N \log N)$ solution consists of checking the cross-sectional triangles at $z = 1, \ldots$ planes for an integer point.

Example 6. Consider the pyramid with

$$\mathbf{M} = \begin{bmatrix} 56 & 54 & 55 \\ 36 & 38 & 31 \\ 28 & 25 & 30 \end{bmatrix} \quad \text{and} \quad \mathbf{g} = \begin{bmatrix} 3/2 \\ 2/5 \\ 1/4 \end{bmatrix}.$$

Transforming the pyramid by $\mathbf{T}(1_{3,1}).\mathbf{T}(1_{3,2})$ yields

$$\mathbf{M}' = \begin{bmatrix} 56 & 54 & 55 \\ 36 & 38 & 31 \\ 120 & 117 & 116 \end{bmatrix} \quad \text{and} \quad \mathbf{g}' = \begin{bmatrix} 3/2 \\ 2/5 \\ 43/20 \end{bmatrix}.$$

The minimum z for which the cross-sectional triangle of pyramid $(\mathbf{M}', \mathbf{g}')$ contains an integer point at $z = 91$; the point is $(x, y) = (43, 27)$. Back-transforming, obtain the solution point in the original domain as $[43, 27, 21]^T$.

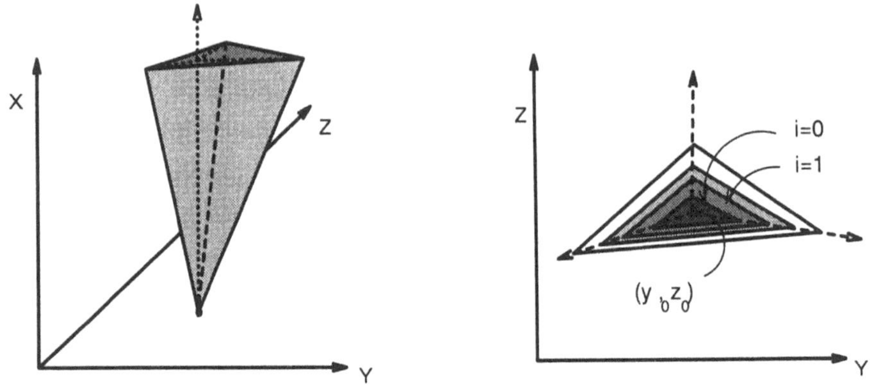

FIGURE 1.4. A *right pyramid* and its projections of cross-sectional triangles on the z-y plane at $x = \lceil x_0 \rceil + i$ for $i = 0, 1, \ldots$. The triangle at a certain value of x encloses all the triangles at lower values of x. The projections of the vectors of the given pyramid join the vertices of the triangles to create three *2-D* sectors on the y-z plane. The given pyramid is divided into three pyramids, each of which project to one of these *2-D* sectors.

Algorithm 2 is applied at most three times in Algorithm 3. This requires $O(\log N)$ arithmetic operations. We apply this algorithm for each $z = d$ plane, where $d \geq 1$ is an integer. There can be at most $O(N)$ such planes. Thus the overall complexity is $O(N \log N)$.

1.5 Right Pyramid

Now we will try to extend the $O(\log N)$ algorithm for the *2-D* problem to *3-D*. As in *2-D*, we will proceed from a simpler problem to a more difficult one. We consider two special cases: the *right pyramid* and the *homogeneous pyramid*, and combine them for the solution of an arbitrary pyramid.

A pyramid enclosing a ray parallel to one of the coordinate axes is called a *right pyramid*. Given a general pyramid, our strategy will be to reduce it to a right pyramid. W.l.o.g. consider a right pyramid that encloses a ray parallel to the x-axis.

> Given a pyramid $(\mathbf{M}, \mathbf{g})$ such that $\mathbf{g}^T = [x_0, y_0, z_0] \neq [0, 0, 0]^T$ and the pyramid encloses the ray $y = y_0, z = z_0, x \geq x_0$, obtain an integer point $\mathbf{p}$ inside the pyramid $(\mathbf{M}, \mathbf{g})$, which is closest to the apex $\mathbf{g}$.

The condition of enclosing a ray parallel to the x-axis is that the cross-sectional triangle at the $x = \lceil x_0 \rceil$ plane includes the point (y_0, z_0). The projections on the y-z plane of cross-sectional triangles at planes $x = \lceil x_0 \rceil + i$ for $i = 1, \ldots,$ enclose the projections of all the cross-sectional triangles on the planes $x = \lceil x_0 \rceil + j$ for $j = 0, \ldots, i - 1$.

The three vectors of the given pyramid project on the y-z plane to create three *2-D* vectors emanating from the point (y_0, z_0) to the corresponding vertices of the projections of the cross-sectional triangles (Figure 1.4).

Now partition the right pyramid into three different pyramids with one edge parallel to the x-axis. The projections of each pair of the three vectors create a single *2-D* sector on the y-z plane corresponding to each of the new pyramids. The $\mathbf{M}$ matrix of each pyramid contains one pair from the three vectors of the given pyramid, and the vector $[1,0,0]^T$, corresponding to the ray $y = y_0, z = z_0, x \geq x_0$. The apex $\mathbf{g}$ of all the pyramids is the same.

Find the closest integer point to the vertex in each of these three pyramids and choose the best among them.

Consider one of the pyramids so created. If the angle of the projected sector is obtuse, we can divide, rotate, and translate the pyramid so that the pyramid is confined to the positive octant. Thus, w.l.o.g. assume that the pyramid is confined to the positive octant with its vertex in the unit cube, $0 \leq g_x, g_y, g_z \leq 1$.

Transform the pyramid by ISTs $\mathbf{T}(1_{1,3})\ \mathbf{T}(1_{1,2})$. This transformation makes the equidistant planes parallel to the y-z plane, and leaves the edge-vector parallel to the x-axis unchanged. Also, since it does not change the y and the z coordinates, the projected sector remains unchanged. After the transformation our objective becomes obtaining the minimum integer x such that the cross-sectional triangle at x contains an integer point.

The projections of the cross-sectional triangles of this pyramid at planes $x = \lceil x_0 \rceil + i$ for $i = 0, 1, \ldots$ have a common vertex at (y_0, z_0) and a set of parallel base-lines cutting the sector at linearly increasing intervals. If there is an integer point inside one of these cross-sectional triangles, then there is an integer point with the same y and z value, inside the pyramid. Its x-value corresponds to the x-value of that cross-sectional triangle. Thus the problem is to find the cross-sectional triangle with minimum x that contains an integer point inside the sector.

Transform the pyramid by a sequence of ISTs so that these base-lines become parallel to the z-axis. These ISTs correspond to the continued-fraction expansion of the slope of the base-lines. These transformations modify only the y and the z rows of the $\mathbf{M}$ matrix. They do not change the containment relationship (Property 4 of the IST).

Now determine the closest integer point inside the sector by Algorithm 2. There is no integer point inside the sector left of this point. Since the base-lines are parallel to the z-axis, it is easy to determine the first triangle that contains this point. Thus, we obtain all the coordinates of the solution point, and then back-transform it to the original pyramid.

Algorithm 4.

Input Matrix $\mathbf{M}$ and vector $\mathbf{g}^T = [x_0, y_0, z_0]$

1. Divide the pyramid in three different contiguous pyramids with the same apex at $\mathbf{g}$ and edges consisting of a new vector $[1,0,0]^T$ and each pair of vectors from $\mathbf{M}$. (If any of the vectors of $\mathbf{M}$ is parallel to the x-$axis$, then there will be only one pyramid)

2. **for** each of the three pyramids **do**

   (a) Transform the pyramid to the positive octant.

   (b) Transform the pyramid by $\mathbf{T}(1_{1,3})\mathbf{T}(1_{1,2})$.

   (c) Project the pyramid on the y-z plane to obtain a sector and the set of parallel base-lines for each integer x.

   (d) Obtain the continued fraction expansion of the slope of the base lines.

   (e) Transform the pyramid by ISTs with off-diagonal terms corresponding to the expansion in the continued fraction, so that base-lines become parallel to the z-axis (or, y-axis).

   (f) Determine the closest integer point to the common vertex, inside the sector, using Algorithm 2.

(g) Obtain the the minimum x for which the cross-sectional triangle contains the obtained integer point.

(h) Back-transform each of the obtained points to the original domain.

3. Choose the closest of the three points so derived.

Note that each of the candidate points obtained in Step 2(h) above maps to a spanning point in the given pyramid, but they can be at different distances from the apex. The closest one is the optimal approximation to the apex.

Example 7. Given

$$\mathbf{M} = \begin{bmatrix} 1 & 9 & 1 \\ 0 & 2 & 8 \\ 0 & 3 & 9 \end{bmatrix} \quad \text{and} \quad \mathbf{g} = \begin{bmatrix} 1/10 \\ 1/5 \\ 2/5 \end{bmatrix}.$$

In this case $\mathbf{a}$ is parallel to the x-axis.
Transform $\mathbf{M}$ by $\mathbf{T}(1_{1,3}).\mathbf{T}(1_{1,2})$:

$$\mathbf{M}' = \begin{bmatrix} 1 & 14 & 18 \\ 0 & 2 & 8 \\ 0 & 3 & 9 \end{bmatrix} \quad \text{and} \quad \mathbf{g}' = \begin{bmatrix} 7/10 \\ 1/5 \\ 2/5 \end{bmatrix}.$$

The slope of the base lines is

$$\frac{3/14 - 9/18}{2/14 - 8/18} = 18/19 = \frac{1}{1 + 1/18}$$

Thus, after two ISTs $\mathbf{T}(-1_{2,3}).\mathbf{T}(-18_{3,2})$,

$$\mathbf{M}'' = \begin{bmatrix} 1 & 14 & 18 \\ 0 & -1 & -1 \\ 0 & 21 & 27 \end{bmatrix} \quad \text{and} \quad \mathbf{g}'' = \begin{bmatrix} 7/10 \\ -1/5 \\ 4 \end{bmatrix}.$$

The corresponding sector on the y-z plane is given by lines $z - 27y = 47/5$ and $z - 21y = 41/5$, intersecting at $(y, z) = (-1/5, 4)$. The closest integer point inside the sector is (-1, 21). The corresponding minimum x value is 13. Back-transforming the point $[13, -1, 21]^T$ in the original domain, obtain $[8, 2, 3]^T$ as the optimal solution point.

Steps 2(e) and 2(f) of this algorithm are $O(\log N)$. The other steps are of constant time. The overall complexity of Algorithm 4 is $O(\log N)$.

Our strategy to obtain a solution point for the *3-D* problem is to transform the original pyramid to a right pyramid. Whether the pyramid includes a ray of certain slope depends only on the $\mathbf{M}$ matrix, and not on the coordinates of the apex $\mathbf{g}$. Since, the reduction will be dictated by $\mathbf{M}$, we will consider a *homogeneous pyramid* with the same $\mathbf{M}$ matrix but with the apex at the origin. Once we compute all the transformations that are necessary to transform $\mathbf{M}$ to a right pyramid, we can apply the same set of transformations to $\mathbf{g}$.

1.6 Homogeneous Pyramid

Consider a pyramid confined to the positive octant, with apex at the origin. If the cross-sectional triangle at the $x = 1$ plane contains an integer point (y', z'), then two ISTs of $\mathbf{T}(-y'_{2,1}).\mathbf{T}(-z'_{3,1})$, will yield a right pyramid that includes the x-axis. Similar transformations may be applied if the cross-sectional triangle in any other one-plane contains an integer point. Thus our interim objective now is to reduce $\mathbf{M}$ in such a way that there is an integer point in one of the cross-sectional triangles of the one-planes. This integer-point will be the *seed* for the transformation to the right-pyramid.

> Given a pyramid $(\mathbf{M}, \mathbf{g})$, with $\mathbf{g}^T = [0, 0, 0]$ and all the elements of $\mathbf{M}$ non-negative, obtain an integer point inside the pyramid with one coordinate being unity.

Using a sequence of ISTs, expand the pyramid as much as possible while still keeping it in the first octant. Each step in this reduction is a multiplication by an IST $\mathbf{T}(-l_{j,k})$.
Algorithm 5.

Input: matrix $\mathbf{M} = [\mathbf{a}, \mathbf{b}, \mathbf{c}]$
repeat until $\mathbf{M}$ remains the same at two consecutive iterations
  for each row $k \in \{x, y, z\}$ of $\mathbf{M}$ **do**
    for rows $j \in \{x, y, z\}$, $j \neq k$ of $\mathbf{M}$ **do**
      Find *maximum* l such that $a_j \geq l.a_k$, $b_j \geq l.b_k$, and $c_j \geq l.c_k$
      $\mathbf{M} \leftarrow \mathbf{T}(-l_{j,k}).\mathbf{M}$

Example 8. Given:

$$\mathbf{M} = \begin{bmatrix} 56 & 54 & 55 \\ 36 & 38 & 31 \\ 28 & 25 & 30 \end{bmatrix} \quad \text{and} \quad \mathbf{g} = \begin{bmatrix} 0 \\ 0 \\ 0 \end{bmatrix}.$$

After four modular transformations: $\mathbf{T}_1(-1_{1,2})$, $\mathbf{T}_2(-1_{2,3})$, $\mathbf{T}_3(-1_{3,1})$, and $\mathbf{T}_4(-1_{1,3})$

$$\mathbf{M}' = \mathbf{T}_4\mathbf{T}_3\mathbf{T}_2\mathbf{T}_1\mathbf{M} = \begin{bmatrix} 12 & 7 & 18 \\ 8 & 13 & 1 \\ 8 & 9 & 6 \end{bmatrix} \quad \text{and} \quad \mathbf{g}' = \begin{bmatrix} 0 \\ 0 \\ 0 \end{bmatrix}.$$

The number of forward ISTs in the reduction of Algorithm 5 (i.e., the number of times $\mathbf{M}$ is updated in Algorithm 5) is bounded by the maximum number of steps in the GCD of any two elements in the given $\mathbf{M}$. Now the number of steps in the GCD algorithm is also bounded by $O(\log N)$, where N is the largest integer element in $\mathbf{M}$. Finding l in Line 5 of Algorithm 5 is constant time if we assume that the integer division is a constant-time operation. All other operations involve integer multiplication and addition, which are constant time. Thus the reduction takes $O(\log N)$ arithmetic operations.

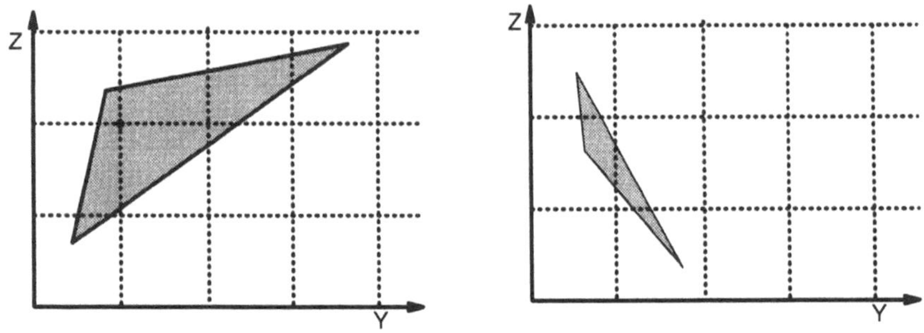

FIGURE 1.5. The terminating cases of Algorithm 5. In Case 1, left, the cross-sectional triangle at the $x = 1$ plane contains an integer point. In Case 2, right, it does not. In both cases it crosses the *one-lines*.

This algorithm terminates when no pair of rows of matrix **M** can be reduced further. Then none of the cross-sectional triangles at any of the one-planes can be translated by any integer amount towards the origin, without leaving the positive quadrant. The triangle at the $x = 1$ plane must cross both the $y = 1$ and $z = 1$ lines (*one-lines* with reference to the $x = 1$ plane). Either this triangle includes an integer point, or it does not (Figure 1.5). (The same holds for the $y = 1$ and $z = 1$ planes.) Algorithm 3 is used to determine whether a triangle on one of the one-planes contains an integer point or not.
Case 1. If the triangle at any of the one-planes contains one or more integer points, choose any one of the points as the seed of the transformation to a right pyramid.
Case 2. If none of the cross-sectional triangles at these one-planes contains an integer point, then determine some value of x, y, or z such that the cross-sectional triangle at that x, y, or z contains an integer point.

Consider the triangle at $x = 1$. The pyramid includes an integer point at $x = k$ if it includes a rational point whose coordinates have a common denominator k at the triangle at $x = 1$. The triangle intersects the lines $y = 1$ and $z = 1$. Now, determine the interval on the $y = 1$ line inside the triangle on the $x = 1$ plane (Figure 1.6). Find the rational number with the smallest denominator within that interval, using the continued fraction expansion of the limits of the interval.

If that rational number on the line $y = 1$ in the plane $x = 1$ happens to be r/s, then the pyramid includes the point $(x, y, z) = (s, s, r)$. Do the same for the other one-lines in each of the cross-sectional triangles in the other one-planes.

Algorithm 6.

Input : The final **M** of Algorithm 5.
for all three *one-planes* **do**

> obtain *cross-sectional* triangle.
> **for** both *one-lines* on the *one-plane* do
> 1. obtain the cross-section of the triangle at that *one-line*.
> 2. obtain the rational number with the smallest denominator inside the interval.
>
> scale up the triangle to obtain the integer point inside the pyramid

choose the point **q** at minimum distance.

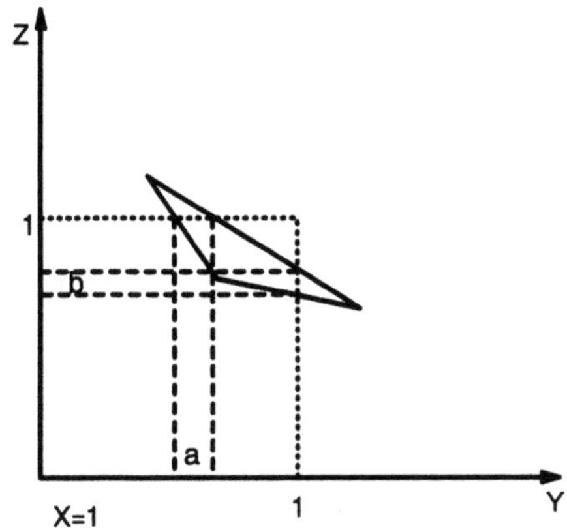

FIGURE 1.6. The intervals of the intersection of the *one-lines* with the cross-sectional triangle at the $x = 1$ plane. Any rational number with denominator k in intervals a or b corresponds to an integer point within the cross-sectional triangle at $x = k$.

Example 9. Consider the final matrix from Example 8

$$\mathbf{M}' = \begin{bmatrix} 12 & 7 & 18 \\ 8 & 13 & 1 \\ 8 & 9 & 6 \end{bmatrix} \quad \text{and} \quad \mathbf{g}' = \begin{bmatrix} 0 \\ 0 \\ 0 \end{bmatrix}.$$

None of the cross-sectional triangles at the three one-planes contains any integer point.
Consider the triangle at $z = 1$ plane. It cuts $y = 1$ at $\frac{739}{500}$ and $\frac{3}{2}$.
Applying Algorithm 6, obtain 3/2, the smallest rational number within that interval. Scaling up, obtain a point $\mathbf{q}^T = [3, 2, 2]$ inside the pyramid.

Since the maximum number of steps in the continued fraction expansion is $O(\log N)$, the number of arithmetic operations in Algorithm 6 is also $O(\log N)$.

Now transform the pyramid so that one of the coordinates of $\mathbf{q}$ becomes 1 after the transformation. This point will be the seed for the transformation to the right pyramid.

Algorithm 7.

Input: integer point $\mathbf{q}(x, y, z)$.
do for each pair of coordinates u and v

while one is not divisible by the other or one of them is not 1 **do**

1. $u \leftarrow u \bmod v$.
2. $v \leftarrow v \bmod u$.

Example 10. Given a point $[15, 26, 15]^T$. After reductions by ISTs $\mathbf{T}(-1_{2,1})$, $\mathbf{T}(-1_{1,2})$, $\mathbf{T}(-2_{2,1})$, and $\mathbf{T}(-1_{1,2})$, the point is reduced to $[1, 3, 15]^T$.

The algorithm terminates because x, y, and z are not mutually divisible. The arithmetic complexity of this algorithm is bounded by the maximum number of steps to obtain the GCD of any pair of the given triplet, i.e., $O(\log N)$.

1.7 Arbitrary Pyramid

Now we come back to the original problem as stated at the beginning of the chapter. We still assume that all the elements of $\mathbf{M}$ are non-negative. All other pyramids can be systematically divided, translated, or rotated to reduce them to this type of pyramid.

> Given a pyramid $(\mathbf{M}, \mathbf{g})$, with $\mathbf{g}$ a rational point, and all the elements of $\mathbf{M}$ non-negative, obtain an approximating integer point that is inside the pyramid.

To reduce the pyramid to a right pyramid, first use Algorithms 5-7 to find an integer point in the cross-sectional triangle at one of the one-planes of the corresponding homogeneous pyramid.

Suppose that the integer point is (y', z') in the $x = 1$ plane. Translate the cross-sectional triangle by $(-y', -z')$ (i.e., apply ISTs $\mathbf{T}(-y'_{2,1})$, $\mathbf{T}(-z'_{3,1})$), so that the projected integer point coincides with the origin (which in this case is the apex) and the pyramid satisfies the condition of a right pyramid.

Now we can turn our attention to the non-zero $\mathbf{g}$. Transform $\mathbf{g}$ by the transformation matrices used, to obtain the apex of the final pyramid. Then obtain the solution integer point by Algorithm 4, and transform the solution point back to the original domain. The entire procedure is summarized below.

Algorithm 8.

Input: pyramid $(\mathbf{M}, \mathbf{g})$.

1. Open up $\mathbf{M}$ by ISTs using Algorithm 5.
2. Determine if a cross-sectional triangle of any of the one-planes contains an integer point by Algorithm 3.
3. If it does, choose arbitrarily an integer point from that triangle and go to Step 6.
4. Else use Algorithm 6 to obtain some triangle (parallel to a coordinate plane) that contains an integer point.
5. Use Algorithm 7 to transform the pyramid so that the above integer point has at least one unity coordinate.
6. Transform the pyramid to a right pyramid.
7. Transform g by the ISTs applied so far.
8. Obtain three integer points by Algorithm 4, after dividing the given pyramid into three mutually adjacent pyramids.
9. Transform back the points to the original domain.
10. Choose the best solution point among the three.

Example 11. Consider the pyramid with

$$\mathbf{M} = \begin{bmatrix} 56 & 54 & 55 \\ 36 & 38 & 31 \\ 28 & 25 & 30 \end{bmatrix} \quad \text{and} \quad \mathbf{g} = \begin{bmatrix} 3/2 \\ 2/5 \\ 1/4 \end{bmatrix}.$$

Notice that the **M** matrix is the same as given in Example 8 and thus, using Algorithm 5, after four modular transformations,

$$\mathbf{M}' = \begin{bmatrix} 12 & 7 & 18 \\ 8 & 13 & 1 \\ 8 & 9 & 6 \end{bmatrix} \quad \text{and} \quad \mathbf{g}' = \begin{bmatrix} 39/20 \\ 3/20 \\ -17/20 \end{bmatrix}.$$

Whence, as in Example 9, use Algorithm 6 to obtain the solution point $[3, 2, 2]^T$. Reducing one more step using Algorithm 7,

$$\mathbf{M}'' = \begin{bmatrix} 1 & -2 & 12 \\ 2 & 13 & 1 \\ 2 & 9 & 6 \end{bmatrix} \quad \text{and} \quad \mathbf{g}'' = \begin{bmatrix} 14/5 \\ 3/20 \\ -17/20 \end{bmatrix}.$$

This contains the integer point $[1, 2, 2]^T$ in the cross-sectional triangle at the $x = 1$ plane. Translate by $(y, z) = (-2, -2)$ on the $x = 1$ plane (i.e., apply ISTs $\mathbf{T}(-2_{2,1})$, $\mathbf{T}(-2_{3,1})$),

$$\mathbf{M}''' = \begin{bmatrix} 1 & -2 & 12 \\ 0 & 17 & -13 \\ 0 & 13 & -18 \end{bmatrix} \quad \text{and} \quad \mathbf{g}''' = \begin{bmatrix} 14/5 \\ -109/20 \\ -129/20 \end{bmatrix}.$$

Now, the pyramid $(\mathbf{M}''', \mathbf{g}''')$ is a right pyramid. The solution-point of this pyramid is $[6, -6, -7]^T$. Back-transforming, obtain the solution point in the original domain as $[43, 27, 21]^T$.

In Algorithm 8, we use Algorithms 3, 4, 5, 6, 7. Each of these algorithms has arithmetic complexity $O(\log N)$. Thus the overall complexity of this algorithm is $O(\log N)$.

1.8 Conclusion

We posed the *2-D* and *3-D constrained integer approximation problems* with linear constraints. We presented $O(\log N)$ solutions for the general *2-D* problem and for the special case of a *right pyramid* in *3-D*. We also presented an $O(N \log N)$ solution for the general *3-D* problem. The above solutions are guaranteed to produce the closest integer point to the apex. Our approach to extend the $O(\log N)$ solution for the *2-D* problem to *3-D* guarantees finding an integer point *inside* the pyramid, but not that it is the *closest* solution point. The hurdles are described below.

In the *2-D* problem, the closest integer point inside a sector confined to the positive quadrant is *unique*. It has the minimum x *and* minimum y distance from the intersection point. In *3-D*, some of the coordinates of the closest integer point to the apex may not be minimal. (For instance, for the closest integer-point $\mathbf{p}$ and another point $\mathbf{q}$ inside the pyramid, $|\mathbf{p} - \mathbf{g}| < |\mathbf{q} - \mathbf{g}|$, but it may happen that $|p_x - g_x| > |q_x - g_x|$.) In the *3-D* problem there can be more than one solution point. These two facts make it more difficult to prove the optimality of the solution point in *3-D*.

The IST of the form $\mathbf{T}(l_{j,k})$ described in this chapter is a generalization of *continued fractions* to dimensions higher than 2. It is also an *automorphism* on the integer lattice. In *2-D* the optimal solution of a sector maps to the optimal solution after the transformation. In higher dimensions this may not be the case.

However, there are certain aspects of the IST which make it attractive for the solution of this kind of problem.

- The forward *integer shear transformation* with a negative off-diagonal element widens the pyramid in the sense that the cross-sectional triangles at the one-planes are expanded to include more lattice-points. This facilitates obtaining an integer point inside such a triangle.

- The integer points map *bijectively* by an IST. This guarantees that the point obtained in the reduced domain corresponds to one of the integer points in the original domain and *vice-versa.*

- The orientation of any point, line, or plane with respect to another point, line, or plane remains unchanged. Thus points in the upper half-space of a plane, remain so after the transformation. This property guarantees that the integer points inside a pyramid remain within the pyramid after the transformation.

- Parallelism is preserved. This leads to the optimal solution for *right pyramids.*

We are investigating whether the IST also has the following properties, which would suffice to prove the optimality of the general solution.

- A *spanning point* in the *3-D* pyramid maps to a spanning point after any IST. Though this is easy to prove for a forward IST with a *negative* off-diagonal element, it is not clear whether it holds for the backward IST with a *positive* off-diagonal element. This property would guarantee that the integer point obtained in the reduced pyramid (which is a spanning point) maps to a spanning point in the original pyramid.

- The optimal solution point maps to the optimal solution-point after any IST. Though this is certainly the case in *2-D*, we have not found any conclusive proof for or against it in *3-D.*

So far, in all the experiments we have performed in 3-D, the $O(\log N)$ algorithm has yielded the optimal solution. As expected, the algorithm runs very fast. For instance, the number of steps in the $O(\log N)$ algorithm takes to solve Example 11 is 8, while the solution obtained by an integer programming package (LINDO) takes 436 steps. Thus, we believe that this is a practical solution for the *3-D* problem.

Acknowledgements:

This work was partially supported by the National Science Foundation Information, Robotics, and Systems grant IRI-8704718; and by a New York State Science and Technology Foundation grant to the Rensselaer Center for Advanced Technology. Maharaj Mukherjee gratefully acknowledges the support of Institute of Applied Mathematics, CNR, Genova, Italy, to attend and present this work at the Eurographics Workshop on Computer Graphics and Mathematics in Santa Margherita, Italy.

1.9 References

[1] D. Dobkin and D. Silver. Recipes for geometry and numerical analysis - part I: An empirical study. In *Proceedings of the ACM Symposium on Computational Geometry*, pages 93 - 105. Champaign-Urbana, Illinois, June 1988.

[2] D. H. Greene and F. F. Yao. Finite resolution computational geometry. In *Proceedings of Twenty-seventh Annual IEEE-FOCS Conference*, pages 143 - 152. Toronto, Canada, October 1986.

[3] L. Guibas, D. Salesin, and J. Stolfi. Epsilon-geometry: Building robust algorithms from imprecise computations. In *Proceedings of the ACM Symposium on Computational Geometry*, pages 208 - 217. Saarbrücken, Germany, June 1989.

[4] C. M. Hoffmann, J. E. Hopcroft, and M. S. Karasick. Robust set operations on polyhedral solids. *IEEE Computer Graphics & Applications*, 9(6):50 - 59, November 1989.

[5] D. E. Knuth. *The Art of Computer Programming - Seminumerical Algorithms (Vol. II)*. Addison Wesley, 1981.

[6] J. C. Lagarias. The computational complexity of simultaneous Diophantine approximation problems. In *Proceedings of Twenty-third Annual IEEE-FOCS Conference*, pages 32 - 39, 1982.

[7] A. K. Lenstra. Polynomial factorization by root approximation. In J. Fitch, editor, *Lecture Notes on Computer Science: Int. Symp. Symbolic and Algebraic Computations*, volume 174, pages 272 - 276. Springer-Verlag, 1984.

[8] A. K. Lenstra, H. W. Lenstra, and L. Lovász. Factoring polynomials with rational coefficients. *Math. Ann.*, 261:515 - 534, 1982.

[9] L. R. Nackman M. S. Karasick, D. Lieber. Efficient Delaunay triangulation using rational arithmetic. IBM Research Report, RC 14455, IBM Research Division, Yorktown Heights, New York, March 1989.

[10] D. W. Matula and P. Kornerup. Foundations of Finite Precision Rational Arithmetic. In *Computing, Suppl 2*, pages 85 - 111. Springer-Verlag, 1980.

[11] S. Mehta, M. Mukherjee, and G. Nagy. Constrained integer approximation to 2-D line intersections. In *Proceedings Second Canadian Conf. on Computational Geometry*, pages 302 - 305. Ottawa, Canada, August 1990.

[12] S. Mehta, M. Mukherjee, and G. Nagy. Constrained integer approximation to planar line intersections. *Information Processing Letters*, 40(3):137–139, November 1991.

[13] V. J. Milenkovic. Verifiable implementations of geometric algorithms using finite precision arithmetic. In D. Kapur and J. L. Mundy, editors, *Geometric Reasoning*, pages 377 - 401. Cambridge, Massachusetts, MIT Press, 1989.

[14] R. E. Moore. *Interval Analysis*. Prentice Hall, Englewood Cliffs, New Jersey, 1966.

[15] M. Segal and C. H. Sequin. Consistent calculations for solid modeling. In *Proceedings of the ACM Symposium on Computational Geometry*, pages 29 - 38. Baltimore, Maryland, June 1985.

[16] J. Vignes. New methods for evaluating the validity of mathematical computations. In *Mathematics and Computers in Simulation*, volume 20, pages 227 - 249, 1978.

[17] J. H. Wilkinson. *Rounding Errors in Algebraic Processes.* Prentice Hall Inc., Englewood Cliffs, New Jersey, 1963.

2 Approximate C^1–blending with Triangular Cubic Patches

Hartmut Prautzsch

ABSTRACT This paper discusses a simple construction of triangular surfaces with prescribed boundary and cross boundary derivative curves. These surfaces are continuous piecewise cubics with arbitrarily small jumps in their first derivatives. Moreover, they form a sequence which converges to a proper C^1–surface which is again piecewise cubic and interpolates the given boundary information.

2.1 Introduction

A particular problem in CAGD and the finite element method is to build a triangular C^1 surface whose boundary and cross boundary derivatives are given. A triangular surface is described by a real or vector valued function $\mathbf{p}$ with a triangular parameter domain τ as shown in Figure 2.1.

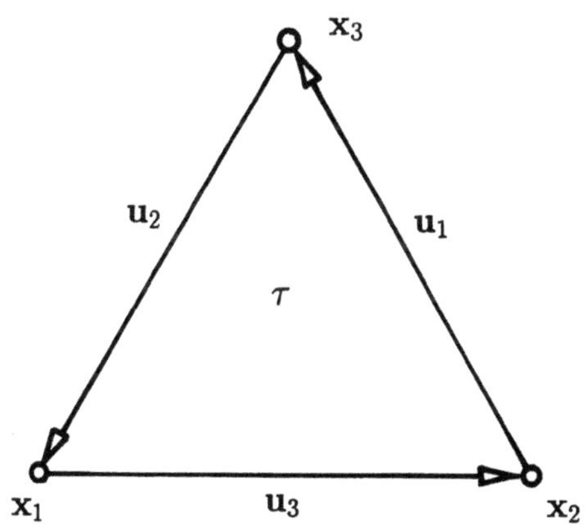

FIGURE 2.1. The triangular parameter domain τ.

On using barycentric coordinates with respect to τ one has that $(u, v, w) \in \tau$ iff $u + v + w = 1$ and $(u, v, w) \geq (0, 0, 0)$. In particular $\mathbf{x}_1 = (1, 0, 0)$, $\mathbf{x}_2 = (0, 1, 0)$, $\mathbf{x}_3 = (0, 0, 1)$ are the vertices of τ and $\mathbf{v}_1 = (0, -1, 1)$, $\mathbf{v}_2 = (1, 0, -1)$, $\mathbf{v}_3 = (-1, 1, 0)$ are the vectors pointing from one vertex of τ to another one.

Let $\mathbf{p}(u, v, w), (u, v, w) \in \tau$, be the triangular patch to be constructed and let $D_i\mathbf{p}$ denote the directional derivative of $\mathbf{p}$ with respect to $\mathbf{v}_i$, $i = 1, 2, 3$. Then $\mathbf{p}(u, v, w)$ and all $D_i\mathbf{p}(u, v, w)$ are initially given on the boundary of τ.

If the three boundary curves of $\mathbf{p}$ are cubic and for each $D_i\mathbf{p}$ quadratic poynomials, then one can hope to be able to construct the otherwise unknown surface patch $\mathbf{p}$ as a cubic. Such a cubic patch $\mathbf{p}$ exists iff the corner twists are compatible, i.e., iff in $\mathbf{x}_1$

$$\lim_{v \to 0} D_2 D_3 \mathbf{p}(1 - v, v, 0) = \lim_{w \to 0} D_3 D_2 \mathbf{p}(1 - w, 0, w)$$

holds and analogous identities in $\mathbf{x}_2$ and $\mathbf{x}_3$.

In case the corner twists are incompatible, there exists a solution **p** which is cubic at least over each of the subtriangles of τ shown in Figure 2.2. This is the Clough–Tocher solution, see [8].

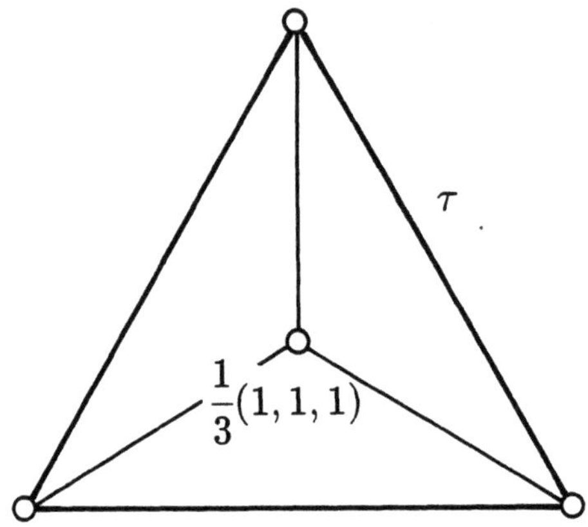

FIGURE 2.2. The Clough–Tocher split of τ.

In case of arbitrary boundary data one has to resort to other constructions, as the ones deviced in [3], [4] and [5]. The solutions described there are rational of relatively high degree.

Here another method derived from [7] is proposed which in some cases is even simpler than the Clough–Tocher solution. Yet, it can also be formulated for arbitrary boundary data. This will be the subject of a future paper. The price to be paid for the simplicity is that the method generates only an approximate solution. However, the approximation can be constructed so as to be arbitrarily close to an exact solution.

Before we present the method in Sections 2.3, 2.4 and 2.5, we recall some facts about the Bernstein–Bézier representation in Section 2.2. Section 2.6 contains some error bounds for the approximate solutions developed in this paper.

2.2 Bézier triangles and subdivision

Every cubic polynomial **b** has a unique *Bézier representation* over τ

$$\mathbf{b}(u,v,w) = \sum_{\mathbf{i}\in\Delta_3} \mathbf{b}_{\mathbf{i}} B_{\mathbf{i}}(u,v,w)$$

where (u,v,w) are the barycentric coordinates with respect to τ,

$$\Delta_m = \{(i,j,k) \in \{0,1,\ldots,m\}^3 \mid i+j+k=m\} \quad ,$$

$$B_{i,j,k}(u,v,w) = \frac{3!}{i!j!k!}u^i v^j w^k \quad ,$$

see, e.g., [1]. The set of the *Bézier* or *control points* $\{\mathbf{b}_{\mathbf{i}} \mid \mathbf{i} \in \Delta_3\}$ is called the *Bézier net* of **b** with respect to τ. Let $\tau_1, \tau_2, \tau_3, \tau_4$ be the four equal triangles decomposing τ as shown in Figure 2.3.

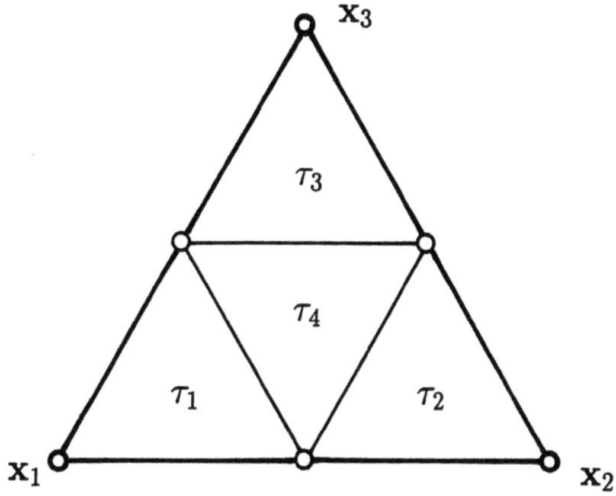

FIGURE 2.3. The uniform decomposition of τ.

Then $\mathbf{b}$ has a "finer" representation with respect to the triangulation $T = \{\tau_1, \tau_2, \tau_3, \tau_4\}$. Namely there are control points $\mathbf{b_i}, \mathbf{i} \in \Delta_6$, such that

$$\mathbf{b}(u,v,w) = \begin{cases} \sum_{\Delta_3} \mathbf{b}_{i+3,j,k} B_{ijk}(2u-1, 2v, 2w) & \text{in } \tau_1 \\ \sum_{\Delta_3} \mathbf{b}_{i,j+3,k} B_{ijk}(2u, 2v-1, 2w) & \text{in } \tau_2 \\ \sum_{\Delta_3} \mathbf{b}_{i,j,k+3} B_{ijk}(2u, 2v, 2w-1) & \text{in } \tau_3 \\ \sum_{\Delta_3} \mathbf{b}_{3-i,3-j,3-k} B_{ijk}(1-2u, 1-2v, 1-2w) & \text{in } \tau_4 \end{cases}$$

Here, the set $\{\mathbf{b_i} \mid \mathbf{i} \in \Delta_6\}$ is called the Bézier net of $\mathbf{b}$ with respect to the triangulation T. There are efficient and numerically stable algorithms, so called *subdivision* or *refinement algorithms*, to compute it [2, 6].

2.3 Initial approximation

From the subdivision technique described in the previous section one can derive an interpolation method which generates a patch $\mathbf{p}$ with a prescribed boundary and cross boundary derivatives, i.e. $\mathbf{p}$ is the interpolant. We will assume here that the three boundary curves of $\mathbf{p}$ are cubics while the boundary curves of $D_i\mathbf{p}$, $i = 1, 2, 3$, are quadratics.

Under these assumptions there exist three unique cubic patches

$$\mathbf{q}_\mu = \sum_{\mathbf{i} \in \Delta_3} \mathbf{b}^\mu_{\mathbf{i}} B_{\mathbf{i}}, \quad \mu = 1, 2, 3,$$

over τ such that $\mathbf{q}_\mu$ interpolates the entire boundary of $\mathbf{p}$ and $D_i\mathbf{q}_\mu$ the boundary of $D_i\mathbf{p}$ over $[\mathbf{x}_i, \mathbf{x}_j]$ for all permutations (i, j, μ) of $(1, 2, 3)$. These conditions imply that

$$\mathbf{b}^1_{\mathbf{i}} = \mathbf{b}^2_{\mathbf{i}} = \mathbf{b}^3_{\mathbf{i}} \quad \text{for all } \mathbf{i} \in \Delta_3 \setminus \{(1,1,1)\},$$

and only if all corner twists are compatible also

$$\mathbf{b}^1_{111} = \mathbf{b}^2_{111} = \mathbf{b}^3_{111}.$$

A simple approximation to the interpolant $\mathbf{p}$ — it will be defined in the sequel — is given by

$$\mathbf{p}_0 = \frac{1}{3}(\mathbf{q}_1 + \mathbf{q}_2 + \mathbf{q}_3).$$

It agrees with $\mathbf{p}$ on the boundary $\partial\tau$ of τ and approximates the cross boundary derivatives of $\mathbf{p}$. Some straight forward calculations show that for any arbitrary norm $\| \; \|$

$$\max_{\partial\tau} \| D_i\mathbf{p}_0 - D_i\mathbf{p} \| \leq \frac{3}{2}K \leq L$$

where $K = \max_{\mu=1}^{3} \| \mathbf{b}_{111}^{\mu} - \frac{1}{3}\sum_{\nu=1}^{3} \mathbf{b}_{111}^{\nu} \|$ and $L = \max\{\| \mathbf{b}_{111}^{\mu} - \mathbf{b}_{111}^{\nu} \| \mid \mu, \nu \in \{1,2,3\}\}$.

2.4 Quasi Bézier nets

The initial approximation can be improved. In explaining how the notion of quasi Bézier nets will be useful. Namely we will call a set $\{\mathbf{b}_{\mathbf{i}} \mid \mathbf{i} \in \Delta_3\}$ a *quasi Bézier net* if $\mathbf{b}_{\mathbf{i}}$, $\mathbf{i} \in \Delta_3 \setminus \{(1,1,1)\}$ is some (control) point and $\mathbf{b}_{111}$ an array of three (control) points. Thus the quasi Bézier net $\{\mathbf{b}_{\mathbf{i}} \mid \mathbf{i} \in \Delta_3\}$,

$$\begin{aligned} \mathbf{b}_{111} &= (\mathbf{b}_{111}^{1}, \mathbf{b}_{111}^{2}, \mathbf{b}_{111}^{3}) \\ \mathbf{b}_{\mathbf{i}} &= \mathbf{b}_{\mathbf{i}}^{1} = \mathbf{b}_{\mathbf{i}}^{2} = \mathbf{b}_{\mathbf{i}}^{3}, \quad \text{for } \mathbf{i} \neq (1,1,1), \end{aligned}$$

represents the interpolation problem of the previous section.

The initial approximation $\mathbf{p}_0$ to this interpolation problem will be called the *medium cubic* associated with this net. For later reference we introduce the notation

$$M\{\mathbf{b}_{\mathbf{i}} \mid \mathbf{i} \in \Delta_3\} := \frac{1}{3}\sum_{\mu=1}^{3} \mathbf{b}_{111}^{\mu} B_{111} + \sum_{\Delta_3 \setminus \{(1,1,1)\}} \mathbf{b}_{\mathbf{i}} B_{\mathbf{i}}.$$

Bézier nets can be refined as described in section 2.2. Based on these ideas we define a refinement process for quasi Bézier nets:

Let $\{\mathbf{b}_{\mathbf{i}}^{\mu} \mid \mathbf{i} \in \Delta_6\}$ be the Bézier nets of $\mathbf{q}_{\mu}$ with respect to T. Then the refinement of the quasi Bézier net $\{\mathbf{b}_{\mathbf{i}} \mid \mathbf{i} \in \Delta_3\}$ is the set $\{\mathbf{b}_{\mathbf{i}} \mid \mathbf{i} \in \Delta_6\}$ as given by the marks in Figure 2.4 and the relation

$$\mathbf{b}_{\mathbf{i}} = \begin{cases} \mathbf{b}_{\mathbf{i}}^{\mu} & \text{if } \mathbf{i}/6 \text{ is marked by } \mu \text{ or a circle} \\ 1/3\sum_{\mu=1}^{3} \mathbf{b}_{\mathbf{i}}^{\mu} & \text{if } \mathbf{i}/6 \text{ is marked by } * \\ (\mathbf{a}_{\mathbf{i}}^{1}, \mathbf{b}_{\mathbf{i}}^{2}, \mathbf{b}_{\mathbf{i}}^{3}) & \text{if } \mathbf{i}/6 \text{ is marked by } \square_1 \\ (\mathbf{b}_{\mathbf{i}}^{1}, \mathbf{a}_{\mathbf{i}}^{2}, \mathbf{b}_{\mathbf{i}}^{3}) & \text{if } \mathbf{i}/6 \text{ is marked by } \square_2 \\ (\mathbf{b}_{\mathbf{i}}^{1}, \mathbf{b}_{\mathbf{i}}^{2}, \mathbf{a}_{\mathbf{i}}^{3}) & \text{if } \mathbf{i}/6 \text{ is marked by } \square_3 \end{cases}$$

where

$$\mathbf{a}_{\mathbf{i}}^{k} = (4\mathbf{b}_{\mathbf{i}}^{i} + 4\mathbf{b}_{\mathbf{i}}^{j} - 2\mathbf{b}_{\mathbf{i}}^{k})/3, \quad \{i,j,k\} = \{1,2,3\}.$$

Figure 2.4 shows the point set $\frac{1}{6}\Delta_6 \subset \tau$ with certain marks. Note that $\mathbf{b}_{\mathbf{i}}^{1} = \mathbf{b}_{\mathbf{i}}^{2} = \mathbf{b}_{\mathbf{i}}^{3}$ whenever $\mathbf{i}/6$ is marked by a circle. Figure 2.5 shows the construction of the $\mathbf{a}$'s. Note that the common subscript $\mathbf{i}$ is omitted.

2.5 Improving the initial approximation

The refined quasi Bézier net $\{\mathbf{b}_{\mathbf{i}} \mid \mathbf{i} \in \Delta_6\}$ contains a Bézier net representing a cubic over τ_4 and three quasi Bézier nets representing an interpolation problem over τ_1, τ_2, and τ_3. Solving the three interpolation problems will generate three surfaces. Together with the cubic over τ_4 they will form one large C^1 surface over τ which also interpolates the original

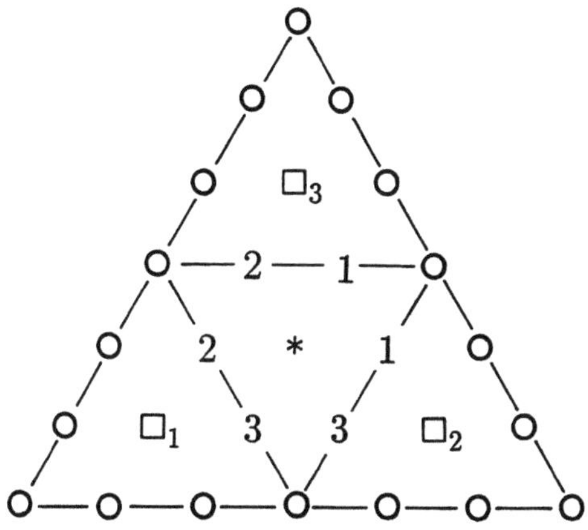

FIGURE 2.4. Refinement of a quasi Bézier net schematically.

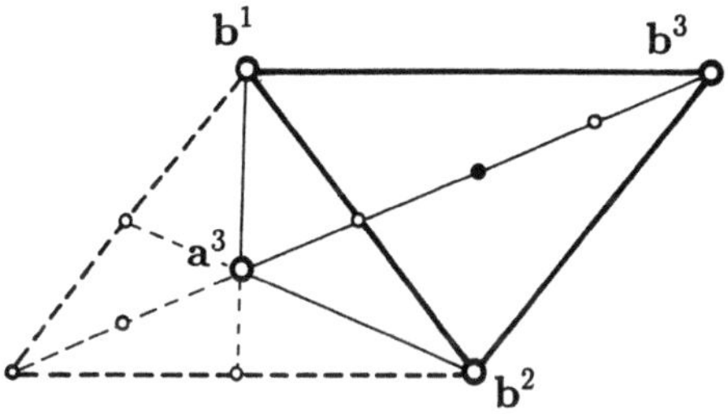

FIGURE 2.5. The construction of the **a** 's.

interpolation problem over τ. All this can be seen using standard results on Bézier nets, see, e.g. [1].

Rather than solving the interpolation problems over τ_1, τ_2, τ_3 exactly one can resort to the medium cubic as in section 2.3. Thus one gets the surface

$$\mathbf{p}_1(u,v,w) = \begin{cases} M\{\mathbf{b}_{i+3,j,k} \mid (i,j,k) \in \Delta_3\}(2u-1,2v,2w) & \text{in } \tau_1 \\ M\{\mathbf{b}_{i,j+3,k} \mid (i,j,k) \in \Delta_3\}(2u,2v-1,2w) & \text{in } \tau_2 \\ M\{\mathbf{b}_{i,j,k+3} \mid (i,j,k) \in \Delta_3\}(2u,2v,2w-1) & \text{in } \tau_3 \\ \sum_{\Delta_3} \mathbf{b}_{3-i,3-j,3-k} B_{ijk}(1-2u,1-2v,1-2w) & \text{in } \tau_4 \end{cases}$$

It solves the original interpolation problem over τ better than $\mathbf{p}_0$ as shown in the next section. Furthermore, the approximation $\mathbf{p}_1$ can be improved over τ_1, τ_2, τ_3 in exactly the same way as $\mathbf{p}_0$ was improved over τ. This yields a surface $\mathbf{p}_2$ which is cubic over each of the subtriangles of τ shown in Figure 2.6.

One has that $\mathbf{p}_2 = \mathbf{p}_1$ over τ_4, and $\mathbf{p}_2$ can be improved over the white triangles in Figure 2.6 in exactly the same way as before. The resulting surface $\mathbf{p}_3$ equals $\mathbf{p}_2$ over all black triangles shown in Figure 2.6. Thus one can construct an infinite sequence of surfaces $\mathbf{p}_0, \mathbf{p}_1, \mathbf{p}_2, \ldots$. The next section will show that this sequence converges to a C^1 surface $\mathbf{p}$ which solves the original interpolation problem exactly.

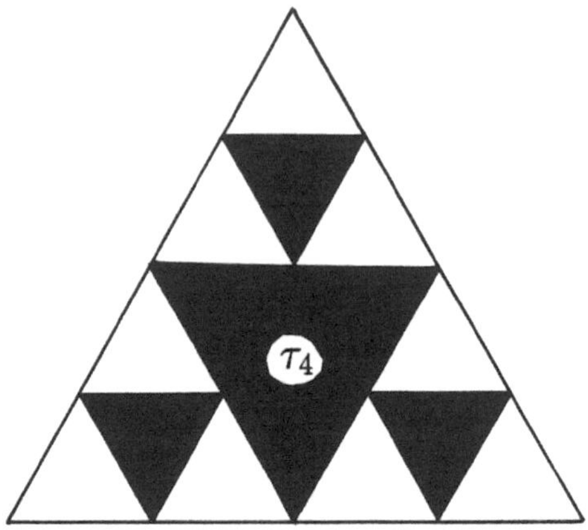

FIGURE 2.6. The parameter domain of $\mathbf{p}_2$.

2.6 Error estimation

Let $\mathbf{b}_{\mathbf{i}}^{\mu}$ denote a Bézier point of $\mathbf{q}_{\mu}$ with respect to τ or respectively T as before. From the subdivision algorithms [2, 6] one can infer that for all $\mathbf{i} \in \Delta_6$ and all $\mu, \nu \in \{1,2,3\}$

$$\mathbf{b}_{\mathbf{i}}^{\mu} - \mathbf{b}_{\mathbf{i}}^{\nu} = \begin{cases} (\mathbf{b}_{111}^{\mu} - \mathbf{b}_{111}^{\nu})/4 & \mathbf{i} \geq (1,1,1) \\ \mathbf{0} & \text{otherwise} \end{cases} .$$

This implies that the norm of the Bézier points of $\mathbf{p}_1 - \mathbf{p}_0$ with respect to each τ_k, $k = 1,2,3$, is bounded by $K/4 \leq L/6$ where

$$\begin{aligned} K &= \max_{\mu=1}^{3} \| \mathbf{b}_{111}^{\mu} - \frac{1}{3}\sum_{\nu=1}^{3} \mathbf{b}_{111}^{\mu} \| \quad \text{and} \\ L &= \max\{\| \mathbf{b}_{111}^{\mu} - \mathbf{b}_{111}^{\nu} \| \mid \mu, \nu \in \{1,2,3\}\} \end{aligned}$$

as before.

Since a polynomial over some triangle lies in the convex hull of its Bézier net with respect to this triangle, one has $\max_\tau \| \mathbf{p}_1 - \mathbf{p}_0 \| \leq K/4$. A better estimate is obtained if the Bézier nets of $\mathbf{p}_0 - \mathbf{p}_1$ are further subdivided. Then one gets

$$\max_{\tau} \| \mathbf{p}_1 - \mathbf{p}_0 \| \leq K/8 \leq L/12.$$

Let $\{\mathbf{c}_{\mathbf{i}} \mid \mathbf{i} \in \Delta_3\}$ be the quasi Bézier net of $\mathbf{p}_1$ with respect to some τ_k, where $\mathbf{c}_{111} = (\mathbf{c}_{111}^1, \mathbf{c}_{111}^2, \mathbf{c}_{111}^3)$. Then

$$\max_{\mu=1}^{3} \| \mathbf{c}_{111}^{\mu} - \frac{1}{3}\sum_{\nu=1}^{3} \mathbf{c}_{111}^{\mu} \| \leq K/4.$$

Thus

$$\max_{\tau} \| \mathbf{p}_2 - \mathbf{p}_1 \| \leq K/32$$

and analogously

$$\max_{\tau} \| \mathbf{p}_{i+1} - \mathbf{p}_i \| \leq \frac{4^{-i}}{8} K,$$

i.e., the sequence $(\mathbf{p}_i)$ converges uniformly to a continuous surface $\mathbf{p}$ over τ where

$$\max_{\tau} \| \mathbf{p} - \mathbf{p}_i \| \leq \frac{4^{-i}}{6} K \leq \frac{4^{-i}}{9} L.$$

It remains to analyze the derivatives. As before the polynomials $D_\mu(\mathbf{p}_{i+1} - \mathbf{p}_i)$ are bounded by the norm of their largest Bézier point, see, e.g., [1] for how to compute the Bézier points of such derivatives. As a result one obtains

$$\max_{\tau} \| D_\mu(\mathbf{p}_{i+1} - \mathbf{p}_i) \| \leq \frac{3}{2^{i+1}} K$$

and

$$\max_{\partial\tau} \| D_\mu(\mathbf{p} - \mathbf{p}_i) \| \leq \frac{3}{2^{i+1}} K \leq \frac{1}{2^i} L.$$

While the derivatives of $\mathbf{p}$ are given on the boundary of τ, it is not clear whether $\mathbf{p}$ is differentiable otherwise. $D_\mu \mathbf{p}_i$ has discontinuities on the boundaries of certain subtriangles of τ. For example, $D_\mu \mathbf{p}_1$ has discontinities only over two edges of τ_4. Let $J_{\mu,i}$ be the norm of the maximum jump of $D_\mu \mathbf{p}_i$. Then it is not difficult to observe that

$$J_{\mu,i} \leq \frac{3}{2^{i+1}} K \leq \frac{L}{2^i}.$$

Hence, standard results of Analysis show that $\mathbf{p}$ is differentiable and that $(D_\mu \mathbf{p}_i)$ converges to $D_\mu \mathbf{p}$.

We also have

$$\max_{\tau} \| D_i(\mathbf{p} - \mathbf{p}_i) \| \leq 3 \cdot 2^{-i} K.$$

Note that the derivatives converge slower than the positional values. The reason is that refinement means a scaling by the factor 2 or 2^i if the refinement procedure is repeated. The scaling factor appears in the argument of the functions and does not matter if positional values are considered. However when computing the derivatives the factor "comes up front".

2.7 References

[1] G. Farin. Triangular Bernstein–Bézier patches. *Computer Aided Geometric Design*, 3:83–127, 1986.

[2] R. Goldman. Subdivision algorithms for Bézier triangles. *Computer Aided Geometric Design*, 15:159–166, 1983.

[3] J.A. Gregory. A C^2 polygonal surface patch. *Computer Aided Geometric Design*, 6:69–75, 1989.

[4] H. Hagen and H. Pottmann. Curvature continuous triangular interpolants. In T. Lyche and L.L. Schumaker, editors, *Mathematical Methods in Computer Aided Geometric Design*, Academic Press, Boston, 1989.

[5] G.M. Nielson. A transfinite, visually continuous triangular interpolant. In G. Farin, editor, *Geometric Modeling: Applications and new trends*, Philadelphia, 1986. SIAM.

[6] H. Prautzsch. *Subdivision algorithms for multivariate splines — a geometric approach.* PhD thesis, Technical Univ. Braunschweig, 1984. in German.

[7] H. Prautzsch. Algorithmic blending. to appear in J. Approx. Theory, 1992.

[8] G. Strang and G. Fix. *An analysis of the finite element method.* Prentice–Hall, Englewood Cliffs, 1973.

3 A Geometric Approach to Bézier Curves

Hans-Peter Seidel

ABSTRACT Using techniques from classical geometry we present a purely geometric approach to Bézier curves and B-splines. The approach is based on the intersection of osculating flats: The osculating 1-flat is simply the tangent line, the osculating 2-flat is the osculating plane, etc. The intersection of osculating flats leads to the so-called *polar form*. We discuss the main properties of the polar form and show how polar forms lead to a simple new labeling scheme for Bézier curves and B-splines.

3.1 Introduction

Polar forms are a classical mathematical tool for the study of polynomials. In the context of Computer Graphics and Computer Aided Design they have first been considered by Paul de Faget de Casteljau [8, 6, 7] at Citroen and by Lyle Ramshaw [11, 10, 12] at DEC. Polar forms have led to considerable simplifications and to some significant new results in the theory of spline curves and surfaces. In particular, they have led to a new multivariate B-spline scheme that allows to construct smooth piecewise polynomial surfaces over arbitrary triangulations of the parameter plane [18, 2, 17].

It is the goal of these notes to present an elementary introduction to polar forms for curves by stressing as much as possible the geometric principles that are underlying the approach. Instead of treating the subject in full generality we have chosen to explicitly discuss both the quadratic case $n = 2$ and the cubic case $n = 3$. It is hoped that this style of presentation makes the underlying ideas more visible than a development in full generality. For additional details on the geometry behind polar forms the reader is referred to [11, 13, 14].

3.2 The quadratic case

3.2.1 Polynomials and polar forms

Consider a quadratic polynomial curve $\mathbf{F} : \mathbb{R} \to \mathbb{R}^2$ in the plane. Let $u, v \in \mathbb{R}$ be arbitrary parameter values, and let $\mathbf{F}(u)$ and $\mathbf{F}(v)$ be the corresponding points on the curve, see Figure 3.1. The *polar form* $\mathbf{f}$ of $\mathbf{F}$ can then be defined geometrically as follows: For $u \neq v$ the point $\mathbf{f}(u, v)$ is the intersection of the tangent line $T_1\mathbf{F}(u)$ to $\mathbf{F}$ at u with the tangent line $T_1\mathbf{F}(v)$ of $\mathbf{F}$ at v, while for $u = v$ the point $\mathbf{f}(u, v) = \mathbf{f}(u, u) = \mathbf{F}(u)$ is simply a point on the curve, i.e., we get

$$\mathbf{f}_{geom} : \mathbb{R}^2 \to \mathbb{R}^2 : (u, v) \mapsto \begin{cases} \mathbf{F}(u) & \text{if } u = v \\ T_1\mathbf{F}(u) \cap T_1\mathbf{F}(v) & \text{if } u \neq v. \end{cases} \tag{3.2.1}$$

What are the properties of the polar form $\mathbf{f}$, and why are polar forms useful?

In order to discuss these questions we would first like to derive an explicit formula for $\mathbf{f}$. We therefore consider the map

$$\mathbf{f}_{pol}(u, v) = 2\,\mathbf{F}(\frac{u+v}{2}) - \frac{1}{2}\,\mathbf{F}(u) - \frac{1}{2}\,\mathbf{F}(v). \tag{3.2.2}$$

Obviously $\mathbf{f}_{pol}$ is symmetric and $\mathbf{f}_{pol}(u, u) = \mathbf{F}(u)$. Suppose we fix the first argument u.

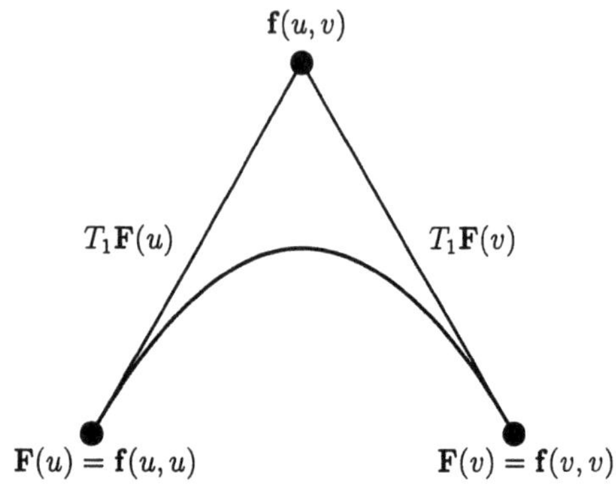

FIGURE 3.1. The polar form of a quadratic polynomial.

How does $\mathbf{f}_{pol}(u,v)$ depend on the second argument v? Setting

$$\mathbf{L}_u(v) = \mathbf{f}_{pol}(u,v),$$

and observing that the first derivative $\mathbf{F}'$ of $\mathbf{F}$ is a degree one polynomial, differentiation of $\mathbf{L}_u(v)$ with respect to v then yields

$$\begin{aligned} \mathbf{L}'_u(v) &= \mathbf{F}'(\frac{u+v}{2}) - \frac{1}{2}\,\mathbf{F}'(v) \\ &= \frac{1}{2}\,\mathbf{F}'(u) + \frac{1}{2}\,\mathbf{F}'(v) - \frac{1}{2}\,\mathbf{F}'(v) \\ &= \frac{1}{2}\,\mathbf{F}'(u), \end{aligned}$$

and it follows that

$$\mathbf{f}_{pol}(u,v) = L_u(v) = \mathbf{F}(u) + \frac{v-u}{2}\,\mathbf{F}'(u). \tag{3.2.3}$$

In particular $\mathbf{f}_{pol}(u,v)$ lies on the tangent line $T_1\mathbf{F}(u)$, and therefore

$$\mathbf{f} = \mathbf{f}_{geom} = \mathbf{f}_{pol}. \tag{3.2.4}$$

Thus we now have both a geometric and an algebraic definition of $\mathbf{f}$.

Furthermore, the above argument shows that the point $\mathbf{f}(u,v)$ moves on the tangent line $T_1\mathbf{F}(u)$ with uniform speed $\mathbf{v} = \mathbf{F}'(u)/2$. Note that 'moving with uniform speed' is equivalent to the condition

$$\mathbf{f}(u, \alpha v_1 + (1-\alpha)v_2) = \alpha\,\mathbf{f}(u,v_1) + (1-\alpha)\,\mathbf{f}(u,v_2). \tag{3.2.5}$$

This means that $\mathbf{f}(u,v)$ is an *affine* function in v. The above results may then be phrased as follows:

- The polar form $\mathbf{f}$ is *symmetric.*
- The polar form $\mathbf{f}$ is *multiaffine*, i.e., $\mathbf{f}$ is affine in each variable if the other is fixed.
- $\mathbf{F}$ is the *diagonal* of $\mathbf{f}$, i.e., $\mathbf{F}(u) = \mathbf{f}(u,u)$.

We also note that the above three properties actually characterize the polar form $\mathbf{f}$: Let $\mathbf{g} : \mathbb{R}^2 \to \mathbb{R}^2$ be any map that satisfies these three conditions. Writing

$$\begin{aligned}
\mathbf{F}(\frac{u+v}{2}) &= \mathbf{g}(\frac{u+v}{2}, \frac{u+v}{2}) \\
&= \frac{1}{4}\,\mathbf{g}(u,u) + \frac{1}{4}\,\mathbf{g}(u,v) + \frac{1}{4}\,\mathbf{g}(v,u) + \frac{1}{4}\mathbf{g}(v,v) \\
&= \frac{1}{4}\,\mathbf{F}(u) + \frac{1}{2}\,\mathbf{g}(u,v) + \frac{1}{4}\,\mathbf{F}(v)
\end{aligned}$$

we see that

$$\mathbf{g}(u,v) = 2\,\mathbf{F}(\frac{u+v}{2}) - \frac{1}{2}\,\mathbf{F}(u) - \frac{1}{2}\,\mathbf{F}(v) = \mathbf{f}(u,v)\,,$$

i.e., $\mathbf{g} = \mathbf{f}$. We summarize our results as follows [7, 11]:

Theorem 3.2.1 (Blossoming Principle) *Given a quadratic polynomial* $\mathbf{F} : \mathbb{R} \to \mathbb{R}^2$ *there exists a unique symmetric multiaffine map* $\mathbf{f} : \mathbb{R}^2 \to \mathbb{R}^2$ *satisfying*

$$\mathbf{f}(u,u) = \mathbf{F}(u). \tag{3.2.6}$$

In this situation $\mathbf{f}$ *is called the multiaffine* polar form *or* blossom *of* $\mathbf{F}$.

We conclude this section by briefly discussing the case where the polynomial curve $\mathbf{F}$ is given in *monomial form*, i.e.,

$$\mathbf{F}(u) = \mathbf{a}_0 + \mathbf{a}_1 u + \mathbf{a}_2 u^2, \quad \mathbf{a}_i \in \mathbb{R}^2. \tag{3.2.7}$$

We claim that the polar form $\mathbf{f}$ is then given as

$$\mathbf{f}(u,v) = \mathbf{a}_0 + \frac{\mathbf{a}_1}{2}(u+v) + \mathbf{a}_2 uv. \tag{3.2.8}$$

Using Theorem 3.2.1 this immediately follows from the fact that the right side of Equation (3.2.8) is symmetric, affine in each argument, and satisfies $\mathbf{f}(u,u) = \mathbf{F}(u)$.

3.2.2 Bézier Curves

Why are polar forms useful? Let us consider again a quadratic polynomial curve $\mathbf{F}$ in the plane. Suppose we wish to represent $\mathbf{F}$ as a Bézier curve over some given interval $\Delta = [r,s]$. What are the Bézier points? Writing

$$u = \frac{u-r}{s-r}s + \frac{s-u}{s-r}r$$

we obtain

$$\begin{aligned}
\mathbf{F}(u) &= \mathbf{f}(u,u) \\
&= \frac{s-u}{s-r}\,\mathbf{f}(u,r) + \frac{u-r}{s-r}\,\mathbf{f}(u,s) \\
&= \left(\frac{s-u}{s-r}\right)^2 \mathbf{f}(r,r) + 2\left(\frac{u-r}{s-r}\right)\left(\frac{s-u}{s-r}\right)\mathbf{f}(r,s) + \left(\frac{u-r}{s-r}\right)^2 \mathbf{f}(s,s).
\end{aligned}$$

But this means that we have represented $\mathbf{F}$ in Bernstein-Bézier form w.r.t. $\Delta = [r,s]$, and it follows that the Bézier points of $\mathbf{F}$ are given as

$$\mathbf{b}_0 = \mathbf{f}(r,r), \;\; \mathbf{b}_1 = \mathbf{f}(r,s) \;\text{ and }\; \mathbf{b}_2 = \mathbf{f}(s,s). \tag{3.2.9}$$

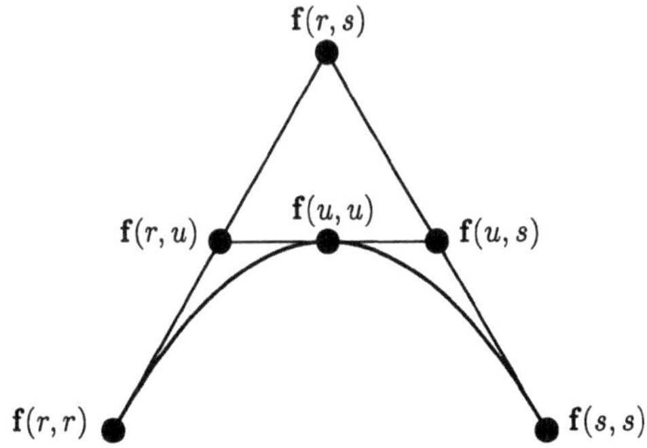

FIGURE 3.2. The de Casteljau Algorithm for a quadratic Bézier curve.

It is then obvious that a quadratic Bézier curve lies in the convex hull of its control points, goes through the end points $\mathbf{b}_0$ and $\mathbf{b}_2$, and is tangent to the Bézier polygon in these points.

How can we evaluate Bézier curves? Consider a quadratic Bézier curve $\mathbf{F}$ over some given interval $\Delta = [r, s]$, and let $u \in \mathbb{R}$. A look at Figure 3.2 convinces us that we can compute the point $\mathbf{F}(u) = \mathbf{f}(u, u)$ from the points $\mathbf{f}(r, u)$ and $\mathbf{f}(u, s)$, which in turn can be computed from $\mathbf{f}(r, r)$ and $\mathbf{f}(r, s)$, and from $\mathbf{f}(r, s)$ and $\mathbf{f}(s, s)$, respectively. It is also clear *how* to compute these points: Writing

$$u = \frac{u-r}{s-r}s + \frac{s-u}{s-r}r \ ,$$

and using the fact that the polar form $\mathbf{f}$ is symmetric and affine in each argument we obtain

$$\mathbf{f}(u, u) = \frac{u-r}{s-r}\,\mathbf{f}(u, s) + \frac{s-u}{s-r}\,\mathbf{f}(r, u).$$

as well as

$$\mathbf{f}(r, u) = \frac{u-r}{s-r}\,\mathbf{f}(r, s) + \frac{s-u}{s-r}\,\mathbf{f}(r, r)$$

and

$$\mathbf{f}(u, s) = \frac{u-r}{s-r}\,\mathbf{f}(s, s) + \frac{s-u}{s-r}\,\mathbf{f}(r, s).$$

We thus obtain the triangular scheme of computation as shown in Figure 3.3. This algorithm was first studied by Paul de Faget de Casteljau [8, 6, 7] and is therefore called *de Casteljau Algorithm*.

Note that the de Casteljau Algorithm offers much more than just evaluation: Suppose that we wish to subdivide a Bézier curve $\mathbf{F}$ over a given interval $\Delta = [s, t]$ at an arbitrary parameter $u \in [r, s]$. What are the new Bézier points of the left and right segments $\mathbf{F}_l$ and $\mathbf{F}_r$ with respect to the subintervals $\Delta_l = [r, u]$ and $\Delta_r = [u, s]$? Equation (3.2.9) tells us that the new Bézier points after subdivision are given as

$$\mathbf{b}_0^l = \mathbf{f}(r, r), \quad \mathbf{b}_1^l = \mathbf{f}(r, u), \quad \mathbf{b}_2^l = \mathbf{f}(u, u) \tag{3.2.10}$$

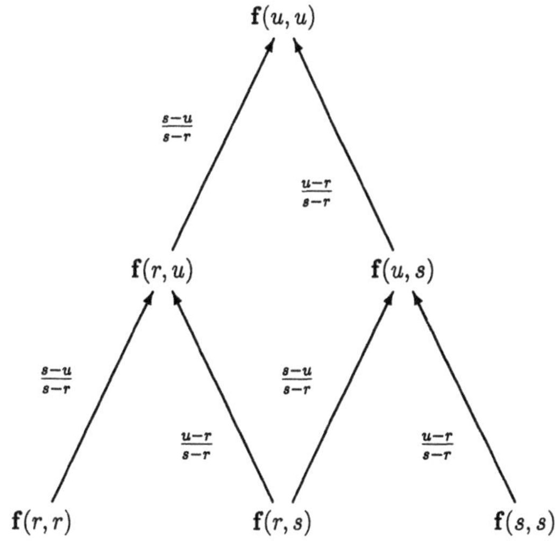

FIGURE 3.3. The de Casteljau Algorithm for a quadratic Bézier curve.

and

$$\mathbf{b}_0^r = \mathbf{f}(u,u), \quad \mathbf{b}_1^r = \mathbf{f}(u,s), \quad \mathbf{b}_2^r = \mathbf{f}(s,s). \tag{3.2.11}$$

Inspection of Figure 3.3 shows that these points are automatically computed during the de Casteljau Algorithm and are stored along the left and right diagonals.

We conclude this section with a brief remark on affine invariance. Let $\tilde{\mathbf{F}} = \phi \circ \mathbf{F}$ be the image of $\mathbf{F}$ under an affine map (e.g., translation, scaling, rotation) ϕ. Checking the conditions of Theorem 3.2.1, we see that the polar form $\tilde{\mathbf{f}}$ of $\tilde{\mathbf{F}}$ is given as $\tilde{\mathbf{f}} = \phi \circ \mathbf{f}$. It follows that the Bézier points $\tilde{\mathbf{b}}_i$ of $\tilde{\mathbf{F}}$ satisfy

$$\tilde{\mathbf{b}}_0 = \tilde{\mathbf{f}}(r,r) = \phi(\mathbf{b}_0), \quad \tilde{\mathbf{b}}_1 = \tilde{\mathbf{f}}(r,s) = \phi(\mathbf{b}_1), \quad \tilde{\mathbf{b}}_2 = \tilde{\mathbf{f}}(s,s) = \phi(\mathbf{b}_2). \tag{3.2.12}$$

This means that the relationship between the curve $\mathbf{F}$ and its Bézier control polygon is invariant under affine maps.

3.2.3 Continuity conditions between adjacent polynomial segments

So far we have only discussed a single polynomial piece. How can we stich several pieces together in a smooth fashion? Let $\mathbf{F} : [r,s] \to \mathbb{R}^2$ and $\mathbf{G} : [s,t] \to \mathbb{R}^2$ be two quadratic polynomial curves in the plane. Using our results about Bézier points we see that

- $\mathbf{F}$ and $\mathbf{G}$ are continuous at s iff $\mathbf{f}(s,s) = \mathbf{g}(s,s)$.

What are the conditions for C^1-continuity? Equation 3.2.3 shows that

$$\mathbf{F}'(s) = 2\,\frac{\mathbf{f}(s,u) - \mathbf{f}(s,s)}{u - s} \quad \text{for all } u \neq s,$$

and similarly for $\mathbf{G}$. Therefore

- $\mathbf{F}$ and $\mathbf{G}$ are C^1-continuous at s iff $\mathbf{f}(s,u) = \mathbf{g}(s,u)$ for all $u \in \mathbb{R}$.

How can we apply the above smoothness conditions to Bézier curves? Let $\mathbf{F}$ and $\mathbf{G}$ be two quadratic Bézier curves over $\Delta_l = [r,s]$ and $\Delta_r = [s,t]$ with Bézier points $\mathbf{b}_i$ and $\mathbf{c}_i$, respectively. Using Equation (3.2.9) we see that

- $\mathbf{F}$ and $\mathbf{G}$ are C^0-continuous at s iff $\mathbf{c}_0 = \mathbf{b}_2$.

Furthermore, if $\mathbf{g}(s,s) = \mathbf{f}(s,s)$, then the condition $\mathbf{g}(s,u) = \mathbf{f}(s,u)$ is equivalent to $\mathbf{g}(s,t) = \mathbf{f}(s,t)$, and we see that

- $\mathbf{F}$ and $\mathbf{G}$ are C^1-continuous at s iff $\mathbf{c}_1$, $\mathbf{c}_0 = \mathbf{b}_2$, and $\mathbf{b}_1$ are collinear and satisfy the ratios $|\mathbf{c}_1 - \mathbf{c}_0| : |\mathbf{b}_2 - \mathbf{b}_1| = (t-s) : (s-r) = \Delta_r : \Delta_l$.

These are the well-known C^1-conditions for Bézier curves.

3.2.4 B-splines

We have seen above that a quadratic polynomial $\mathbf{F}$ and its polar form $\mathbf{f}$ is completely defined by its Bézier points $\mathbf{b}_0 = \mathbf{f}(r,r)$, $\mathbf{b}_1 = \mathbf{f}(r,s)$, and $\mathbf{b}_2 = \mathbf{f}(s,s)$. Let

$$q \leq r < s \leq t$$

be a non-decreasing sequence of real numbers. We wish to show that $\mathbf{F}$ can equally well be defined by its *de Boor points*

$$\mathbf{d}_0 = \mathbf{f}(q,r), \quad \mathbf{d}_1 = \mathbf{f}(r,s), \quad \text{and } \mathbf{d}_2 = \mathbf{f}(s,t). \tag{3.2.13}$$

Representing u as an affine combination w.r.t $[r,s]$, $[q,s]$, and $[r,t]$, a similar argument as above shows

$$\begin{aligned}\mathbf{F}(u) &= \mathbf{f}(u,u)\\ &= \left(\frac{s-u}{s-r}\right)\left(\frac{s-u}{s-q}\right)\mathbf{f}(q,r)\\ &\quad +\{\left(\frac{s-u}{s-r}\right)\left(\frac{u-q}{s-q}\right) + \left(\frac{u-r}{s-r}\right)\left(\frac{t-u}{t-r}\right)\}\mathbf{f}(r,s)\\ &\quad + \left(\frac{u-r}{s-r}\right)\left(\frac{u-r}{t-r}\right)\mathbf{f}(s,t),\end{aligned}$$

and $\mathbf{F}$ is in fact completely determined by the points $\mathbf{f}(q,r)$, $\mathbf{f}(r,s)$, and $\mathbf{f}(s,t)$. Using the properties of Theorem 3.2.1 it is also clear that we can evaluate the curve at an arbitrary parameter $u \in [r,s]$ by the triangular scheme shown in Figure 3.4. Inspection shows that this algorithm is identical to the evaluation of a quadratic B-spline segment with respect to the knot sequence $...q,r,s,t...$ by the *de Boor Algorithm*. Therefore our curve $\mathbf{F}$ is a B-spline segment with B-spline control points $\mathbf{d}_0 = \mathbf{f}(q,r)$, $\mathbf{d}_1 = \mathbf{f}(r,s)$, and $\mathbf{d}_2 = \mathbf{f}(s,t)$.

Let $\mathbf{G}$ be another curve segment over the adjacent interval $[s,t]$, and assume that $\mathbf{G}$ is also given by its de Boor points

$$\mathbf{e}_0 = \mathbf{g}(r,s), \quad \mathbf{e}_1 = \mathbf{g}(s,t), \quad \text{and } \mathbf{e}_2 = \mathbf{g}(t,x) \tag{3.2.14}$$

with respect to the knot sequence

$$r \leq s < t \leq x.$$

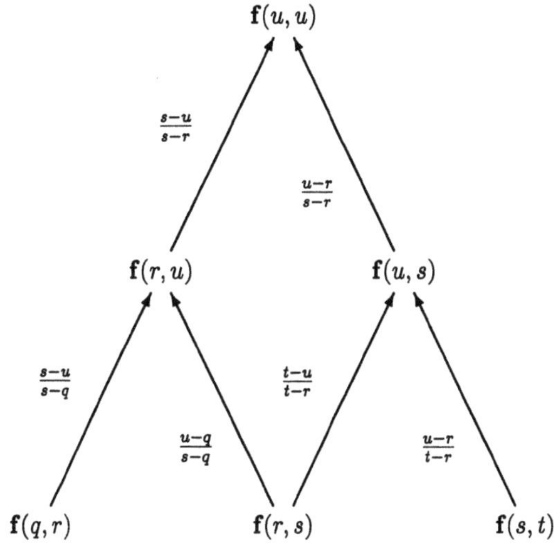

FIGURE 3.4. The de Boor Algorithm for a quadratic B-spline segment.

Furthermore, we also assume that

$$\mathbf{e}_0 = \mathbf{d}_1 \quad \text{and } \mathbf{e}_1 = \mathbf{d}_2. \tag{3.2.15}$$

We claim that $\mathbf{F}$ and $\mathbf{G}$ are automatically C^1-continuous at s under these circumstances: Since

$$\mathbf{f}(r,s) = \mathbf{g}(r,s) \quad \text{and } \mathbf{f}(s,t) = \mathbf{g}(s,t), \tag{3.2.16}$$

hence $\mathbf{f}(s,u) = \mathbf{g}(s,u)$ for all $u \in \mathbb{R}$, this follows immediately from the C^1-conditions in the previous section. The resulting computational scheme for both segments $\mathbf{F}$ and $\mathbf{G}$ is shown in Figure 3.5. Of course, if s is a knot of multiplicity 2, then the order of continuity decreases by one. In this situation $\mathbf{F}$ and $\mathbf{G}$ will still be continuous, but no longer differentiable.

3.3 The cubic case

3.3.1 Polynomials and polar forms

Things work pretty much the same for cubics. Let $F : \mathbb{R} \to \mathbb{R}^3$ be a cubic polynomial curve in $\mathbb{R}^3$. In view of Theorem 3.2.1 we are looking for a unique symmetric 3-affine map $\mathbf{f} : \mathbb{R}^3 \to \mathbb{R}^3$ satisfying $\mathbf{f}(u,u,u) = \mathbf{F}(u)$. Suppose that $\mathbf{f}$ does in fact exist. Writing

$$\begin{aligned}
&\mathbf{F}(\frac{u+v}{2}) \\
&= \mathbf{f}(\frac{u+v}{2}, \frac{u+v}{2}, \frac{u+v}{2}) \\
&= \frac{1}{8}\left(\mathbf{f}(u,u,u) + 3\,\mathbf{f}(u,u,v) + 3\,\mathbf{f}(u,v,v) + \mathbf{f}(v,v,v)\right)
\end{aligned}$$

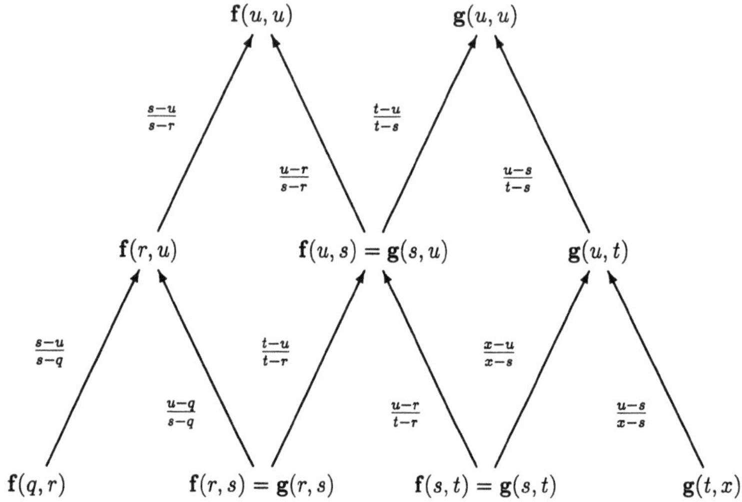

FIGURE 3.5. Overlapping de Boor schemes for two adjacent B-spline segments **F** and **G**.

we see that **f** satisfies

$$3\ (\mathbf{f}(u,u,v)+\mathbf{f}(u,v,v)) = 8\ \mathbf{F}(\frac{u+v}{2}) - \mathbf{F}(u) - \mathbf{F}(v),$$

and using

$$\begin{aligned}&\mathbf{F}(\frac{u+v+w}{3}) = \\ &\quad \frac{1}{27}\ (\mathbf{F}(u)+\mathbf{F}(v)+\mathbf{F}(w)) + \frac{1}{9}\ (\mathbf{f}(u,u,v)+\mathbf{f}(u,u,w) \\ &\quad + \mathbf{f}(u,v,v)+\mathbf{f}(v,v,w)+\mathbf{f}(u,w,w)+\mathbf{f}(v,w,w)) \\ &\quad + \frac{6}{27}\ \mathbf{f}(u,v,w)\end{aligned}$$

we get

$$\begin{aligned}&\mathbf{f}(u,v,w) = \\ &\quad \frac{1}{6}\ \{\mathbf{F}(u)+\mathbf{F}(v)+\mathbf{F}(w) - 8\ (\mathbf{F}(\frac{u+v}{2})+\mathbf{F}(\frac{v+w}{2}) \\ &\quad + \mathbf{F}(\frac{w+u}{2})) + 27\ \mathbf{F}(\frac{u+v+w}{3})\}\end{aligned} \tag{3.3.1}$$

which is the generalization of Formula (3.2.2) from quadratics to cubics. Conversely, a little bit of computation shows that **f** as defined by (3.3.1) satisfies

$$\frac{\partial \mathbf{f}(u,v,w)}{\partial u} = \mathbf{f}'(v,w), \tag{3.3.2}$$

i.e., **f** is *affine* in u. Since **f** is also *symmetric*, and satisfies $\mathbf{f}(u,u,u) = \mathbf{F}(u)$, we obtain the following [7, 11]:

Theorem 3.3.1 (Blossoming Principle) *Given a cubic polynomial* $\mathbf{F} : \mathbb{R} \to \mathbb{R}^3$ *there exists a unique symmetric multiaffine map* $\mathbf{f} : \mathbb{R}^3 \to \mathbb{R}^3$ *satisfying*

$$\mathbf{f}(u, u, u) = \mathbf{F}(u). \tag{3.3.3}$$

In this situation $\mathbf{f}$ *is called the multiaffine* polar form *or* blossom *of* $\mathbf{F}$.

What can we say about the geometry behind polar forms? Since we are now operating in $\mathbf{R}^3$, there isn't much hope to successfully intersect lines. But how about planes instead? High school geometry tells us that three planes in $\mathbf{R}^3$ in general position always intersect in a single point, and that the same is true for a single plane and a line. But what planes should we pick? Let $\mathbf{F}(u)$ be an arbitrary point on the cubic polynomial curve $\mathbf{F} : \mathbf{R}^3 \to \mathbf{R}^3$. The plane that best approximates the curve at the point $\mathbf{F}(u)$ is the *osculating plane*

$$T_2\mathbf{F}(u) = \{\mathbf{F}(u) + \lambda\mathbf{F}'(u) + \mu\mathbf{F}''(u) \mid \lambda, \mu \in \mathbf{R}\}. \tag{3.3.4}$$

If $\mathbf{F}$ is non-planar, its osculating planes are in general position, and we thus obtain the following well-defined map

$$\mathbf{f}_{geom} : \mathbf{R}^3 \to \mathbf{R}^3 : (u, v, w) \mapsto \begin{cases} \mathbf{F}(u) & \text{if } u = v = w \\ T_1\mathbf{F}(u) \cap T_2\mathbf{F}(w) & \text{if } u = v \neq w \\ T_2\mathbf{F}(u) \cap T_2\mathbf{F}(v) \cap T_2\mathbf{F}(w) & \text{if } u \neq v \neq w \neq u. \end{cases} \tag{3.3.5}$$

We claim that

$$\mathbf{f}_{geom} = \mathbf{f}. \tag{3.3.6}$$

Since

$$\mathbf{F}'(u) = 3\,\frac{\mathbf{f}(u, u, v) - \mathbf{f}(u, u, u)}{v - u} \tag{3.3.7}$$

and

$$\mathbf{F}''(u) = 6\,\frac{\mathbf{f}(u, v, v) - 2\,\mathbf{f}(u, u, v) + \mathbf{f}(u, u, , u)}{(v - u)^2}, \tag{3.3.8}$$

both derivative vectors lie in the plane $T = \{\mathbf{f}(u, v, w) \mid v, w \in \mathbf{R}\}$, hence $T_2\mathbf{F}(u) \subseteq T$, and therefore $T_2\mathbf{F}(u) = T$. In particular, $\mathbf{f}(u, v, w) \in T_2\mathbf{F}(u)$. Furthermore, Formula (3.3.7) also shows that $\mathbf{f}(u, u, w)$ lies on the tangent line $T_1\mathbf{F}(u)$ and we see that in fact $\mathbf{f}_{geom} = \mathbf{f}$. Thus we again have both an algebraic and a geometric interpretation of the polar form $\mathbf{f}$. Note, however, that this geometric interpretation breaks down if $\mathbf{F}$ is planar, since all osculating planes agree in this situation.

We conclude this section by again looking specifically at *monomials*. For a monomial curve

$$\mathbf{F}(u) = \mathbf{a}_0 + \mathbf{a}_1 u + \mathbf{a}_2 u^2 + \mathbf{a}_3 u^3, \quad \mathbf{a}_i \in \mathbf{R}^2. \tag{3.3.9}$$

the polar form $\mathbf{f}$ is given as

$$\mathbf{f}(u, v, w) = \mathbf{a}_0 + \frac{\mathbf{a}_1}{3}(u + v + w) + \mathbf{a}_2(uv + vw + wu) + \mathbf{a}_3 uvw. \tag{3.3.10}$$

This is a direct generalization of the case $n = 2$.

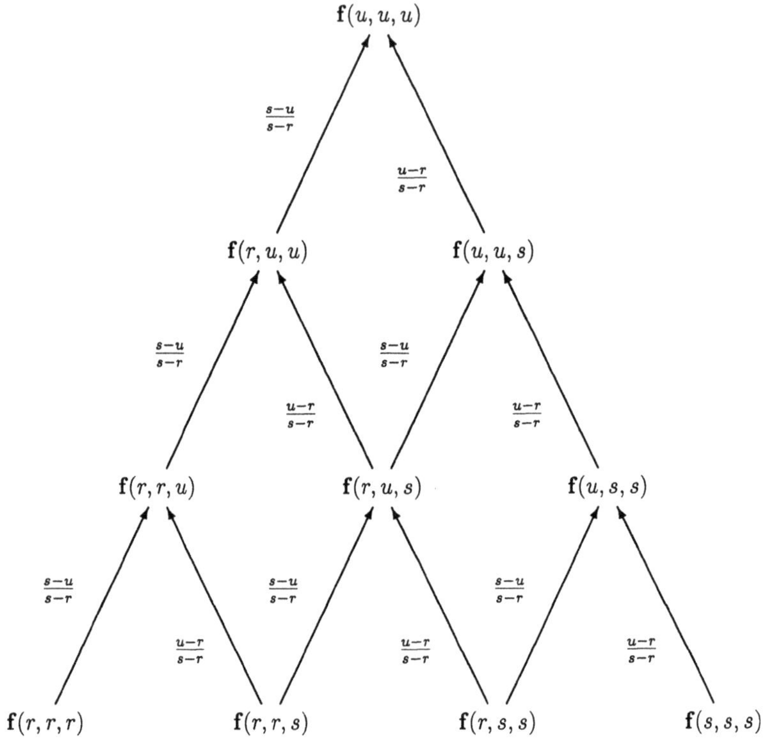

FIGURE 3.6. The de Casteljau Algorithm for a cubic Bézier curve.

3.3.2 Bézier curves, continuity conditions, and B-splines

Once existence and uniqueness of the polar form are established, our treatment of Bézier curves and B-splines from the preceding section can be adopted almost word by word. Again writing

$$u = \frac{u-r}{s-r}s + \frac{s-u}{s-r}r$$

we obtain

$$\begin{aligned} \mathbf{F}(u) &= \mathbf{f}(u,u,u) \\ &= \left(\frac{s-u}{s-r}\right)^3 \mathbf{f}(r,r,r) + 3\left(\frac{u-r}{s-r}\right)\left(\frac{s-u}{s-r}\right)^2 \mathbf{f}(r,r,s) \\ &\quad +3\left(\frac{u-r}{s-r}\right)^2\left(\frac{s-u}{s-r}\right)\mathbf{f}(r,s,s) + \left(\frac{u-r}{s-r}\right)^3 \mathbf{f}(s,s,s). \end{aligned}$$

and therefore

$$\mathbf{b}_0 = \mathbf{f}(r,r,r),\ \mathbf{b}_1 = \mathbf{f}(r,r,s),\ \mathbf{b}_2 = \mathbf{f}(r,s,s),\ \text{ and } \mathbf{b}_3 = \mathbf{f}(s,s,s). \tag{3.3.11}$$

This result then implies all standard properties of Bézier curves, including affine invariance, the convex hull property, end point interpolation and tangency, the de Casteljau

Algorithm, and subdivision [8, 6, 7]. The de Casteljau Algorithm for a cubic Bézier curve is illustrated by Figure 3.6.

It is also easy to phrase the continuity conditions between adjacent polynomial segments in terms of polar forms. Let $\mathbf{F} : [r, s] \to \mathbb{R}^3$ and $\mathbf{G} : [s, t] \to \mathbb{R}^3$ be two cubic polynomial curves in $\mathbb{R}^3$. Using the above results we see that

- $\mathbf{F}$ and $\mathbf{G}$ are continuous at s iff $\mathbf{f}(s, s, s) = \mathbf{g}(s, s, s)$,

and using the derivative formulae (3.3.7), (3.3.8) we see that

- $\mathbf{F}$ and $\mathbf{G}$ are C^1-continuous at s iff $\mathbf{f}(s, s, u) = \mathbf{g}(s, s, u)$ for $u \in \mathbb{R}$,

and that

- $\mathbf{F}$ and $\mathbf{G}$ are C^2-continuous at s iff $\mathbf{f}(s, u, v) = \mathbf{g}(s, u, v)$ for all $u, v \in \mathbb{R}$.

For Bézier curves this immediately translates into the following:

- $\mathbf{F}$ and $\mathbf{G}$ are C^0-continuous at s iff $\mathbf{c}_0 = \mathbf{b}_3$.
- $\mathbf{F}$ and $\mathbf{G}$ are C^1-continuous at s iff $\mathbf{c}_1$, $\mathbf{c}_0 = \mathbf{b}_3$, and $\mathbf{b}_2$ are collinear and satisfy the ratios $|\mathbf{c}_1 - \mathbf{c}_0| : |\mathbf{b}_3 - \mathbf{b}_2| = (t - s) : (s - r) = \Delta_r : \Delta_l$.
- $\mathbf{F}$ and $\mathbf{G}$ are C^2-continuous at s iff the lines $\mathbf{b}_1\mathbf{b}_2$ and $\mathbf{c}_1\mathbf{c}_2$ are coplanar and intersect in a point $\mathbf{d}$, and satisfy the ratios $|\mathbf{d} - \mathbf{b}_2| : |\mathbf{b}_2 - \mathbf{b}_1| = |\mathbf{c}_2 - \mathbf{c}_1| : |\mathbf{c}_1 - \mathbf{d}| = (t - s) : (s - r) = \Delta_r : \Delta_l$.

Finally, all these results can also be applied to B-splines. Let

$$p \leq q \leq r < s \leq t \leq x$$

be a non-decreasing sequence of knots. A similar expansion as in the quadratic case shows that the *de Boor* points of $\mathbf{F}$ w.r.t. the given knot sequence are given as

$$\mathbf{d}_0 = \mathbf{f}(p, q, r), \quad \mathbf{d}_1 = \mathbf{f}(q, r, s), \quad \mathbf{d}_2 = \mathbf{f}(r, s, t) \text{ and } \mathbf{d}_3 = \mathbf{f}(s, t, x). \tag{3.3.12}$$

From this we can immediately recover the de Boor Algorithm (see Fig. 3.7). Finally, inspection shows that two adjacent B-spline segments $\mathbf{F} : [r, s] \to \mathbb{R}^3$ and $\mathbf{G} : [s, t] \to \mathbb{R}^3$ at a simple knot s satisfy

$$\mathbf{f}(s, s, u) = \mathbf{g}(s, s, u) \text{ and } \mathbf{f}(s, u, v) = \mathbf{g}(s, u, v) \tag{3.3.13}$$

for arbitrary $u, v \in \mathbb{R}$. Using the above continuity conditions we thus see that cubic B-splines are in fact C^2-continuous at a simple knot s. Again, this order of continuity will decrease in the presence of knot multiplicities: If s is a knot of multiplicity μ, a cubic B-spline will be $C^{3-\mu}$-continuous at s.

Acknowledgements:

This work has been partly supported by the Natural Sciences and Engineering Research Council of Canada through Personal Operating Grant OGP0105573 and Srategic Operating Grant STR0040527.

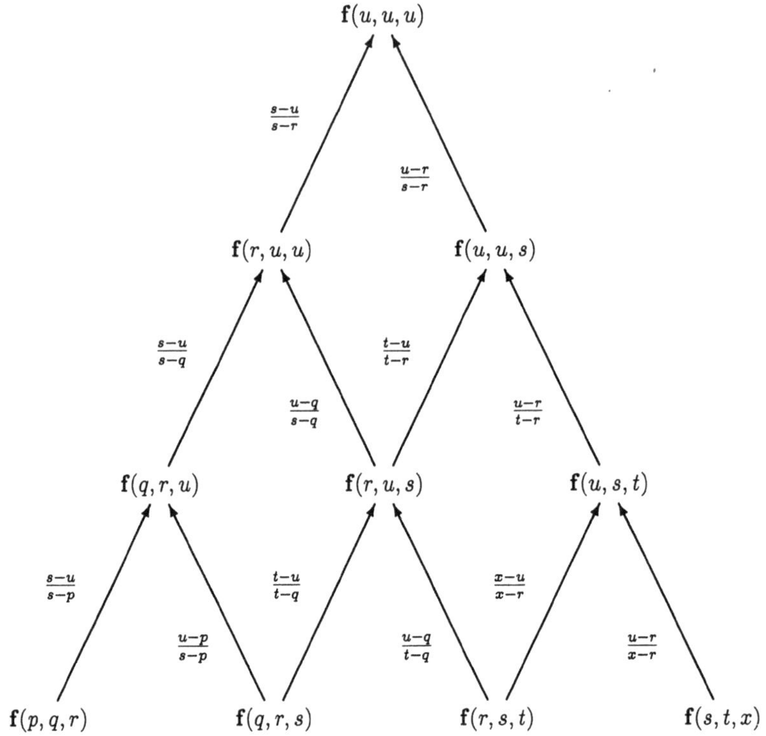

FIGURE 3.7. The de Boor Algorithm for a cubic B-spline segment.

3.4 References

[1] W. Boehm, G. Farin, and J. Kahmann. A survey of curve and surface methods in CAGD. *Computer-Aided Geom. Design*, 1:1–60, 1984.

[2] W. Dahmen, C.A. Micchelli and H.-P. Seidel. *Blossoming Begets B-Splines Built Better by B-Patches.* Research Report RC 16261(#72182), IBM Research Division, Yorktown Heights, 1990. To appear in Math. Comp..

[3] C. de Boor. B-form basics. In G. Farin, editor, *Geometric Modeling, Algorithms and New Trends*, 131–148, SIAM, 1987.

[4] C. de Boor. On calculating with B-splines. *J. Approx. Th.*, 6:50–62, 1972.

[5] C. de Boor. *A Practical Guide to Splines.* Springer, New York, 1978.

[6] P. de Casteljau. *Courbes et surfaces à pôles.* Technical Report, André Citroen, Paris, 1963.

[7] P. de Casteljau. *Formes à Pôles.* Hermes, Paris, 1985.

[8] P. de Casteljau. *Outillages méthodes calcul.* Technical Report, André Citroen, Paris, 1959.

[9] R.N. Goldman. Blossoming and knot insertion algorithms for B-spline curves. *Computer-Aided Geom. Design*, 7:69–81, 1990.

[10] L. Ramshaw. Béziers and B-splines as multiaffine maps. In *Theoretical Foundations of Computer Graphics and CAD*, 757–776, Springer, 1988.

[11] L. Ramshaw. *Blossoming: A connect-the-dots approach to splines.* Technical Report, Digital Systems Research Center, Palo Alto, 1987.

[12] L. Ramshaw. Blossoms are polar forms. *Computer-Aided Geom. Design*, 6:323–358, 1989.

[13] L. Ramshaw. The polar forms of polynomial curves. In *Blossoming: The New Polar-Form Approach to Spline Curves and Surfaces, SIGGRAPH '91 Course Notes #26*, 1.1–1.29, ACM SIGGRAPH, 1991.

[14] H.-P. Seidel. *Geometric Constructions and Knot Inscrtion for Geometrically Continuous Spline Curves of Arbitrary Degree.* Technical Report CS-90-24, Dept. of Computer Science, University of Waterloo, 1990.

[15] H.-P. Seidel. Knot insertion from a blossoming point of view. *Computer-Aided Geom. Design*, 5:81–86, 1988.

[16] H.-P. Seidel. A new multiaffine approach to B-splines. *Computer-Aided Geom. Design*, 6:23–32, 1989.

[17] H.-P. Seidel. Polar forms and triangular B-Spline surfaces. In *Blossoming: The New Polar-Form Approach to Spline Curves and Surfaces, SIGGRAPH '91 Course Notes #26*, 8.1–8.52, ACM SIGGRAPH, 1991.

[18] H.-P. Seidel. Symmetric recursive algorithms for surfaces: B-patches and the de Boor algorithm for polynomials over triangles. *Constr. Approx.*, 7:257–279, 1991.

4 Representing and Modeling of Cyclide Patches using NURBS

Xiaolin Zhou

ABSTRACT
In this paper we introduce a method to represent cyclide patches using NURBS. This representation enables us to implement cyclide patches in a NURBS–based interactive environment and to enhance the flexibility of modeling with cyclide patches.

4.1 Introduction

Cyclides have attracted the interest of researchers in recent years. They have been proposed for the purpose of piecewise surface design [12] and for the variable radius blending [3, 14]. The geometric properties of cyclides have been reviewed in [2, 4, 14].

In this paper, we discuss the representation and modeling of cyclide patches using NURBS. NURBS possess a number of superior properties compared to other traditional representations [21, 20, 13]. We try to obtain a implementation of cyclide patches as a basis for NURBS–based modelers and to enhance the flexibility of modeling with cyclides.

This paper is organized as follows: Section 2 introduces the construction and properties of cyclide patches. In Section 3, we give a rough idea of the blossoming principle, which is used in Section 4 to find the control points of a NURBS representation of cyclide patches. In Section 5 we discuss some useful properties of cyclide patches and the possibility of enhancing the flexibility of cyclide patches by using their NURBS representations. In the last section, we summarize the results and indicate directions for future researchs.

4.2 The construction and properties of cyclide patches

We start from the principle patches introduced by Martin [11]. We believe that this construction is more suitable for designers because it allows them to have more control over the patches. The material of this section is based on [11, 6, 19]; the reason for including it here is to make this article self–contained.

Remembering from differential geometry that the intersection of a surface with a plane containing the surface normal at a point is a plane curve, the curvature of this curve at this point is the normal curvature of the surface in the direction of the plane. At any point on a surface there is infinite number of such planes, each of which will give a different value for the normal curvature. The normal curvature takes on minimum and maximum values for a pair of directions which are orthogonal. These are called the principal directions at this point. The normal curvatures in these directions are called the principal curvatures. On any surface we can draw curves whose tangents are always in one of the principal directions. Such curves are called lines of curvature. There are two families of thes curves and they intersect orthogonally (except for umbilical points, where they need to be treated specially). A four–sided patch whose boundary curves are lines of curvature is called a principal patch. If all the lines of curvature on a principal patch are circular arcs, this patch is called a cyclide patch. This type of surfaces was originally discovered by Dupin [8]. Martin et al have investigated the properties of a cyclide patch and constructed it whilst considering its four boundary curves.

One important property of a cyclide patch is that its four corners lie on a single circle.

This means that if we take two adjacent sides of the patch as being given, the fourth corner of the patch is constrained to move around on the circle defined by the three given corners, and there is only one degree of freedom (see Figure 4.1). Martin [11] has shown that 7 parameters are needed to define a cyclide patch. They are the geodesic curvature g, the normal curvature n and the chord length l of sides 1 and 2 as well as a shape parameter P, where P ranges from 0 to 1 and determines the position of the fourth corner on the circle (see Figure 4.2).

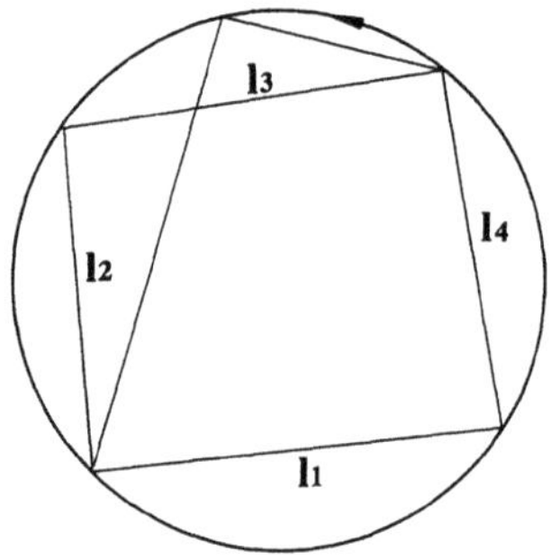

FIGURE 4.1. The four corners of a cyclide patch lie on a circle.

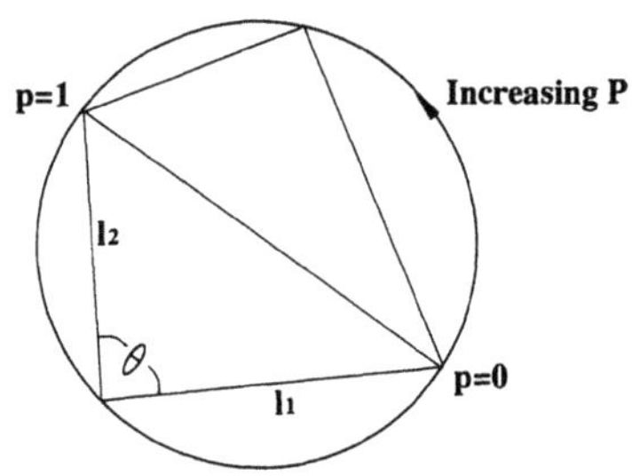

FIGURE 4.2. The parameters P and θ

Considering side 1, we have:

$$\begin{aligned} g_1 &= \kappa_1 cos\phi_1 \\ n_1 &= \kappa_1 sin\phi_1 \\ l_1 &= \frac{2sin\frac{\psi_1}{2}}{\kappa_1} \end{aligned}$$

where κ_1 is the curvature of side 1, ϕ_1 is the angle between the space curve binormal and the surface normal. The same is also valid for side 2.

Let ψ_1 and ψ_2 be the angles between the end tangents of sides 1 and 2 respectively (see Figure 4.3). Taking μ and λ as the fractional arc lengths along sides 1 and 2 respectively, de Pont [6] has defined two parameters u and v:

$$u = \frac{tan\mu\frac{\psi_1}{2}}{tan\frac{\psi_1}{2}}$$
$$v = \frac{tan\lambda\frac{\psi_2}{2}}{tan\frac{\psi_2}{2}}$$

It is obvious that $0 \le u \le 1$ and $0 \le v \le 1$.

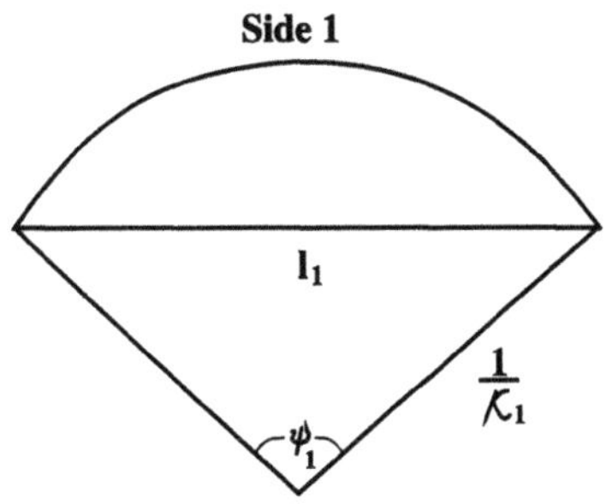

FIGURE 4.3. Some parameters of side 1.

Using u and v as variables, a cyclide patch in the location shown in Figure 4.4 can be described by the following parametric form:

$$\begin{aligned} x(u,v) &= \frac{a_1u(A_2l_1 - B_2l_2) - b_2v^2(A_1l_2 + B_1l_1)}{A_1A_2 + B_1B_2} \\ y(u,v) &= \frac{b_1u^2(A_2l_1 - B_2l_2) + a_2v(A_1l_2 + B_1l_1)}{A_1A_2 + B_1B_2} \\ z(u,v) &= \frac{c_1u^2(A_2l_1 - B_2l_2) + c_2v^2(A_1l_2 + B_1l_1)}{A_1A_2 + B_1B_2} \end{aligned} \tag{4.1}$$

where

$$\begin{aligned} b_i &= cos\phi_i sin\frac{\psi_i}{2} = \frac{g_il_i}{2} \\ c_i &= sin\phi_i sin\frac{\psi_i}{2} = \frac{n_il_i}{2} \\ a_i &= (1 - b_i^2 - c_i^2)^{1/2} \qquad i = 1,2 \end{aligned} \tag{4.2}$$

are the parameters of sides 1 and 2 respectively. A_i, B_i $i = 1, 2$ can be calculated:

$$\begin{aligned} A_1 &= a_1^2 + (b_1^2 + c_1^2)u^2 \\ B_1 &= pu^2 + 2b_2a_1u(1-u) \\ A_2 &= a_2^2 + (b_2^2 + c_2^2)v^2 \\ B_2 &= qv^2 + 2b_1a_2v(1-v) \end{aligned} \tag{4.3}$$

In order to calculate p and q, we define a parameter θ. The geometric meaning of θ can be seen on Figure 4.2 and can be calculated:

$$cos\theta = -a_1b_2 + b_1a_2 + c_1c_2 \tag{4.4}$$

p and q can be obtained:

$$\begin{aligned} p &= \frac{-l_1l_2 + 2P(l_1l2 - l_2^2cos\theta)}{(1-P)l_1^2 + Pl_2^2} \\ q &= -\frac{-l_1l_2 + 2(1-P)(l_1l_2 - l_1^2cos\theta)}{(1-P)l_1^2 + Pl_2^2} \end{aligned} \tag{4.5}$$

Formula (4.1) enables us to construct a cyclide patch by controlling the curvatures of side 1 and side 2. The parameters g_i and n_i ($i = 1, 2$) have clear geometric meanings. Figure 4.4 and Figure 4.5 show two cyclide patches defined be formula (4.1). They have different $g1$ and $n1$. Other parameters of both patches are equal.

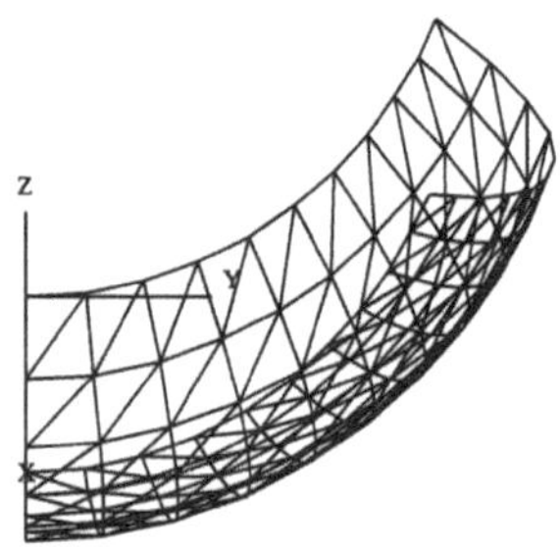

FIGURE 4.4. A cyclide patch with $g1 = 0$

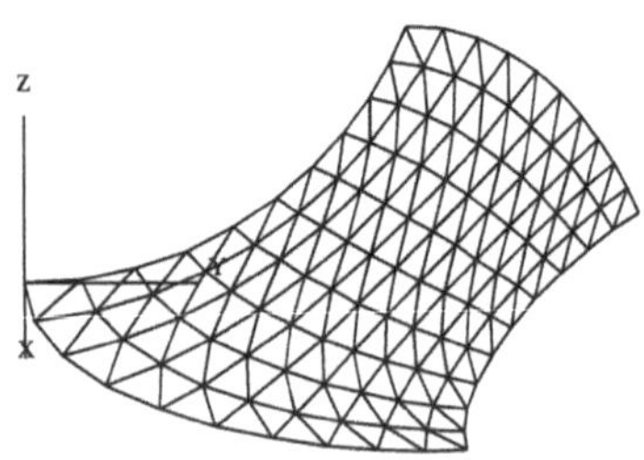

FIGURE 4.5. A cyclide patch with $n1 = 0$

In the following we consider the parametric form given in 4.1, for which we try to find a proper NURBS representation.

4.3 NURBS and the blossoming principle

Since the introduction of polar forms and blossoming in computer graphics and CAD by de Casteljau [5] and Ramshaw [15] and [16], many researchers have used these ideas to prove known results in a more effective way and to achieve new results in the area of interactive curve and surface design [1, 10, 18, 17]. Here we try to exploit its potentials in practical applications.

This new approach is based on the well known fact that polynomials of degree n and symmetric n–affine mappings are equivalent to each other.

Let F be a nth–degree polynomial of variable u. Then $f = f(u_1, ..., u_n)$ is called polar form or blossom of $F(u)$, if the following conditions are fulfilled:

- f is affine in each variable (n-affine).
- f does not change its value under any permutation of its variables (symmetry).
- $f(u, ..., u) = F(u)$

A B–Spline curve is defined as

$$F(u) = \sum_{i=0}^{m} d_i \cdot N_i^n(u) \qquad n \leq m \tag{4.6}$$

where $N(u)$ is the normalized basis function of degree n over the knot vector $T = (t_0, ..., t_{m+n+1}), t_i \leq t_{i+1}$. We know from [18] that the de Boor points (also called control points) d_i of the B-Spline curve can be expressed in terms of blossoming. Let F_j be the restriction of F to the interval $I_j := [t_j, t_j + 1)$ and f_j be the blossom of F_j. The de Boor points d_l are given by the formula [18]

$$d_l = f_j(t_{l+1}, ..., t_{l+n}) \tag{4.7}$$

where $j - n \leq l \leq j$.

The same property is valid for NURBS as well. We consider here only the case of NURBS surfaces.

Let $F(u, v)$ be a NURBS surface which is defined as:

$$F(u, v) = \sum_{i=0}^{m}\sum_{j=0}^{q} R_{i,j}^{n,p} \cdot d_{ij} = \sum_{i=0}^{m}\sum_{j=0}^{q} \frac{N_i^n N_j^p(u, v) \cdot w_{ij} \cdot d_{ij}}{\sum_{i=0}^{m}\sum_{j=0}^{q} N_i^p N_j^n(u, v) w_{ij}} \tag{4.8}$$

where R is the bivariate basis function of degree $(n + p)$ over the knot vectors $T = (t_0, ..., t_{m+n+1})$ in the u–direction and $S = (s_0, ..., s_{q+p+1})$ in the v–direction respectively. w_{ij} are the weights.

Let F_{ij} be the restriction of F on the area of $[t_i, t_{i+1}) \times [s_j, s_{j+1})$ and f_{ij} the blossom of F_{ij}. The de Boor points d_{hr} can be computed by the following formula [22]:

$$d_{hr} = f_{ij}(t_{h+1}, \ldots, t_{h+n}; s_{r+1}, \ldots, s_{r+p}) \tag{4.9}$$

where $i - n \leq h \leq i$, $j - p \leq r \leq j$ with $t_i \leq t_{i+1}, s_j \leq s_{j+1}$.

One can see, using the results above, that the de Boor points of a polynomial F can be computed from its blossom f; this leads to the NURBS representation of F.

4.4 Representing the cyclide patches through NURBS

In this section we show how to obtain the NURBS representation of cyclide patches. The method can be shortly summarized as follows: we begin with the parametric form (4.1) of the cyclide patches and choose the appropriate knot vectors. With the help of the blossom, thė de Boor points will be evaluated and we can finally get the NURBS representation of the cyclide patches.

The corresponding homogeneous representations of (4.1) are given by:

$$\begin{aligned}
x(u,v) &= a_1u(A_2l_1 - B_2l_2) - b_2v^2(A_1l_2 + B_1l_1)\\
y(u,v) &= b_1u^2(A_2l_1 - B_2l_2) + a_2v(A_1l_2 + B_1l_1)\\
z(u,v) &= c_1u^2(A_2l_1 - B_2l_2) + c_2v^2(A_1l_2 + B_1l_1)\\
h(u,v) &= A_1A_2 + B_1B_2
\end{aligned}$$

where $0 \le u \le 1$ and $0 \le v \le 1$

From the homogeneous representations, we can obtain the blossoms directly. The blossom of $x(u,v)$ is given:

$$\begin{aligned}
f_x(u_1,u_2;v_1,v_2) &= \frac{1}{2}a_1(A_2(u_1,u_2;v_1,v_2)l_1 - B_2(u_1,u_2;v_1,v_2)l_2)(u_1+u_2)\\
&\quad -b_2(A_1(u_1,u_2;v_1,v_2)l_2 + B_1(u_1,u_2;v_1,v_2)l_1)v_1v_2
\end{aligned}$$

With the same method we can also get the blossoms of $y(u,v)$, $z(u,v)$ and $h(u,v)$:

$$\begin{aligned}
f_y(u_1,u_2;v_1,v_2) &= b_1(A_2(u_1,u_2;v_1,v_2)l_1 - B_2(u_1,u_2;v_1,v_2)l_2)u_1u_2\\
&\quad +\frac{a_2}{2}(A_1(u_1,u_2;v_1,v_2)l_2 + B_1(u_1,u_2;v_1,v_2)l_1)(v_1+v_2)\\
f_z(u_1,u_2;v_1,v_2) &= c_1(A_2(u_1,u_2;v_1,v_2)l_1 - B_2(u_1,u_2;v_1,v_2)l_2)u_1u_2\\
&\quad +c_2(A_1(u_1,u_2;v_1,v_2)l_2 + B_1(u_1,u_2;v_1,v_2)l_1)v_1v_2\\
f_h(u_1,u_2;v_1,v_2) &= A_1(u_1,u_2;v_1,v_2)A_2(u_1,u_2;v_1,v_2)\\
&\quad +B_1(u_1,u_2;v_1,v_2)B_2(u_1,u_2;v_1,v_2)
\end{aligned}$$

where

$$\begin{aligned}
A_1(u_1,u_2;v_1,v_2) &= a_1^2 + (b_1^2 + c_1^2)(u_1+u_2)\\
B_1(u_1,u_2;v_1,v_2) &= (p - 2b_2a_1)u_1u_2 + b_2a_1(u_1+u_2)\\
A_2(u_1,u_2;v_1,v_2) &= a_2^2 + (b_2^2 + c_2^2)(u_1+u_2)\\
B_2(u_1,u_2;v_1,v_2) &= (q - 2b_1a_2)v_1v_2 + b_1a_2(v_1+v_2)
\end{aligned}$$

Let $d_{ij} = (x_{ij}, y_{ij}, z_{ij}, h_{ij})$ be the de Boor points in homogeneous coordinates. By the use of the simple knot vector $T = (0,0,0,1,1,1)$ in the u–direction and $S = (0,0,0,1,1,1)$ in the v–direction the de Boor points (here identical with the Bézier points) can be computed as follows:

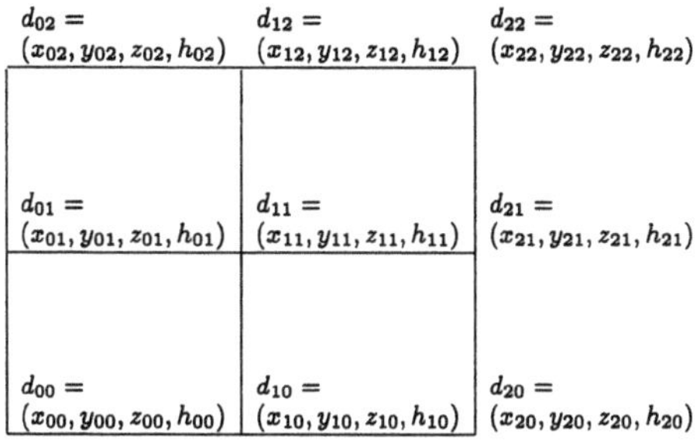

FIGURE 4.6. The control net of the NURBS representation of a cyclide patch

$$
\begin{aligned}
x_{00} &= f_x(0,0;0,0) = 0 \\
x_{10} &= f_x(0,1;0,0) = \frac{1}{2} a_1 a_2^2 l_1 \\
x_{20} &= f_x(1,1;0,0) = a_1 a_2^2 l_1 \\
x_{01} &= f_x(0,0;0,1) = 0 \\
x_{11} &= f_x(0,1;0,1) = \frac{1}{2} a_1 a_2 (a_2 l_1 - b_1 l_2) \\
x_{21} &= f_x(1,1;0,1) = a_1 a_2 (a_2 l_1 - b_1 l_2) \\
x_{02} &= f_x(0,0;1,1) = -a_1^2 b_2 l_2 \\
x_{12} &= f_x(0,1;1,1) = \frac{a_1}{2}(l_1 - q l_2) - a_1 b_2 (a_1 l_2 + b_2 l_1) \\
x_{22} &= f_x(1,1;1,1) = a_1 (l_1 - q l_2) - b_2 (l_2 + p l_1)
\end{aligned}
$$

With blossoming applied analogously to $y(u,v), z(u,v)$ and $h(u,v)$ we have the control net of a cyclide patch shown in Figure 4.6 with the following results:

$$
\begin{aligned}
d_{00} &= (0,\ 0,\ 0,\ a_1^2 a_2^2) \\
d_{10} &= (\frac{1}{2} a_1 a_2^2 l_1,\ 0,\ 0,\ a_1^2 a_2^2) \\
d_{20} &= (a_1 a_2^2 l_1,\ b_1 a_2^2 l_1,\ c_1 a_2^2 l_1,\ a_2^2) \\
d_{01} &= (0,\ \frac{1}{2} a_1^2 a_2 l_2,\ 0,\ a_1^2 a_2^2) \\
d_{11} &= (\frac{1}{2} a_1 a_2 (a_2 l_1 - b_1 l_2),\ \frac{1}{2} a_1 a_2 (a_1 l_2 + b_2),\ 0,\ a_1 a_2 (a_1 a_2 + b_1 b_2)) \\
d_{21} &= (a_1 a_2 (a_2 l_1 - b_1 l_2),\ \frac{a_1}{2}(l_2 + p l_1) + b_1 a_2 (a_2 l_1 - b_1 l_2)), \\
&\quad c_1 a_2 (a_2 l_1 - b_1 l_2),\ a_2 (a_2 + p b_1) \\
d_{02} &= (-a_1^2 b_2 l_2,\ a_1^2 a_2 l_2,\ a_1^2 c_2 l_2,\ a_1^2) \\
d_{12} &= (\frac{a_1}{2}(l_1 - q l_2) - a_1 b_2 (a_1 l_2 + b_2 l_1)),\ a_1 a_2 (a_1 l_2 + b_2), \\
&\quad a_1 c_2 (a_1 l_2 + b_2 l_1),\ a_1 (a_1 + q b_2) \\
d_{22} &= (a_1 (l_1 - q l_2) - b_2 (l_2 + p l_1),\ b_1 (l_1 - q l_2) + a_2 (p l_1 + l_2), \\
&\quad c_1 (l_1 - q l_2) + c_2 (p l_1 + l_2), \\
&\quad 1 + pq) \qquad (4.10)
\end{aligned}
$$

With the knot vectors T and S chosen above as well as the control net, we have obtained a NURBS formulation which represents exactly the cyclide patch. This formulation is actually a rational Bézier patch (see Figure 4.7).

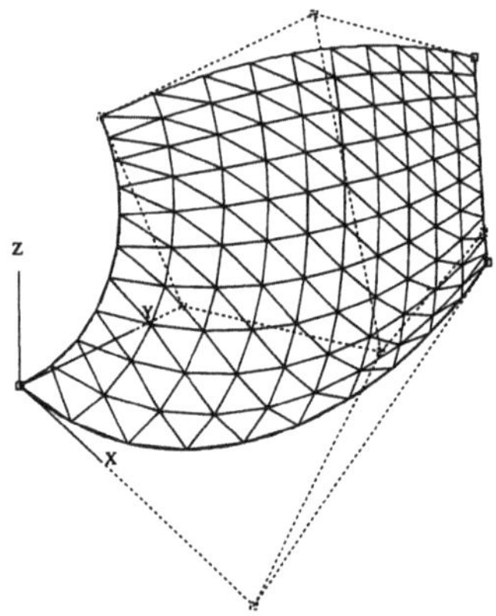

FIGURE 4.7. The NURBS representation of a cyclide patch with its control points.

4.5 Some Practical Considerations

In practice it is desirable to have a global representation of a larger part of surfaces. Quadrics have been used as such a representation. But, in general, they are not flexible enough. The type of cyclide surfaces has more degrees of freedom and includes the natural quadrics (plane, circular cyclinder, cone, sphere) and the torus. With the formulation above NURBS we can get a uniform and compact internal representation of such objects for a NURBS-based modeler. Figure 4.8 to Figure 4.12 show some examples.

An important property of a cyclide patch is that its offset surface is also a cyclide patch. Here we try to get the formulation of the offset surface of the cyclide patch given in (4.1). Let $F_o(u,v)$ be such an offset surface with the distance d to the original patch $F(u,v)$ given in (4.1). We have

$$F_o(u,v) = F(u,v) + d\vec{N} \tag{4.11}$$

where $\vec{N}$ is the unit surface normal of $F(u,v)$. From (4.1) $\vec{N}$ can be calculated:

$$\vec{N}(u,v) = \begin{pmatrix} \frac{2}{h(u,v)}(-a_1u(c_1A_2 - c_2B_2) + b_2v^2(c_2A_1 + c_1B_1) \\ \frac{2}{h(u,v)}(-b_1u^2(c_1A_2 - c_2B_2) - a_2v(c_2A_1 + c_1B_1) \\ 1 + \frac{2}{h(u,v)}(-c_1u^2(c_1A_2 - c_2B_2) - c_2v^2(c_2A_1 + c_1B_1) \end{pmatrix}^T \tag{4.12}$$

Substituting 4.1 and 4.12 into 4.11 we have

$$\begin{aligned}
x_o(u,v) &= \frac{a_1u(A_2(l_1-2dc_1)-B_2(l_2-2dc_2))-b_2v^2(A_1(l_2-2dc_2)+B_1(l_1-2dc_1))}{A_1A_2+B_1B_2} \\
y_o(u,v) &= \frac{b_1u^2(A_2(l_1-2dc_1)-B_2(l_2-2dc_2))+a_2v(A_1(l_2-2dc_2)+B_1(l_1-2dc_1))}{A_1A_2+B_1B_2} \\
z_o(u,v) &= \frac{c_1u^2(A_2(l_1-2dc_1)-B_2(l_2-2dc_2))+c_2v^2(A_1(l_2-2dc_2)+B_1(l_1-2dc_1))}{A_1A_2+B_1B_2}+d
\end{aligned} \tag{4.13}$$

Let $l_{oi} = l_i(1-2dc_i) = l_i(1-dn_i)$, $i = 1, 2$. From (4.13) it can be seen that if we want to calculate an offset surface of a cyclide patch given in (4.1) we only need to replace l_1, l_2 by $l_{o1} = l_1(1-dn_1)$ and $l_{o2} = l_2(1-dn_2)$ in 4.1 respectively, then add d to z. The NURBS representation of the offset surface can be obtained by replacing l_1, l_2 through l_{o1} and l_{o2} in (4.10) as well as adding h_{ij} to z_{ij} $i,j = 0,2$ in (4.10) (see Figure. 4.13).

Another consideration is that the weights at the four corners of a patch h_{00}, h_{20}, h_{02}, h_{22} are dependent on the values of the parameters by which the patch is defined. It is desirable to have four equal weights, otherwise the parameterization could be very non–uniform with respect to the arc length isoparametric lines. Given that the weights at the corners are equal, we can get from h_{00}, h_{20}, h_{02} in (4.10):

$$a_1 = a_2 = 1$$

Together with (4.3) this yields:

$$b_1 = c_1 = b_2 = c_2 = 0$$

This means that the curvature of the side 1 and side 2 must be equal to zero. Also, from (4.4) we have:

$$cos\theta = -a_1b_2 + b_1a_2 + c_1c_2 = 0$$

This means that angle between the chord vectors of the side 1 and side 2 must be $\pi/2$. Further:

$$h_{22} = 1 + (p - a_1b_2)(q - b_1a_2) = 1 + pq$$

Let $h_{22} = 1$, then p or q must be zero. This implies that the four chords l_1, l_2, l_3, l_4 must form a trapezium [12] and because $\theta = \pi/2$, this trapezium is actually a rectangle. The cyclide patch is therefore a rectangular plane (see Figure 4.8). The uniformity of the parameterization can be improved, if needed, by a rational linear transformation as suggested in [9].

It is well known that a NURBS surface can be modeled by evaluating its control points and weights. This is also true for the NURBS representation of cyclide patches. The algorithms for evaluating NURBS can be used directly for cyclide patches. Figure 4.14 shows a cyclide patch whose weights h_{10}, h_{01}, h_{21}, h_{12} are all set to be zero. With the NURBS representation we have more possibilities to model the cyclide patches.

It is possible to get smooth connections between cyclide patches from the NURBS point of view (see Figure 4.15). It seems to be useful if we approximate a surface, for example, a blending surface, by using piecewise cyclide patches.

It is also possible to obtain smooth connections between cyclide patches and other free–form surfaces by controlling the NURBS control points (see Figure 4.16). In fact, this construction could be done automatically in each NURBS–based modeler if the cyclide patches were implemented as NURBS surfaces.

4.6 Conclusion

We have introduced a method to obtain a NURBS representation of cyclide patches and its possible applications for surface modeling and blending. It enables us to integrate the cyclide patches into a NURBS–based modeler conveniently. Another goal of this work is to try to exploit the potentials of the blossoming principle in practical use. In general, we focus on the applications rather than the mathematical foundations. With the same method as given above we have also obtained another NURBS representation of a whole cyclide with different parameterizations [23]. Together with the representation presented in this paper we have the possibility to choose adequate parameterizations of cyclides for different applications based on NURBS. Ongoing work currently includes building a user–friendly interactive environment to deal with cyclides; the investigation of combining the generalized cyclide patches [7] and NURBS surfaces for use in more general situations and, finally, the use of cyclides and cyclide patches as piecewise algebraic surfaces.

Acknowledgements:

The author wishes to express his special thanks to Professor W. Straßer for his direction and valuable advices in doing this work.

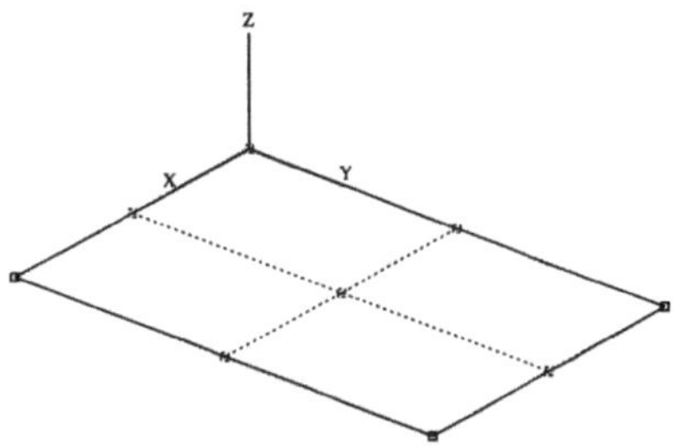

FIGURE 4.8. A rectangular plane as a cyclide patch.

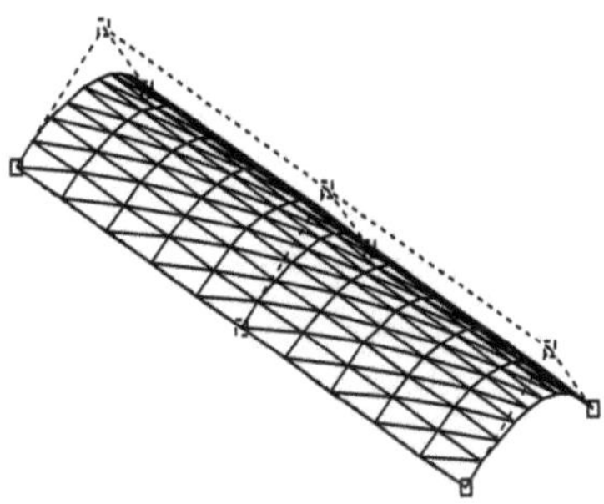

FIGURE 4.9. Part of a circular cylinder as a cyclide patch.

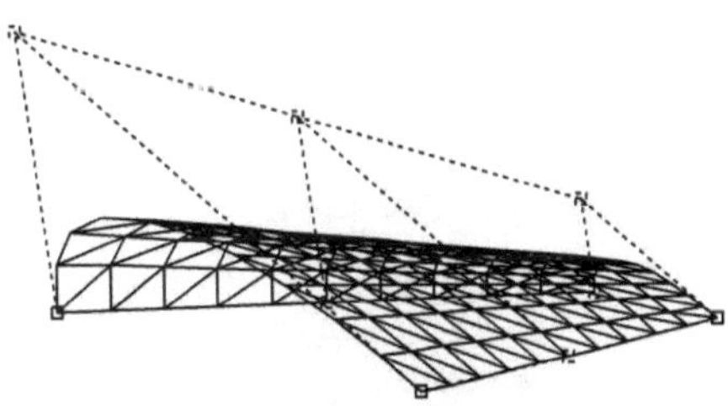

FIGURE 4.10. Part of a cone as a cyclide patch.

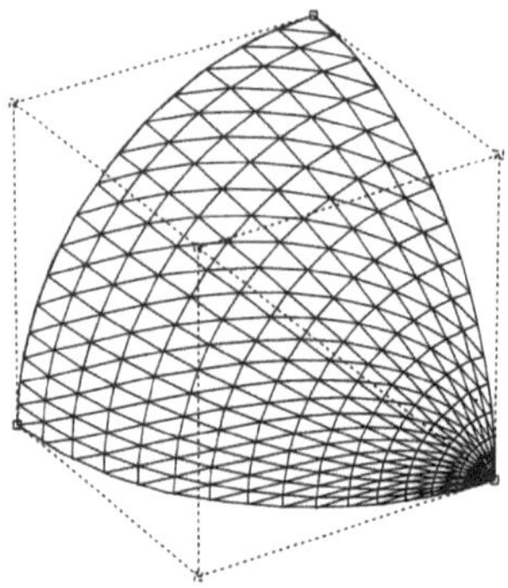

FIGURE 4.11. Part of a sphere as a cyclide patch.

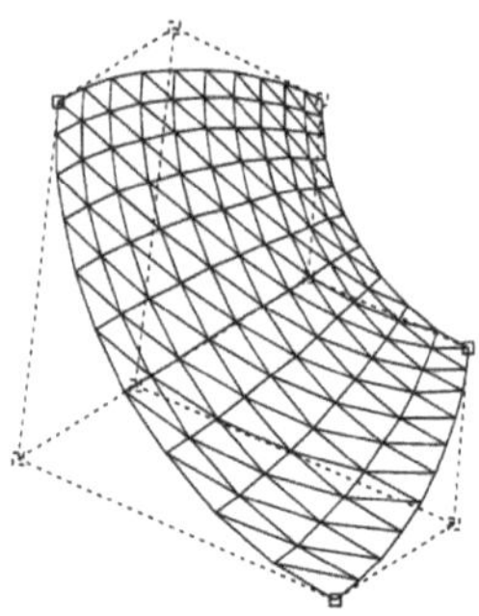

FIGURE 4.12. Part of a torus as a cyclide patch.

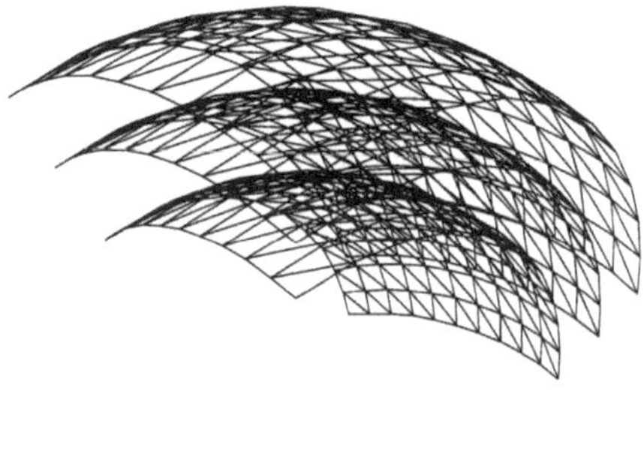

FIGURE 4.13. The offset patches of a cyclide patch.

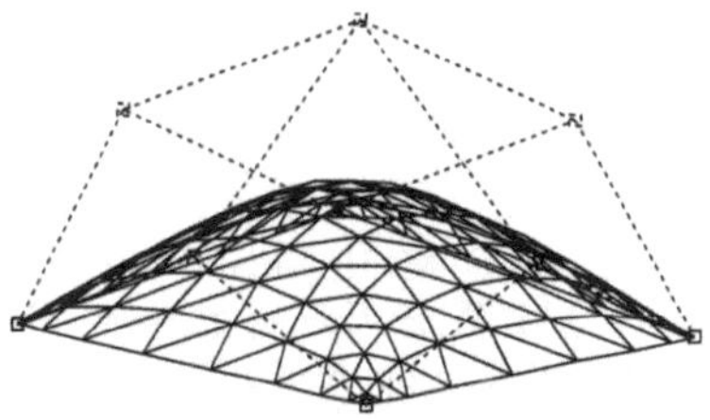

FIGURE 4.14. A cyclide patch for which some weights of its control points are modified.

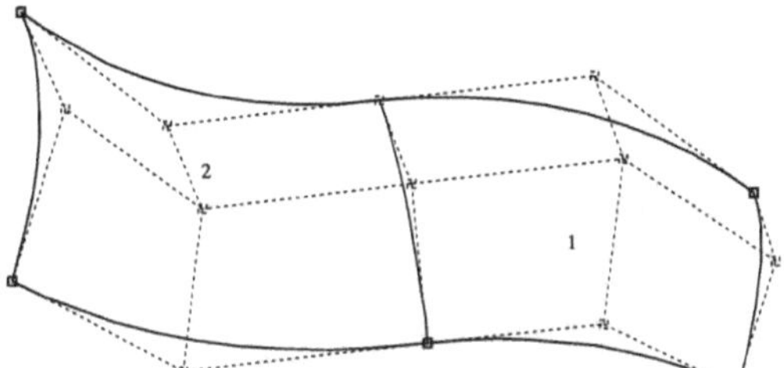

FIGURE 4.15. Two cyclide patches which are connected with C^1 continuity.

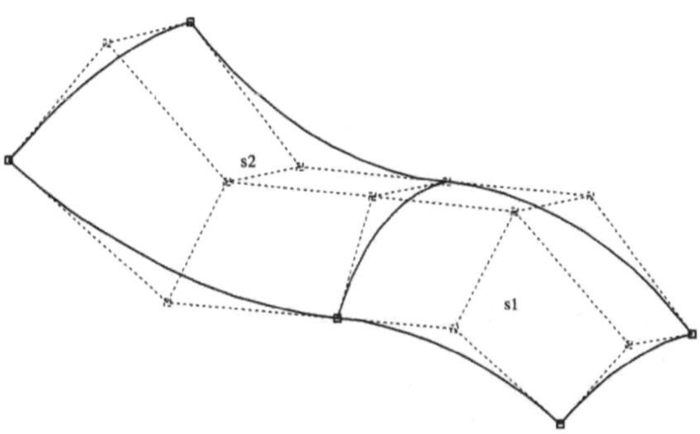

FIGURE 4.16. A cyclide patch and a NUBRS patch which are connected with C^1 continuity.

4.7 References

[1] W. Boehm. On de Boor–like Algorithms and Blossoming. *Computer Aided Geometric Design*, 5:pp. 71–79, 1988.

[2] W. Boehm. On cyclides in geometric modeling,. *Computer Aided Geometric Design*, 7:pp. 243–256, 1990.

[3] V. Chandru, D. Dutta, and C. M. Hoffmann. Variable radius blending using Dupin cyclides. In K. Preiss M. J. Wozny, J. U. Turner, editor, *NSF-IFIP Working Conference on Geometric Modeling*, pp. 39–60. North-Holland, 1988.

[4] V. Chandru, D. Dutta, and C. M. Hoffmann. On the geometry of dupin cyclides. *The Visual Computer*, (5):pp. 277–290, 1989.

[5] P. de Casteljau. *Formes à Pôles*. Hermes, Paris, 1985.

[6] de Pont. *Essays on the Cyclide Patch*. PhD thesis, Cambrige University, Cambrige, UK, 1984.

[7] W. L. F. Degen. Generalized cyclides for use in CAGD. In *the IMA Conference Proceedings Series, to appear*. Oxford Press, 1991.

[8] C. P. Dupin. *Applications de Géométrie et de Méchanique*. Bachelier Paris, 1822.

[9] I. Faux and M. Pratt. *Computational Geometry for Design and Manufacture*. Ellis Horwood, Chichester, 1979.

[10] R.N. Goldman. Blossoming and Knot insertion algorithms for B–Spline curves. *Computer Aided Geometric Design*, 6:pp. 57–64, 1989.

[11] R. R. Martin. *Principle Patches for Computational Geometry*. PhD thesis, Cambrige University, Cambrige, UK, 1982.

[12] W. Nutbourne and R. Martin. *Differential Geometry Applied to Curve and Surface Design*. Ellis Horwood, 1988.

[13] L. Piegl. On Nurbs: A Survey. *IEEE Computer Graphics & Applications*, 1:55–71, January 1991.

[14] M. J. Pratt. Cyclides in CAGD. *CAGD*, 7:221–242, 1990.

[15] L. Ramshaw. *Blossoming: A Connect-The-Dots Approach to Splines*. Digital Systems Research Center, Palo Alto, 1987.

[16] L. Ramshaw. Béziers and B-Splines as Multiaffine Maps. In *Theoretical Foundations of Computer Graphics and CAD*, pp. 757–776. Springer, 1988.

[17] H.-P. Seidel. Computing B-Spline Control Points. In Wolfgang Straßer and H.-P. Seidel, editors, *Theory and Practice of Geometric Modelling*. Springer, 1989.

[18] H.-P. Seidel. A New Approach to B-Splines. *Computer Aided Geometric Design*, 6:pp. 23–32, 1989.

[19] T. J. Sharrock. *Surface Design with Cyclide Patches*. PhD thesis, Cambrige University, Cambridge, UK, 1985.

[20] V. Skytt T. Dokken, A. M. Ytrehus. The Role Of NURBS in Geometric Modeling and CAD/CAM. In *International Symposium on Advanced Geometric Modeling For Engineering Applications, Berlin(West)*, 1989.

[21] W. Tiller. Geometric Modeling using Non-Uniform Rational B-splines: Mathematical Techniques. In *Siggraph 86 Tutorial Presentation*, 1986.

[22] X. Zhou. Parametrisierung der Flächen zweiten Grades und Konvertierung in die NURBS–Tensorproduktdarstellung. Technical Report WSI 89-1, Universität Tübingen, Tübingen, Germany, 1989.

[23] X. Zhou and W. Straßer. A NURBS Representation for Cyclides. In T. L. Kunii, editor, *Modeling in Computer Graphics, Proc. of the IFIP 5.10 Working Conference*, pp. 77–92. Springer-Verlag, 1991.

5 Local Parametrization of Space Curves at Singular Points

Maria E. Alonso
Teo Mora
Gianfranco Niesi
Mario Raimondo

ABSTRACT
We propose a symbolic computation algorithm for computing local parametrization of analytic branches and real analytic branches of a curve in n-dimensional space, which is defined by implicit polynomial equations. The algorithm can be used in space curve tracing near a singular point, as an alternative to symbolic computations based on resolutions of singularities.

5.1 Introduction

We discuss an algorithmical approach to the problem of authomatic parametrization of a curve in n-dimensional space, which is defined by implicit polynomial equations. The focus of our interest is the structure of the curve near singular points.

Let us be given a curve $\Gamma \subset \mathbf{C}^n$ defined by s polynomials $P_1, \ldots, P_s$ in $\mathbf{C}[X_1, \ldots, X_n]$ and let $(\alpha_1, \ldots, \alpha_n) \in \Gamma$ be a singular point of the curve. We assume that Γ is not contained in the hyperplane $X_1 - \alpha_1 = 0$; there is no loss of generality in doing this, since otherwise by the substitution $X_1 = \alpha_1$ in $P_1, \ldots, P_s$ we can reduce to a problem in one variable less.

It is then known that the analytic branches of Γ have a parametrization:

$$X_1 - \alpha_1 = t^{\alpha_1}, X_2 - \alpha_2 = f_2(t), \ldots, X_n - \alpha_n = f_n(t)$$

with $\alpha_1 \in \mathbf{N} - 0, f_2, \ldots, f_n$ non-invertible power series in $\mathbf{C}[[t]]$. Our aim is to explicitly "compute" such parametrizations.

More exactly for each analytic branch we intend to compute integers $a_1, \ldots, a_n$, polynomials $T_2(t), \ldots, T_n(t)$, polynomials $S_2(t, X_2, \ldots, X_n), \ldots, S_n(t, X_2, \ldots, X_n)$ s.t.

1) the Jacobian $(\frac{\delta S_i}{\delta X_j})_{ij}$ is non-zero at the origin.

2) denoting $Q_2(t), \ldots, Q_n(t)$ the unique formal power series s.t.

$$\forall i S_i(t, Q_2(t), \ldots, Q_n(t)) = 0$$

a parametrization of the analytic branch is given by:

$$X_1 - \alpha_1 = t^{a_1}, X_2 - \alpha_2 = T_2(t) + t^{a_2} Q_2(u), \ldots, X_n - \alpha_n = T_n(t) + t^{a_n} Q_n(u)$$

Since the problem is a local problem, we can assume without loss of generality that the singular point is the origin, reducing ourselves to this case by means of a translation.

While the problem is a typical problem in Computer Algebra and we are going to use fairly standard symbolic computation techniques, we understand that the problem could have interesting applications in geometric modeling. There, space curve tracing is needed and numerical techniques fail near singular points; for curve tracing near a singular point, symbolic computations based on resolutions of singularities have been proposed ([9], [5]). The approach we are developing leads to symbolic tracing of approximations of Puiseux expansions for each branch; we hope it could provide a useful alternative for curve tracing near singularities. An approach for plane curves, which is similar to the one we propose here has been given in[6].

Local parametrization of a curve has been studied in a paper by Maurer ([13]); his approach is not completely algorithmical, but it could be made such by recurring to standard basis computations by means of the Tangent Cone Algorithm [14]. The algorithm requires however too many standard basis computations to result in an efficient algorithm. There is a much older (it goes back to 1912) computational approach to the same problem due to Mac Millan [12]; it is algorithmic only under some assumptions of regularity which represent a major restriction to its applicability; whenever these regularity conditions are not satisfied, Mac Millan recours to an impressive and clever set of tricks, which however fail to produce a complete algorithm.

Our proposal, while mainly based on Mac Millan's ideas, makes a complete algorithm out of them, by making use of the Tangent Cone Algorithm and other symbolic computation techniques. For another recent symbolic approach to the same problem see [2]

Our method requires extensive recourse to solving systems of polynomial equations with finitely many solutions and dealing with the arithmetics of algebraic numbers. We discuss a possible approach to this problems based on the most recent state of the art. There is no current implementation of our algorithm, but we will present some promising experimentations performed by means of CoCoA 1.5.3.

5.2 Branches of a Curve at a Singular Point

We begin by recalling some algebraic facts and fixing some notations.

A *Puiseux series* in X over a field K is a formal expression:

$$P(X) = \sum_{i=0}^{\infty} c_i X^{\nu_i/\nu}$$

with $c_i \in K$, $c_0 \neq 0$, $\nu_i \in \mathbf{Z}$, $\nu \in \mathbf{N}$, $\nu_0 < \nu_1 < \ldots$, $\gcd(\nu, \nu_1, \ldots, \nu_i, \ldots) = 1$.

Remark that there are at most finitely many negative exponents ν_i/ν in the Puiseux series $P(X)$, while there can be finitely or infinitely many positive exponents having non-zero coefficients. The value ν_0/ν is called the *order* of P, $\operatorname{ord}(P) = \nu_0/\nu$, while ν is called its *index*, $\operatorname{ind}(P) = \nu$, order and index of the zero series being undefined.

By the usual definition of sum and product, the set of all Puiseux series is turned into an integral domain; what is less obvious, is that for each non-zero Puiseux series $P(X)$, there is a unique Puiseux series $P^{-1}(X)$ such that $P(X)P^{-1}(X) = 1$; for such a series, necessarily, $\operatorname{ord}(P^{-1}) = -\operatorname{ord}(P)$; moreover if the exponents of P are all integers, the same is true for those of P^{-1}. As a consequence the set of all Puiseux series in X over K is a field, which we denote by $K((X))_{\text{Puis}}$. The subset of $K((X))_{\text{Puis}}$ consisting of those series $P(X)$ such that $\operatorname{ord}(P) \geq 0$ (i.e. none among the exponents with non zero coefficient is negative) is closed under sums and products and is therefore an integral domain which we will denote as $K[[X]]_{\text{Puis}}$. Units of $K[[X]]_{\text{Puis}}$ are then the elements of order zero.

The most important property of Puiseux series is given by Puiseux Theorem which states that, if K is algebraically closed, then the field $K((X))_{\text{Puis}}$ is algebraically closed.

Since a polynomial can be canonically identified with a (finite) Puiseux series with non-negative order and integer exponents, so that $K[X] \subset K[[X]]_{\text{Puis}}$, one concludes that $K(X) \subset K((X))_{\text{Puis}}$, so that, by Puiseux Theorem, $\mathbf{K}((X))_{\text{Puis}}$ contains the algebraic closure of $K(X)$, where $\mathbf{K}$ is the algebraic closure of K.

The Puiseux series $P(X) = \sum_{i=0}^{\infty} c_i X^{\nu_i/\nu}$ with index 1, i.e. with integer exponents, are easily seen to be both a subfield of $K((X))_{\text{Puis}}$ and the field of fractions of $K[[X]]$; it will be denoted by $K((X))$.

Let $\Gamma \subset \mathbf{C}^n$ be a curve defined by

$$F_1(X_1, \ldots, X_n) = 0, \ldots, F_s(X_1, \ldots, X_n) = 0$$

where F_i is a polynomial in $K[X_1, \ldots, X_n]$, the field K is some finite extension of $\mathbf{Q}$ and $\mathbf{K} \subset \mathbf{C}$ denotes its algebraic closure; let $I \subset K[X_1, \ldots, X_n]$ be the ideal generated by $F_1, \ldots, F_s$ and let us denote $K[x_1, \ldots, x_n] := K[X_1, \ldots, X_n]/I$ the coordinate ring of Γ. Let us also consider the ideal $I^e \subset K(X_1)[X_2, \ldots, X_n]$ generated by $F_1, \ldots, F_s$ in the larger ring $K(X_1)[X_2, \ldots, X_n]$.

Let us assume that none of the irreducible components of Γ is contained in some hyperplane $X_1 = \alpha$; this implies that x_1 is not algebraic over K. Under this assumption, the ideal I^e (i.e. the system of equations $F_1 = \ldots = F_s = 0$) has only finitely many solutions in the $(n-1)$-dimensional affine space over the algebraical closure of $K(X_1)$, which is contained in $\mathbf{K}((X_1))_{\text{Puis}}$. Then there are finitely many Puiseux series $P_{jk}(X_1) \in \mathbf{K}((X_1))_{\text{Puis}}$, $j = 1 \ldots, r$, $k = 2 \ldots n$ such that:

$$F_i(X_1, P_{j2}(X_1), \ldots, P_{jn}(X_1)) = 0 \quad \text{for all } i, j$$

each of the $(n-1)$-tuples $(P_{j2}(X_1), \ldots, P_{jn}(X_1))$ being a solution of the system. Moreover each $P_{jk}(X_1)$ converges in a neighborhood of $X_1 = 0$. It is important to remark that, for each such solution, there is a finite algebraic extension K_j of K such that $P_{jk}(X_1) \in K_j((X_1))_{\text{Puis}}$.

There are therefore, a fortiori, at most finitely many solutions $(P_{j2}(X_1), \ldots, P_{jn}(X_1))$ such that $\operatorname{ord}(P_{jk}) > 0$ for all k. We will call them *the solutions centered at the origin.*

Let

$$P_2(X_1) := \sum_{i=0}^{\infty} c_{i2} X^{\nu_{i2}/\nu_2}, \ldots, P_n(X_1) := \sum_{i=0}^{\infty} c_{in} X^{\nu_{in}/\nu_n}$$

be one of these solutions; let $\nu := \operatorname{lcm}(\nu_2, \ldots, \nu_n)$, let $\mu_{ij} := \nu_{ij}\nu/\nu_j$ and let

$$R_j(t) := \sum_{i=0}^{\infty} c_{ij} t^{\mu_{ij}} \in \mathbf{K}[[t]] .$$

Then $P_j \in \mathbf{K}[[X_1^{1/\nu}]]$ and it is the image of R_j under the isomorphism between $\mathbf{K}[[X_1^{1/\nu}]]$ and $\mathbf{K}[[t]]$ given by the identification $t = X_1^{1/\nu}$.

There are ν automorphisms of $\mathbf{K}((t))$ which leave fixed $X_1 = t^\nu$ and so $\mathbf{K}(X_1)$; they are given by $t \mapsto \zeta t$, where ζ is a νth-root of unity, $\zeta^\nu = 1$. As a consequence, if

$$Q_2(X_1) := \sum_{i=0}^{\infty} c_{i2} \zeta^{\mu_{i2}} X^{\nu_{i2}/\nu_2}, \ldots, Q_n(X_1) := \sum_{i=0}^{\infty} c_{in} \zeta^{\mu_{in}} X^{\nu_{in}/\nu_n}$$

then $(Q_2, \ldots, Q_n)$ is another solution of the system. In other words, solutions centered at the origin of a system of equations $F_1 = \ldots = F_s = 0$, $F_i \in K(X_1)[X_2, \ldots, X_n]$ can be divided in cycles; each cycle contains, for some ν, exactly ν solutions, related each other as specified above. Since any of these solutions, say $(P_2, \ldots, P_n)$, is such that each P_i is convergent, then there is an open disk $U_\varepsilon := \{t \in \mathbf{C} : |t| < \varepsilon\}$ such that for each $t \in U_\varepsilon$, $(t^\nu, R_2(t), \ldots, R_n(t))$ is a point of Γ.

Clearly if ζ is a primitive ν^{th}-root of unity, then the sets $\{(t^\nu, R_2(\zeta^\lambda t), \ldots, R_n(\zeta^\lambda t)) : t \in U_\varepsilon\}_{\lambda=1,\ldots,\nu}$ are the same. This motivates the following definition.

Definition 1 An *analytic branch* of Γ (at the origin) is a cycle of solutions $(P_2, \ldots, P_n)$ with $\operatorname{ord}(P_i) > 0$ for all i, of the system $F_1 = \ldots = F_s = 0$. The *order* of branch is the length ν of the corresponding cycle, i.e. the least common multiple of the indexes of the P_i's. The n-tuple $(t^\nu, R_2(t), \ldots, R_n(t))$ is called *a parametrization* of the branch.
Assume moreover that for each neighborhood U of zero such that, for all i, the series $R_i(t)$ is convergent for each $t \in U$, one has $\{(t^\nu, R_2(t), \ldots, R_n(t)) : t \in U\} \cap \mathbf{R}^n \neq (0, \ldots, 0)$. Then the corresponding analytic branch is called *real*.

Clearly if a solution $(P_2, \ldots, P_n)$ is such that $P_i \in \mathbf{R}[[X_1]]_{\text{Puis}}$ for all i, then the corresponding branch is real. The converse is however false as it is shown by the following trivial example.

Let $F(X,Y) = X + Y^2 \in \mathbf{Q}[X,Y]$. The equation $F = 0$ has two solutions centered at the origin in $\mathbf{C}[[X]]_{\text{Puis}}$, which are $Y = iX^{1/2}$ and $Y = -iX^{1/2}$; they form a cycle of length 2, being related by the transformation $t \mapsto -t$, where $t = X^{1/2}$ and therefore they give a single branch $\{(t_2, it) : t \in \mathbf{C}\}$. This branch is real as it is easily realized by the transformation $t \mapsto -iu$, which gives a parametrization of the branch as $\{(-u^2, u) : u \in \mathbf{C}\}$, which satisfies $\{(-u^2, u) : u \in \mathbf{C}\} \cap \mathbf{R}^2 = \{(-u^2, u) : u \in \mathbf{R}\}$.

However each real analytic branch has either a parametrization $(t^\nu, R_2(t), \ldots, R_n(t))$ or a parametrization $(-t^\nu, R_2(t), \ldots, R_n(t))$ where $R_i(t) \in \mathbf{R}[[t]]$ for all i.

Recall, finally, that the origin is a simple point for Γ if and only if there is only an analytic branch of Γ at the origin, this branch being of order 1, i.e. if and only if there is a single solution centered at the origin of the system $F_1 = \ldots = F_s = 0$. Otherwise the origin is singular.

5.3 Local Parametrization at the Origin

Let $\Gamma \subset \mathbf{C}^n$ be a curve defined by s polynomials $F_1(X_1, \ldots, X_n), \ldots, F_s(X_1, \ldots, X_n)$ in $K[X_1, \ldots, X_n]$, where K is a finite algebraic extension of the rationals. We do not require the curve to be irreducible, but only that its irreducible components are all simple. Under this assumption the curve has only finitely many singular points.

Let then $(\alpha_1, \ldots, \alpha_n) \in \Gamma$ be a singular point of the curve. We assume that none of the irreducible components of Γ passing through $(\alpha_1, \ldots, \alpha_n)$ is contained in the hyperplane $X_1 - \alpha_1 = 0$. We refer to Appendix A to discuss how these properties can be tested, how to deal in case the last assumption fails, how to compute singular points.

By the translation $\phi : K[X_1, \ldots, X_n] \to K[X_1, \ldots, X_n]$ defined by $\phi(X_i) = X_i - \alpha_i$ we can assume w.l.o.g. the singular point to be the origin. As remarked above, since the origin is singular, there is more than one solution centered at the origin of the system $F_1 = \ldots = F_s = 0$, corresponding to either several branches or to a single branch of order greater than 1.

By back-translating, each such solution $(P_2(X_1), \ldots, P_n(X_1))$ will give rise to the corresponding solution $(\alpha_2 + P_2(X_1 - \alpha_1), \ldots, \alpha_n + P_n(X_1 - \alpha_1))$ "centered at $(\alpha_1, \ldots, \alpha_n)$", each branch will give rise to a branch, and if $(\alpha_1, \ldots, \alpha_n) \in \mathbf{R}^n$ each real branch to a real branch.

Our aim is to "compute" all branches (and all real branches) centered at the singular points of Γ, by "computing" an element $(\alpha_2 + P_2(X_1 - \alpha_1), \ldots, \alpha_n + P_n(X_1 - \alpha_1))$ in each cycle giving a branch, or, equivalently, a parametrization

$$(\alpha_1 \pm t^\nu, \alpha_2 + R_2(t), \ldots, \alpha_n + R_n(t))$$

such that $R_i(t) \in \mathbf{R}[[t]]$ if the branch is real.

It is clear that our problem is a local problem, since, if the singular points of Γ are known, by means of a translation, we need only to be able to "compute" solutions centered at the origin. For technical reasons, we need to formulate the problem in local terms, by considering the behaviour of Γ only near the origin. In the following we will therefore assume to be given polynomials $F_1, \ldots, F_s$, which:

1. define locally a curve, i.e. all irreducible components of the variety
$$\Gamma := \{(\alpha_1, \ldots, \alpha_n) \in \mathbf{K}^n : F_1(\alpha_1, \ldots, \alpha_n) = \ldots = F_s(\alpha_1, \ldots, \alpha_n) = 0\}$$
passing through the origin are curves;
2. all irreducible components of Γ passing through the origin are simple;
3. none of the irreducible components of Γ passing through the origin is contained in the hyperplane $X_1 = 0$.

If these assumptions are satisfied we will say that $F_1, \ldots, F_s$ locally define an *admissible curve*.

We need obviously to clarify what we mean by computing a Puiseux series. Assume that we are given $n-1$ polynomials $S_2(u, X_2, \ldots, X_n), \ldots, S_n(u, X_2, \ldots, X_n)$ such that the Jacobian $(\partial S_i/\partial X_j)_{ij}$ is non-zero at the origin. Then by the Implicit Function Theorem, there are unique formal power series $Q_2(u), \ldots, Q_n(u)$ s.t. $S_i(u, Q_2(u), \ldots, Q_n(u)) = 0$ for all i; moreover any approximation of these series can be explicitly obtained by back substitution in the S_i.

Therefore we say that a solution centered at the origin $(P_2(X_1), \ldots, P_n(X_1))$ of the system $F_1 = \ldots = F_s = 0$ is given if we are given integers $a_1, \ldots, a_n$, polynomials $T_2(u), \ldots, T_n(u)$, polynomials $S_2(u, X_2, \ldots, X_n), \ldots, S_n(u, X_2, \ldots, X_n)$ such that:

1. the Jacobian $(\partial S_i/\partial X_j)_{ij}$ is non-zero at the origin.
2. denoting $Q_2(u), \ldots, Q_n(u)$ the unique formal power series such that, for all i, $S_i(u, Q_2(u), \ldots, Q_n(u)) = 0$ and $U_i(u) := T_i(u) + u^{u_i} Q_i(u)$ then $P_i(X_1) = U_i(X_1^{1/a_1})$.

5.4 Standard Bases and Multiplicities

Let us fix a non-zero vector of integer non-negative weights $(a_1, \ldots, a_n)$. Let us consider the morphism $\psi : K[X_1, \ldots, X_n] \to K[X_1, \ldots, X_n, t]$ defined by $\psi(X_i) = t^{a_i} X_i$. Then if $F(X_1, \ldots, X_n) \in K[X_1, \ldots, X_n]$, one has:
$$\psi(F) = F(t^{a_1} X_1, \ldots, t^{a_n} X_n) = t^d G(X_1, \ldots, X_n) + t^{d+1} H(t, X_1, \ldots, X_n).$$
The polynomial $G(X_1, \ldots, X_n)$ is said to be the *initial form* of $F(X_1, \ldots, X_n)$ (w.r.t. the weights $(a_1, \ldots, a_n)$), and we put $\text{in}(F) = G$.

An alternative way of describing it is as follows: assign to the variable X_i the pseudodegree a_i and extend this notion of pseudodegree to the whole polynomial ring. Then each non-zero polynomial $F(X_1, \ldots, X_n)$ can be written uniquely as a finite sum of non-zero pseudohomogeneous forms of different pseudodegrees; $\text{in}(F)$ is the form of least pseudodegree in this representation.

Let I be an ideal in $K[X_1, \ldots, X_n]$. Let us denote by $\text{in}(I)$ the ideal in $K[X_1, \ldots, X_n]$ generated by $\{\text{in}(F) : F \in I\}$. It is easy to verify that $\text{in}(I)$ is pseudohomogeneous (i.e. if a polynomial belongs to $\text{in}(I)$, then each pseudohomogeneous form in its representation belongs to in (I) too) and that a pseudohomogeneous element in in (I) is the initial form of some polynomial in I.

In general it is false that if $F_1, \ldots, F_s$ is a basis of I, then $(\text{in}(F_1), \ldots, \text{in}(F_s))$ is a basis of in (I).

To show an easy example consider the ring $K[X_1, X_2, X_3]$ and the ideal $I = (X_1^2 - X_2^3, X_1^2 - X_3^4)$. With weights $a_i = 1$, the notion of pseudodegree coincides with the one of usual degree and the initial form of a polynomial is the homogeneous form of least degree in its expansion. So we have:

$$\text{in}(X_1^2 - X_2^3) = X_1^2, \quad \text{in}(X_1^2 - X_3^4) = X_1^2, \quad \text{in}(X_2^3 - X_3^4) = X_2^3 .$$

Since $X_2^3 - X_3^4 \in I$, the form X_2^3 belong to in (I), but it does not belong to $(X_1^2) = (\text{in}(X_1^2 - X_2^3), \text{in}(X_1^2 - X_3^4))$.

Definition 2 A *standard basis* of an ideal I (w.r.t. $(a_1, \ldots, a_n)$) is a set of elements $\{G_1, \ldots, G_t\}$ of I such that in (I) is generated by in $(G_1), \ldots,$ in (G_t).

It is important to remark that standard bases are a local tool, intended to describe the structure of a variety near the origin.

First we have to phrase our geometry in algebraic terms; if I is an ideal of the ring $K[X_1, \ldots, X_n]$, then the set of points in $\mathbf{K}^n$ which satisfy all the polynomials in I are (by definition) an algebraic variety V, which can be splitted as the union of irreducible varieties; there is more than one ideal whose set of zeroes is the algebraic variety V; it is however possible to define a notion of multiplicity, so that each ideal corresponds uniquely to a union of irreducible varieties each of them with a prescribed multiplicity.

The dimension of an ideal will denote the maximal dimension of the irreducible components of the algebraic variety defined by the ideal. The local dimension of an ideal will denote the maximal dimension of the irreducible components passing through the origin of the algebraic variety defined by the ideal.

As already remarked in the previous paragraph, the object of our study will be an ideal locally defining a curve; the reason why we need this relaxing is that the ideal generated by a standard basis of I can be different from I; however the irreducible components passing through the origin of the variety defined by the two ideals are the same and have the same multiplicity.

In what follows we will restrict the weights a_i to be all positive, and w.l.o.g. we will assume $\gcd(a_i) = 1$. Under this assumption a standard basis of I satisfies the following properties:

1) the ideal in (I) is pseudohomogeneous; therefore if $(\eta_1, \ldots, \eta_n)$ is a zero of in (I) then, for all t, $(t^{a_1}\eta_1, \ldots, t^{a_n}\eta_n)$ is also a zero of in (I).

2) The dimension of in (I) is the local dimension of I.

3) If I locally defines a curve, the algebraic variety defined by in (I) is the composition of irreducible curves through the origin; in this case we can conclude that in (I) defines the union of the curves whose generic points are $(t^{a_1}\eta_1, \ldots, t^{a_n}\eta_n)$, where $(\eta_1, \ldots, \eta_n)$ runs among the zeroes of in (I); in particular those curves which do not lie in the plane $X_1 = 0$ are uniquely determined by the zeroes $(\eta_1, \ldots, \eta_n)$ of in (I), with $\eta_1 = 1$.

4) A curve $(t^{a_1}\eta_1, \ldots, t^{a_n}\eta_n)$ satisfies the equations of in (I) if and only if it is tangent at the origin to the curve defined by I.

When parametrizing a plane curve $F(X, Y) = 0$, solutions centered at the origin are then solutions of an univariate equation in $\mathbf{K}((X))_{\text{Puis}}[Y]$. Multiplicities of roots of a univariate equation, as well know, are related with the vanishing of derivatives. In our context we will have to deal with a notion of multiplicity for zeroes of systems of multivariate equations. Without entering in details here and referring the reader to the treatment of the problem in [8], we simply remark that in this case the multiplicity of a zero can be characterized in terms of the vanishing of linear combinations of partial higher derivatives. In particular we will need the following result:

Let L be a field of characteristic 0, let $H_1, \ldots, H_s \in L[X_1, \ldots, X_n]$, be such that the system $H_1 = \ldots = H_s = 0$ has only finitely many solutions, and let $(\omega_1, \ldots, \omega_n) \in L^n$ be a zero of the system. Then $(\omega_1, \ldots, \omega_n)$ is a multiple zero if and only if there are $\zeta_1, \ldots, \zeta_n \in L$ such that, denoting: $\Delta(H) := \zeta_1 \partial H/\partial X_1 + \ldots + \zeta_n \partial H/\partial X_n$ for $H \in L[X_1, \ldots, X_n]$, then $\Delta(H_i)(\omega_1, \ldots, \omega_n) = 0$ for all i.

Standard bases are a useful tool in studying multiplicities because of the following results:

Let I be an ideal of dimension 0, i.e. which has only finitely many zeroes in $\mathbf{K}^n$; then $\operatorname{in}(I)$ is a proper ideal if and only if the origin is among the zeroes of I; in this case the multiplicity of the origin as a zero of I is the same as its multiplicity as the only zero of $\operatorname{in}(I)$.

Let us pursue further this point: an ideal I has dimension 0 if and only if in the ring $K[x_1, \ldots, x_n] := K[X_1, \ldots, X_n]/I$ each x_i is algebraic over K, if and only if $K[x_1, \ldots, x_n]$ is a finite dimensional K-vector space.

It is possible to show that the multiplicity of the origin as the zero of $\operatorname{in}(I)$ is exactly the K-dimension of $K[X_1, \ldots, X_n]/\operatorname{in}(I)$.

Furthermore we can count multiplicities of zeroes of I other than the origin as follows: let $\alpha := (\alpha_1, \ldots, \alpha_n)$ be a zero of I of multiplicity h and let $\phi_\alpha : K[X_1, \ldots, X_n] \to K[X_1, \ldots, X_n]$ be the morphism defined by $\phi_\alpha(X_i) = X_i - \alpha_i$; then the origin is a zero for $\phi_\alpha(I)$ of multiplicity h and this multiplicity can be computed by means of standard bases. Finally the sum of the multiplicities of the zeroes of a 0-dimensional ideal I is equal to the K-dimension of $K[X_1, \ldots, X_n]/I$ and is called the *multiplicity* of I.

The notion of multiplicity is insufficient for some of our purposes and we need to give a stronger notion; let J be a 0-dimensional ideal and let $\alpha := (\alpha_1, \ldots, \alpha_n)$, $\beta := (\beta_1, \ldots, \beta_n)$ be two zeroes of J. We will say that α and β are *equivalent zeroes* of J if $\operatorname{in}(\phi_\alpha(J)) = \operatorname{in}(\phi_\beta(J))$; by what we remarked above, this implies that α and β have the same multiplicity, but is not equivalent to it.

If α is a zero of I and β is a zero of J (where I and J are 0-dimensional), we will also say that α (as a zero of I) and β (as a zero of J) are equivalent if $\operatorname{in}(\phi_\alpha(I)) = \operatorname{in}(\phi_\beta(J))$.

Standard bases are an important tool in the algorithm we are describing here, so it is important to remark that there is an algorithm which, given any set of generators of an ideal, allows to compute a standard basis $(G_1, \ldots, G_t)$ of the ideal; moreover, for any prescribed term-ordering $<$, the algorithm returns a standard basis $(G_1, \ldots, G_t)$ such that $(\operatorname{in}(G_1), \ldots, \operatorname{in}(G_t))$ are a Gröbner basis w.r.t. $<$.

5.5 The Main Transformation

Let $I = (F_1, \ldots, F_s)$ locally define an admissible curve Γ. First of all we need some more notation; if $(P_2, \ldots, P_n)$ is a solution centered at the origin of I, then

$$P_2(X_1) = \varepsilon_2 X_1^{\nu_2} + Q_2(X_1), \ldots, P_n(X_1) = \varepsilon_n X_1^{\nu_n} + Q_n(X_1)$$

with $\varepsilon_i \neq 0$, $\operatorname{ord}(Q_i) > 0$ for all i; let $\nu := \min(\nu_i)$; $\eta_i := \varepsilon_i$ if $\nu_i = \nu$, $\eta_i := 0$ if $\nu_i > \nu$. Then we say that:

$(\nu_2, \ldots, \nu_n)$ is the *vector of initial exponents* of $(P_2, \ldots, P_n)$

$(\varepsilon_2 X_1^{\nu_2}, \ldots, \varepsilon_n X_1^{\nu_n})$ is the *monomial approximation* of $(P_2, \ldots, P_n)$

ν is the *initial exponent* of $(P_2, \ldots, P_n)$

$(\eta_2 X_1^{\nu}, \ldots, \eta_n X_1^{\nu})$ is the *initial approximation* of $(P_2, \ldots, P_n)$.

Let us fix weights $(a, b, \ldots, b)$ and let $\nu := b/a$. Let $\eta := (\eta_2, \ldots, \eta_n) \in L^{n-1}$, where L is a finite algebraic extension of K. Let $\psi_\eta : L[X_1, \ldots, X_n] \to L[t, X_2, \ldots, X_n]$ be the morphism defined by:

$$X_1 = t^a,\ X_2 = (\eta_2 + X_2)t^b, \ldots, X_n = (\eta_n + X_n)t^b.$$

For a polynomial $F(X_1, \ldots, X_n) \in K[X_1, \ldots, X_n]$, denote

$$G(X_2, \ldots, X_n) := \mathrm{in}\,(F)(1, X_2, \ldots, X_n)$$

and $d(F)$ the pseudodegree of $\mathrm{in}\,(F)$. Then:

$$\psi_\eta(F) = t^{d(F)} G(\eta_2 + X_2, \ldots, \eta_n + X_n) + t^{d(F)+1} H(t, \eta_2 + X_2, \ldots, \eta_n + X_n)$$

so that $\psi_\eta(F)$ is divisible by $t^{d(F)}$, but not by a higher power of t.

Lemma 1 *Let* $F \in K[X_1, \ldots, X_n]$ *and let* $P_2(X_1), \ldots, P_n(X_1) \in \mathbf{K}[[X_1]]_{\mathrm{Puis}}$ *be such that:*

a) $F(X_1, P_2(X_1), \ldots, P_n(X_1)) = 0$

b) $\mathrm{ord}\,(P_i) \geq \nu$ *for all* i*, so that*

c) $P_i(X_1) := \eta_i X_1^{\nu} + X_1^{\nu} Q_i(X_1)$ *for all* i *with* $Q_i(X_1) \in \mathbf{K}[[X_1]]_{\mathrm{Puis}}$ *and* $\mathrm{ord}\,(Q_i) > 0$.

Denote $\eta := (\eta_2, \ldots, \eta_n)$ *and* $R(t, X_2, \ldots, X_n) := \psi_\eta(F)/t^{d(F)}$. *Then:*

1. $\mathrm{in}\,(F)(1, \eta_2, \ldots, \eta_n) = 0$;

2. $R(t, Q_2(t^a), \ldots, Q_n(t^a)) = 0$.

Proof: 1) One has:

$$F(X_1, P_2(X), \ldots, P_n(X)) = 0$$

and so, by the transformation $X_1 \to t^a$,

$$F(t^a, \eta_2 t^b + t^b Q_2(t^a), \ldots, \eta_n t^b + t^b Q_n(t^a)) = 0.$$

Decomposing the equality according to the different powers of t, one then obtains that $t^{d(F)} G(\eta_2, \ldots, \eta_n) = 0$.

2) One has:

$$t^{d(F)} R(t, X_2, \ldots, X_n) = \psi(F) = F(t^a, (\eta_2 + X_2)t^b, \ldots, (\eta_n + X_n)t^b)$$

and

$$P_i(t^a) = \eta_i t^b + t^b Q_i(t^a).$$

Then

$$t^{d(F)} R(t, Q_2(t^a), \ldots, Q_n(t^a)) = F(t^a, P_2(t^a), \ldots, P_n(t^a)) = 0.$$

Lemma 2 *Let* $F \in K[X_1,\dots,X_n]$, $\eta := (\eta_2,\dots,\eta_n) \in \mathbf{K}^{n-1}$, $R(t,X_2,\dots,X_n) := \psi_\eta(F)/t^{d(F)}$, $Q_2(t),\dots,Q_n(t) \in \mathbf{K}[[t]]_{\text{Puis}}$, $\operatorname{ord}(Q_i) > 0$ *such that* $R(t,Q_2(t),\dots,Q_n(t)) = 0$. *Let* $P_i(X_1) = \eta_i X_1^\nu + X_1^\nu Q_i(X_1^{1/a})$. *Then*

$$F(X_1,P_2(X_1),\dots,P_n(X_1)) = 0.$$

Proof: Since:

$$t^{d(F)}R(t,X_2,\dots,X_n) = \psi(F) = F(t^a,(\eta_2+X_2)t^b,\dots,(\eta_n+X_n)t^b)$$

$$P_i(t^a) = \eta_i t^b + t^b Q_i(t^a)$$

then:

$$F(t^a,P_2(t^a),\dots,P_n(t^a)) = t^{d(F)}R(t,Q_2(t^a),\dots,Q_n(t^a)) = 0.$$

Theorem 1 *Let* $F_1,\dots,F_s$ *be a standard basis of* I *for the weights* $(a,b,\dots b)$, $\nu := b/a$. *Let* $G_i(X_2,\dots,X_n) := \operatorname{in}(F_i)(1,X_2,\dots,X_n)$ *and let* d_i *be the pseudodegree of* $\operatorname{in}(F_i)$. *Then:*

1. *The system* $G_1 = \dots = G_s = 0$ *has either no solutions or finitely many solutions only.*

2. *There is a solution of* $F_1 = \dots = F_s = 0$ *centered at the origin with initial approximation*

$$X_2 = \eta_2 X_1^\nu,\dots,X_n = \eta_n X_1^\nu$$

if and only if $(\eta_2,\dots,\eta_n)$ *is a zero of* $G_1 = \dots = G_s = 0$.

Let now $\bar\eta := (\eta_2,\dots,\eta_n)$ *be a zero of* $G_1 = \dots = G_s = 0$ *and* $R_i(t,X_2,\dots,X_n) := \psi_\eta(F_i)/t^{d_i}$. *Then:*

3. $R_1,\dots,R_s$ *locally define an admissible curve (the* $\bar\eta$*-transformation of* Γ*).*

4. *The solutions of the system* $F_1 = \dots = F_s = 0$ *with initial approximation* $X_2 = \eta_2 X_1^\nu,\dots,X_n = \eta_n X_1^\nu$ *are* $(P_2(X_1),\dots,P_n(X_1))$, *with* $P_i(X_1) = \eta_i X_1^\nu + X_1^\nu Q_i(X_1)$ *and* $(Q_2(t^a),\dots,Q_n(t^a))$ *a solution centered at the origin of* $R_1 = \dots = R_s = 0$.

Proof: 1) The ideal $\operatorname{in}(I) = (\operatorname{in}(F_1),\dots,\operatorname{in}(F_s))$ has dimension 1, so its solutions are finitely many curves $(t^a\eta_1, t^b\eta_2,\dots,t^b\eta_n)$. The solutions of $G_1 = \dots = G_s = 0$ are the points which satisfy $\operatorname{in}(F_1) = \dots = \operatorname{in}(F_s) = 0$ and moreover $X_1 = 1$. Therefore there is a solution of $G_1 = \dots = G_s = 0$, for each curve $(t^{a_1}\eta_1,\dots,t^{a_n}\eta_n)$, which is not in the hyperplane $X_1 = 0$.

2) If $(P_2(X_1),\dots,P_n(X_1))$ is a solution of $F_1 = \dots = F_s = 0$ centered at the origin with initial approximation

$$X_2 = \eta_2 X_1^\nu,\dots,X_n = \eta_n X_1^\nu$$

then $(\eta_2,\dots,\eta_n)$ is a zero of $G_1 = \dots = G_s = 0$, as a consequence of Lemma 1. Conversely, since $(F_1,\dots,F_s)$ is a standard basis, if $(\eta_2,\dots,\eta_n)$ is a zero of $G_1 = \dots = G_s = 0$, then $(t^a, t^b\eta_2,\dots,t^b\eta_n)$ is tangent to $F_1 = \dots = F_s = 0$, and so there is a solution of $F_1 = \dots = F_s = 0$ centered at the origin with initial approximation

$$X_2 = \eta_2 X_1^\nu,\dots,X_n = \eta_n X_1^\nu.$$

3) Since $F_1, \ldots, F_s$ locally define a curve and so have finitely many solutions centered at the origin with initial approximation

$$X_2 = \eta_2 X_1^{\nu}, \ldots, X_n = \eta_n X_1^{\nu}$$

Lemmata 1 and 2 prove that $R_1 = \ldots = R_s = 0$ has finitely many zeroes over $\mathbf{K}[[t]]_{\text{Puis}}$, i.e. $R_1, \ldots, R_s$ locally define a curve. If $R_1, \ldots, R_s$ had components in the hyperplane $t = 0$, then the system $R_1 = \ldots = R_s = t = 0$ would have infinitely many solutions in $\mathbf{K}^n$. However this system is equivalent to $G_1 = \ldots = G_s = t = 0$ which has only finitely many solutions. Finally we must prove that all the solutions centered at the origin of $R_1 = \ldots R_s = 0$ are simple. Let $\zeta_2(t), \ldots, \zeta_n(t) \in \mathbf{K}((t))_{\text{Puis}}$ and denote

$$\Delta(H) := \zeta_2 \partial H/\partial X_2 + \ldots + \zeta_n \partial H/\partial X_n$$

for $H \in \mathbf{K}[t, X_2, \ldots, X_n]$.

Let $\omega_i(X_1) := \zeta_i(X_1^{1/a})$ and denote

$$\Delta(F) := \omega_2 \partial F/\partial X_2 + \ldots + \omega_n \partial F/\partial X_n$$

for $F \in \mathbf{K}[X_1, X_2, \ldots, X_n]$.

A direct verification shows that for all i and j:

$$t^{d_i} \partial R_i/\partial X_j = t^b \psi_\eta(\partial F_i/\partial X_j)$$

so that for all i

$$t^{d_i} \Delta(R_i) = t^b \psi_\eta(\Delta(F_i)).$$

By Lemmata 1 and 2 $(Q_2(t), \ldots, Q_n(t))$ is a zero of $\Delta(R_1) = \ldots = \Delta(R_s) = 0$, if and only if $(P_2(X_1), \ldots, P_n(X_1))$ is a zero of $\Delta(F_1) = \ldots = \Delta(F_s) = 0$, i.e. $(Q_2(t), \ldots, Q_n(t))$ is a multiple zero of $R_1 = \ldots = R_s = 0$ if and only if $(P_2(X_1), \ldots, P_n(X_1))$ is a multiple zero of $F_1 = \ldots = F_s = 0$. Since the solutions centered at the origin of $F_1 = \ldots = F_s = 0$ are all simple, the same is necessarily true for those of $R_1 = \ldots = R_s = 0$.

4) It is an immediate consequence of Lemmata 1 and 2.

It is useful to have some more insight into the conjugacy classes of the roots of the system $G_1 = \ldots = G_s = 0$. Let $\pi : K[X_1, \ldots, X_n] \to K[X_2, \ldots, X_n]$ be the projection given by $\pi(X_1) = 1$, $\pi(X_i) = X_i$. Let ζ be a primitive a^{th}-root of unity, $\zeta^a = 1$.

Let $J = \text{in}(I) = (\text{in}(F_1), \ldots, \text{in}(F_s))$, so that $\pi(J) = (G_1, \ldots, G_s)$. Let $\bar{\eta} := (\eta_2, \ldots, \eta_n)$ be a zero of $\pi(J)$. Then $(t^a, \eta_2 t^b, \ldots, \eta_n t^b)$ is the generic zero of an irreducible component of the variety defined by J. Since $(t^a, \eta_2 \zeta^{ib} t^b, \ldots, \eta_n \zeta^{ib} t^b)$ is also a generic zero of the same component, $\bar{\eta}_i := (\eta_2 \zeta^i, \ldots, \eta_n \zeta^i)$ is a zero of $\pi(J)$. It is possible to show that the $\bar{\eta}_i$'s are equivalent zeroes of $\pi(J)$.

Finally denote $\tau_i : \mathrm{K}[t, X_2, \ldots, X_n] \to \mathrm{K}[t, X_2, \ldots, X_n]$ the morphism such that $\tau_i(t) = \zeta^i t$ and $\tau_i(X_j) = \zeta^{-i} X_j$. One has $\psi_{\bar{\eta}_i} = \tau_i \psi_\eta$ so that the the ideals of the $\bar{\eta}_i$-transformations of Γ are isomorphic. In conclusion:

Proposition 1 *Let $\bar{\eta} := (\eta_2, \ldots, \eta_n)$ be a zero of $G_1 = \ldots G_s = 0$, and let $\bar{\eta}_i := (\eta_2 \zeta^i, \ldots, \eta_n \zeta^i)$, where ζ is a primitive a^{th}-root of unity. Then:*

1. *$\bar{\eta}_i$ is a zero of $G_1 = \ldots G_s = 0$.*
2. *The $\bar{\eta}_i$'s are equivalent zeroes of $G_1 = \ldots = G_s = 0$.*
3. *The ideals of the $\bar{\eta}_i$-transformations of Γ are isomorphic.*

5.6 Counting the Number of Roots

Let $F_1, \ldots, F_s$ locally define an admissible curve Γ, so that the system $F_1 = \ldots = F_s = 0$ has actually solutions centered at the origin, i.e. solutions $(P_2, \ldots, P_n)$ such that $ord(P_i) > 0$ for all i. Our first aim is to count them, which we will achieve by using different but equivalent notions of multiplicity.

First of all we recall that the multiplicity of intersection of the hyperplane $X_1 = 0$ with an analytic branch at the origin is the order of the branch, and so the number of solutions whose cycle gives the branch itself. Therefore the multiplicity of intersection at the origin of the hyperplane $X_1 = 0$ with the curve Γ is exactly the number of solutions centered at the origin of the system $F_1 = \ldots = F_s = 0$. The same number is however also the multiplicity of the origin as a solution of the system $X_1 = F_1 = \ldots = F_s = 0$. Since no component of the curve is contained in the hyperplane $X_1 = 0$, this new system is such to have only finitely many points in $\mathbf{K}^n$ as solutions and therefore the multiplicity of the origin as a zero of it is the multiplicity of any standard basis of the ideal $(X_1, F_1, \ldots, F_s) \subset K[X_1, \ldots, X_n]$ or equivalently of the ideal $(F_1(0, X_2, \ldots, X_n), \ldots, F_s(0, X_2, \ldots, X_n)) \subset K[X_2, \ldots, X_n]$. Hence we have proved:

Lemma 3 *Let $F_1, \ldots, F_s$ locally define an admissible curve Γ. Then the system*

$$F_1(0, X_2, \ldots, X_n) = \ldots = F_s(0, X_2, \ldots, X_n) = 0$$

has finitely many solutions. The multiplicity of the origin as a solution of this system is the number of solutions of $F_1 = \ldots = F_s = 0$ centered at the origin.

Remark that since we assume Γ to be admissible, and so in particular with no multiple component, the multiplicity at the origin of the system above counts the number of *distinct* solutions centered at the origin.

Lemma 4 *Let $F_1, \ldots, F_s$ locally define an admissible curve Γ and moreover be a standard basis for the weights $(a, b, \ldots, b)$. Let $G_i(X_2, \ldots, X_n) = \mathrm{in}\,(F_i)(1, X_2, \ldots, X_n)$ and d_i the pseudodegree of $\mathrm{in}\,(F_i)$. Let $(\eta_2, \ldots, \eta_n)$ be a zero of $G_1 = \ldots = G_s = 0$ with multiplicity h and let*

$$R_i(t, X_2, \ldots, X_n) := F_i(t^a, t^b(\eta_2 + X_2), \ldots, t^b(\eta_n + X_n))/t^{d_i}.$$

Then $R_1 = \ldots = R_s = 0$ has exactly h solutions centered at the origin (again, because of Theorem 1, the h solutions are simple and so are all distinct).

Proof: Remark that $G_i(\eta_2 + X_2, \ldots, \eta_2 + X_n) = R_i(0, X_2, \ldots, X_n)$. Since $(\eta_2, \ldots, \eta_n)$ is a zero of $G_1 = \ldots = G_s = 0$ with multiplicity h, the origin is a solution of the system $R_1(0, X_2, \ldots, X_n) = \ldots = R_s(0, X_2, \ldots, X_n) = 0$, with multiplicity h. By Lemma 3, then, $R_1 = \ldots = R_s = 0$ has exactly h solutions centered at the origin.

Theorem 2 *Let $F_1, \ldots, F_s$ locally define an admissible curve Γ and moreover be a standard basis for the weights $(a, b, \ldots, b)$. Let $G_i(X_2, \ldots, X_n) = \mathrm{in}\,(F_i)(1, X_2, \ldots, X_n)$. Then:*

1. *the multiplicity of the ideal $(G_1, \ldots, G_s)$ is the number of solutions centered at the origin with initial exponent $\mu \geq b/a$;*

2. *the multiplicity of the origin as a zero of $(G_1, \ldots, G_s)$ is the number of solutions centered at the origin with initial exponent $\mu > b/a$;*

3. if $(\eta_2, \ldots, \eta_n)$ *is a zero of* $(G_1, \ldots, G_s)$, *different from the origin, with multiplicity* h, *then* h *is the number of solutions centered at the origin with initial approximation* $X_2 = \eta_2 X_1^{b/a}, \ldots, X_n = \eta_n X_1^{b/a}$.

Proof: the result is an immediate consequence of Theorem 1 and Lemma 4.

Proposition 2 *Let* $F_1, \ldots, F_s$ *locally define an admissible curve* Γ *and moreover be a standard basis for the weights* $(a, b, \ldots, b)$. *Let* $G_i(X_2, \ldots, X_n) = \mathrm{in}(F_i)(1, X_2, \ldots, X_n)$, d_i *the pseudodegree of* $\mathrm{in}(F_i)$. *Let* $(\eta_2, \ldots, \eta_n)$ *be a simple zero of* $G_1 = \ldots = G_s = 0$, $L := K(\eta_2, \ldots, \eta_n)$. *Let* $R_i(t, X_2, \ldots, X_n) = F_i(t^a, t^b(\eta_2 + X_2), \ldots, t^b(\eta_n + X_n))/t^{d_i}$. *Then the Jacobian matrix* $(\partial R_i / \partial X_j)_{ij}$ *has maximal rank at the origin. As a consequence there are* $n-1$ *linear combinations of the* R_i*'s,* $S_2, \ldots, S_n$ *such that*

1. $S_i(t, X_2, \ldots, X_n) = X_i + T_i(t, X_2, \ldots, X_n)$ *with* $T_i \in (t) + (X_2, \ldots, X_n)^2$;
2. *there are unique power series* $Q_j(t) \in L[[t]]$ *such that* $S_i(t, Q_2(t), \ldots, Q_n(t)) = 0$;
3. $(Q_2, \ldots, Q_n)$ *is the unique solution centered at the origin of* $R_1 = \ldots = R_s = 0$.

Proof: First of all remark that $R_i(0, X_2, \ldots, X_n) = G_i(\eta_2 + X_2, \ldots, \eta_n + X_n)$, so that $\partial R_i / \partial X_j(0, X_2, \ldots, X_n) = \partial G_i / \partial X_j(\eta_2 + X_2, \ldots, \eta_n + X_n)$. Therefore the Jacobian matrix $(\partial R_i / \partial X_j)_{ij}$ has maximal rank at the origin if and only if the Jacobian matrix $(\partial G_i / \partial X_j)_{ij}$ has maximal rank at $(\eta_2, \ldots, \eta_n)$. If the latter matrix has not maximal rank, then there are $c_2, \ldots, c_n \in L$, not all zero such that $c_2 \partial G_i / \partial X_2 + \ldots + c_n \partial G_i / \partial X_n$ vanishes at $(\eta_2, \ldots, \eta_n)$ for all i, against the assumption that $(\eta_2, \ldots, \eta_n)$ is a simple zero of $G_1 = \ldots = G_s = 0$. The other statement is an elementary consequence of the Implicit Function Theorem.

5.7 Finding Initial Exponents

A basic step of the algorithm we are developing is finding the set of the initial exponents of the solutions centered at the origin of $F_1 = \ldots = F_s = 0$. In his paper [13], Maurer proposed a necessary condition for a vector $(\nu_2, \ldots, \nu_n)$ to be a vector of initial exponents.

Assume that $(\varepsilon_2 X_1^{\nu_2}, \ldots, \varepsilon_n X_1^{\nu_n})$ is the monomial approximation of a solution centered at the origin, so that $\varepsilon_i \neq 0$ for all i. Let $(a, b_2, \ldots, b_n)$ be integer weights with $\nu_i = b_i / a$. By an argument similar to that of Lemma 1, one proves that $(1, \varepsilon_2, \ldots, \varepsilon_n)$ must be a zero of $\mathrm{in}(F)$, for each polynomial F in I, where $\mathrm{in}(F)$ is computed with respect to the weights $(a, b_2, \ldots, b_n)$. Since $\varepsilon_i \neq 0$ for all i, then, for each polynomial F in I, $\mathrm{in}(F)$ cannot consist of a monomial, so that $(a, b_2, \ldots, b_n)$ must be orthogonal to an edge of the Newton polyhedron of F, for each F in I.

Because of this result, Maurer suggests the following approach: take $n-1$ among the F_i; compute their Newton polyhedra; for each possible $(n-1)$-tuple of edges, one for each polyhedron, compute "the" orthogonal direction. The list of vectors of exponents so found contains then all possible vectors of initial exponents.

An obvious improvement over this approach is to test this list with the Newton polyhedra of the other basis elements and of any other polynomial in the ideal produced during the algorithm, discarding those vectors which are not orthogonal to at least one edge of the polyhedron. Call *Maurer vectors* the resulting vectors. Also, if $(\nu_2, \ldots, \nu_n)$ is a Maurer vector, call $\nu := min(\nu_i)$ a *Maurer exponent.*

There are however two problems with Maurer's approach.

First of all it is possible that for some choice of an edge for each of s different Newton polyhedra (even for s larger than $n-1$), the s edges so chosen span a vector space of dimension less than $n-1$ so that there are infinitely many directions orthogonal to all of them and so infinitely many Maurer vectors. A quite inefficient remedy to this is based on the fact that there are bounds on the integers a, b_i such that $(b_2/a, \ldots, b_n/a)$ are vectors of initial exponents.

A second problem is however that the list of Maurer vectors is quite too large; a standard basis computation with weights $(a, b_2, \ldots, b_n)$ with $b_i/a = \nu_i$ is necessary in order to find (if there exist) monomial approximations $(\varepsilon_2 X_1^{\nu_2}, \ldots, \varepsilon_n X_1^{\nu_n})$, by means of an obvious generalization of Theorem 1; since standard basis computations are the most time-expensive part of a local parametrization algorithm, one cannot afford to compute too many of them.

A different proposal which searches for initial exponents and initial approximations only is contained in MacMillan's paper [12].

Let $\pi : \mathbf{N}^n \to \mathbf{N}^2$ given by $\pi(a_1, \ldots, a_n) = (a_1, a_2 + \ldots + a_n)$; consider the Newton polygons of the projections by π of the Newton polyhedra of the F_i's. MacMillan proves that, **if** a certain condition is verified, ν is the initial exponent of a solution centered at the origin, if and only if $-1/\nu$ is the slope of an edge of at least one of these Newton polygons. Call each such value ν a *MacMillan exponent.*

MacMillan's condition, which he states in terms of resultant theory, can be formulated in more modern language and then it states that $F_1, \ldots, F_s$ is a standard basis for each weight $(a, b \ldots, b)$ such that b/a is a MacMillan exponent. In general such a condition, as remarked by MacMillan himself, does not hold, so that MacMillan's list does not produce all initial exponents. There is however an advantage (and this remark again is due to MacMillan) in working with initial exponents instead of vectors of initial exponents, which is given by the counting argument of Theorem 2.

If $\nu_1 = b/a$ and $\nu_2 = d/c$, $\nu_1 < \nu_2$, are consecutive values in the list of MacMillan's exponent, standard basis computations w.r.t. the weights $(a, b, \ldots, b)$ and $(c, d, \ldots d)$ allow to compute the number of solutions with initial exponent $\nu > \nu_1$ and the number of solutions with initial exponent $\nu \geq \nu_2$; if the two numbers are different, their difference gives the number of solutions with initial exponent ν, $\nu_1 < \nu < \nu_2$. In this case, we can make use of Maurer exponents in the interval (ν_1, ν_2); the same counting argument allows to perform some kind of binary search in the list of Maurer exponents, so decreasing the number of standard basis computations required.

Properties of standard bases suggest to use the following heuristics instead.

For each polynomial in the standard basis of $(F_1, \ldots, F_s)$ w.r.t. the weights $(a, b, \ldots b)$ with $b/a = \nu_1$, compute its Newton polyhedron, and the Newton polygon of its projection by π; from the list of values ν, $\nu_1 < \nu < \nu_2$, such that $-1/\nu$ is the slope of an edge of one of these polygons, (if not empty) select the one of minimal value.

In the same way, for each polynomial in the standard basis of $(F_1, \ldots, F_s)$ w.r.t. the weights $(c, d, \ldots, d)$ with $d/c = \nu_2$, compute its Newton polyhedron, and the Newton polygon of its projection by π; from the list of values ν, $\nu_1 < \nu < \nu_2$, such that $-1/\nu$ is the slope of an edge of one of these polygons, (if not empty) select the one of maximal value.

Only in case this procedure fails, i.e. both lists are empty, perform one step of binary search in the list of Maurer exponents in the interval (ν_1, ν_2).

5.8 Finding all Solutions with a given Initial Exponent

Let us now fix a positive rational ν and let us show how to compute all solutions of $F_1 = \ldots = F_s = 0$ with initial exponent ν. Let $a, b \in \mathbf{N}$ be such that $b/a = \nu$, $\gcd(a,b) = 1$.

1. Compute a standard basis $(H_1, \ldots, H_t)$ of the ideal $(F_1, \ldots, F_s)$ w.r.t. the weights $(a, b, \ldots, b)$.
2. Let $G_i := \text{in}(H_i)(1, X_2, \ldots, X_n)$ and d_i be the pseudodegree of $\text{in}(H_i)$.
3. Compute the multiplicity μ of the ideal $J_0 = (G_1, \ldots, G_t)$.
4. Compute the ideal $J := (G_1, \ldots, G_t) : (X_2^{\mu}, \ldots, X_n^{\mu})$.
5. Compute its multiplicity μ_1 and let $\mu_0 := \mu - \mu_1$.
6. Compute ideals $J_1, \ldots, J_u$ such that:

   i) the union of the roots of the J_i's gives the roots of J:

   $$\{(\eta_2, \ldots, \eta_n) : F(\eta_2, \ldots, \eta_n)\ \forall F \in J\} = \cup_i \{(\eta_2, \ldots, \eta_n) : F(\eta_2, \ldots, \eta_n)\ \forall F \in J_i\}$$

   ii) if $(\eta_2, \ldots, \eta_n)$ and $(\varepsilon_2, \ldots, \varepsilon_n)$ are roots of a same ideal J_i then they are equivalent zeroes of J;

   iii) the roots of J_1 are the simple roots of J.
7. For each root $(\eta_2, \ldots, \eta_n)$ of J_1:
   (a) compute $R_i(t, X_2, \ldots, X_n) := H_i(t^a, t^b(\eta_2 + X_2), \ldots, t^b(\eta_n + X_n))/t^{d_i}$;
   (b) compute $S_2, \ldots, S_n$ satisfying the conditions of Proposition 2;
   (c) return $[(t^a, \eta_2 t^b, \ldots, \eta_n t^b);\ t^b;\ (S_2, \ldots, S_n)]$.
8. For each root $(\eta_2, \ldots, \eta_n)$ of J_i, $i > 1$:
   (a) compute $R_i(t, X_2, \ldots, X_n) := H_i(t^a, t^b(\eta_2 + X_2), \ldots, t^b(\eta_n + X_n))/t^{d_i}$
   (b) compute all solutions of $R_1(t, X_2, \ldots, X_n) = \ldots = R_t(t, X_2, \ldots, X_n) = 0$, i.e. compute all

   $$[(u^c, T_2(u), \ldots, T_n(u));\ u^d;\ (S_2(u, X_2, \ldots, X_n), \ldots, S_n(u, X_2, \ldots, X_n))]$$

   such that , denoting $Q_i(u)$ the unique formal power series such that, for all i, $S_i(u, Q_2(u), \ldots, Q_n(u)) = 0$ and putting:

   $$U_i(u) := T_i(u) + u^d Q_i(u),$$

   one has:

   $$R_1(u^c, U_2(u), \ldots, U_n(u)) = \ldots = R_t(u^c, U_2(u), \ldots, U_n(u)) = 0.$$

(c) For each solution

$$[(u^c, T_2(u), \dots, T_n(u));\ u^d;\ (S_2(u, X_2, \dots, X_n), \dots, S_n(u, X_2, \dots, X_n))]$$

return

$$[(u^{ac}, \eta_2 u^{bc} + u^{bc}T_2(u), \dots, \eta_n u^{bc} + u^{bc}T_n(u));\ u^{bc+d};$$
$$(S_2(u, X_2, \dots, X_n), \dots, S_n(u, X_2, \dots, X_n))]$$

so that, denoting $Q_i(u)$ the unique formal power series such that, for all i, $S_i(u, Q_2(u), \dots, Q_n(u)) = 0$ and putting:

$$V_i(u) := T_i(u) + u^d Q_i(u),$$
$$U_i(u) := \eta_i u^{bc} + u^{bc}V_i(u) = \eta_i u^{bc} + u^{bc}T_i(u) + u^{bc+d}Q_i(u)$$

one has:

$$R_1(u^c, V_2(u), \dots, V_n(u)) = \dots = R_t(u^c, V_2(u), \dots, V_n(u)) = 0$$

and

$$H_1(u^{ac}, U_2(u), \dots, U_n(u)) = \dots = H_t(u^{ac}, U_2(u), \dots, U_n(u)) = 0$$

Let us comment on each step of this algorithm.

1. The standard basis computation can be performed by means of the Tangent Cone Algorithm [14], which has the advantage of returning a Gröbner basis of the ideal $J_0 = (G_1, \dots, G_t)$. Since the standard basis computation is the most time-consuming part of the Algorithm, we will discuss in an Appendix possible ways of improving this step of the algorithm, by truncating the standard basis computation.

2. The ideal J_0 has at most finitely many roots and its roots are the coefficients of the initial approximations of the solutions centered at the origin with initial exponent ν.

3. The multiplicity of $J_0 = (G_1, \dots, G_t)$ can be easily computed if $(G_1, \dots, G_t)$ is a Gröbner basis and it gives the sum of the multiplicities of the roots of J_0, i.e. the number of solutions centered at the origin with initial exponent $\nu_1 \geq \nu$.

4. The roots of the ideal $J := (G_1, \dots, G_t) : (X_2^\mu, \dots, X_n^\mu)$ are exactly the same as the non-zero roots of the ideal J_0 and they have the same multiplicity in the two ideals. A Gröbner basis for the ideal J can be obtained from the Gröbner basis $(G_1, \dots, G_t)$ either by several Gröbner basis computations (there are different schemes to do that [16]; the technique is not efficient but has the advantage to be easily available in any system with the Gröbner basis algorithm) or by a linear algebra algorithm [11, 15] which requires $n^2\mu^3$ arithmetical operations in the field K (this last approach is much more efficient, but we don't know of any existing implementation).

5. The knowledge of a Gröbner basis of J allows to compute easily its multiplicity. By Theorem 2, one then has that μ_1 is the number of solutions centered at the origin with initial exponent ν, while μ_0 is the number of solutions centered at the origin with initial exponent $\nu_1 > \nu$. The knowledge of μ_1 and μ_0 is required for the computation of initial exponents as sketched in the previous paragraph.

6. We agree that the formulation of this step is quite odd. An easier formulation would be: *compute all roots of* J. The reason why we have chosen this formulation is that, while we need to separate roots which are not equivalent, there is no need to actually compute them, as it will be made more clear in section 5.11, to which we refer, postponing the discussion of this point.

7. Since the roots of J_1 are the simple roots of J we can apply Proposition 2. Because of it, we know that there are unique formal power series $Q_i(t)$ such that $R_j(t, Q_2(t), \ldots, Q_n(t)) = 0$ for all j and that $(t^a, \eta_2 t^b + t^b Q_2(t), \ldots, \eta_2 t^b + t^b Q_2(t))$ corresponds to the unique solution centered at the origin with initial approximation $(\eta_2 t^\nu, \ldots \eta_n t^\nu)$ by the substitution $t = X^{1/a}$. As it was specified in section 5.3, the returned information is what we intend by "computing" a solution centered at the origin, and it is sufficient to compute polynomial approximations of any order.

8. Here we apply instead Theorem 1 which implies that

$$R_1(u^c, V_2(u), \ldots, V_n(u)) = \ldots = R_t(u^c, V_2(u), \ldots, V_n(u)) = 0$$

if and only if

$$H_1(u^{ac}, U_2(u), \ldots, U_n(u)) = \ldots = H_t(u^{ac}, U_2(u), \ldots, U_n(u)) = 0.$$

5.9 Termination of the Algorithm

We have yet to prove termination of the algorithm; the only possibility for the algorithm to continue forever is that the recursive call in Step 8.b) is performed infinitely many times. We would have therefore an infinite sequence of:

- admissible curves $\Gamma_1, \ldots, \Gamma_r, \ldots$
- polynomial sets $\{F_{i1}, \ldots, F_{is_i}\}$
- exponents $\nu_i = b_i/a_i$
- points $\bar{\eta}_i = (\eta_{2i}, \ldots, \eta_{ni}) \in \mathbf{K}^{n-1}$
- integers h_i

related as follows:

1. $\{F_{i1}, \ldots, F_{is_i}\}$ is a standard basis of Γ_i w.r.t. the weights $(a_i, b_i, \ldots, b_i)$.
2. $(1, \eta_{2i}, \ldots, \eta_{ni})$ is a zero of $\text{in}(F_{i1}) = \ldots = \text{in}(F_{is_i}) = 0$ of multiplicity h_i.
3. Γ_{i+1} Jis generated by $\psi_{\bar{\eta}_i}(F_{ij})/t^{d(F_{ij})}$ and has exactly h_i solutions centered at the origin.

We can then make the following remarks:

1. the sequence of the h_i's is non increasing, so it must stabilize to a common minimal value h;
2. moreover $h > 1$ (otherwise termination is assured);
3. if $a_i > 1$ then $(1, \zeta\eta_{2i}, \ldots, \zeta\eta_{ni})$ is a zero of $\text{in}(F_{i1}) = \ldots = \text{in}(F_{is_i})$ for each ζ such that $\zeta^{a_i} = 1$;

4. therefore if $a_i > 1$ then $h_{i-1} \geq a_i h_i > h_i$;

5. so there is N such that for $i \geq N$ we have $a_i = 1$, $h_i = h > 1$.

The admissible curves Γ_i have therefore h distinct solutions centered at the origin,

$$(P_{2ij}(X_1), \ldots, P_{nij}(X_1))_{j=1\ldots h}$$

with $P_{lij}(X_1) \in \mathbf{K}[[X_1]]$. Moreover

$$P_{lij}(X_1) = \eta_{li} X_1^{b_i} J + X_1^{b_i} P_{l(i+1)j}(X_1)$$

for each l, i, j. Therefore if we set $c_N := b_N$, $c_i := c_{i-1} + b_i$ for all $i > N$ one has that $\sum_{i=N}^{M} \eta_{li} X_1^{c_i}$ is an approximation of P_{lNj} of order c_M for all j, M, against the assumption that the solutions of Γ_N are distinct.

5.10 Finding Real Analytic Branches

Our final aim is however not to find all solutions centered at the origin, and not even all analytic branches (which would require choosing a solution for each cycle), but to find the real analytic branches. To do so we make use of the following result:

Proposition 3 *Let $(P_2(X_1), \ldots, P_n(X_1))$ be a solution centered at the origin of the system $F_1 = \ldots = F_s = 0$, with*

$$P_2(X_1) = \sum c_{i2} X_1^{\nu_{i2}/\nu}, \ldots, P_n(X_1) = \sum c_{in} X_1^{\nu_{in}/\nu}$$

with $c_{ij} \neq 0$ $\nu, \nu_{ij} \in \mathbf{N}$, $\gcd(\nu_{ij}, \nu) = 1$, so that if ζ is a primitive ν-th root of 1, the ν solutions in the cycle of $(P_2, \ldots, P_n)$ are

$$P_{2j}(X_1) = \sum c_{i2} \zeta^{j\nu_{i2}} X_1^{\nu_{i2}/\nu}, \ldots, P_{nj}(X_1) = \sum c_{in} \zeta^{j\nu_{in}} X_1^{\nu_{in}/\nu}.$$

If ν is odd, consider the ν parametrizations of the branch

$$X_1 = t^\nu, X_2 = \sum c_{i2} \zeta^{j\nu_{i2}} t^{\nu_{i2}}, \ldots, X_n = \sum c_{in} \zeta^{j\nu_{in}} t^{\nu_{in}}$$

There is at most one such parametrization, which is "real" in the sense that all coefficients $c_{ik} \zeta^{j\nu_{ik}}$ are real, and there is exactly one if and only if the branch is real.

If ν is even let ξ be a primitive 2ν-root of 1 and consider the 2ν parametrizations of the branch:

$$X_1 = t^\nu, X_2 = \sum c_{i2} \xi^{j\nu_{i2}} t^{\nu_{i2}}, \ldots, X_n = \sum c_{in} \xi^{j\nu_{in}} t^{\nu_{in}} \quad j \text{ even}$$

$$X_1 = -t^\nu, X_2 = \sum c_{i2} \xi^{j\nu_{i2}} t^{\nu_{i2}}, \ldots, X_n = \sum c_{in} \xi^{j\nu_{in}} t^{\nu_{in}} \quad j \text{ odd}$$

There are either none or two real parametrizations; there are two if and only if the branch is real. In this case they are transformed into each other by the substitution $t \mapsto -t$.

In order to consider parametrizations of the kind $(-t^\nu, \sum c_{i2} t^{\nu_{i2}}, \ldots, \sum c_{in} t^{\nu_{in}})$ we have to modify accordingly the results of section 5.5.

Let $I = (F_1, \ldots, F_s)$ locally define an admissible curve Γ. Let us fix weights $(a, b, \ldots b)$, where a is even and b is odd, and let $\nu := b/a$. Let us assume moreover that $(F_1, \ldots, F_s)$ is a standard basis of I for the weights $(a, b, \ldots b)$. Let ξ be a $2a^{th}$-primitive root of unity.

Let $\bar{\eta} := (\eta_2, \ldots, \eta_n) \in \mathbf{K}^{n-1}$, let $L := K(\eta_2, \ldots, \eta_n, \xi)$ and let $\psi_\eta : L[X_1, \ldots, X_n] \to L[t, X_2, \ldots, X_n]$ be the morphism defined by the transformation

$$X_1 = t^a,\ X_2 = (\eta_2 + X_2)t^b,\ \ldots,\ X_n = (\eta_n + X_n)t^b$$

so that, for $F(X_1, \ldots, X_n) \in K[X_1, \ldots, X_n]$, denoting $d(F)$ the pseudodegree of $\mathrm{in}\,(F)$, one has:

$$\psi_\eta(F) = t^{d(F)}\,\mathrm{in}\,(F)(1, \eta_2 + X_2, \ldots, \eta_n + X_n) + t^{d(F)+1} H(t, \eta_2 + X_2, \ldots, \eta_n + X_n).$$

Let $\rho : L[t, X_2, \ldots, X_n] \to L[t, X_2, \ldots, X_n]$ be the morphism such that

$$\rho(t) = \xi t, \quad \rho(X_i) = \xi^{-b} X_i \quad \text{for all } i$$

so that $\rho\psi_\eta : L[X_1, \ldots, X_n] \to L[t, X_2, \ldots, X_n]$ is given by the transformation:

$$X_1 = -t^a,\ X_2 = (\xi^b\eta_2 + X_2)t^b,\ \ldots,\ X_n = (\xi^b\eta_n + X_n)t^b$$

and

$$\rho\psi_\eta(F) = t^{d(F)}\,\mathrm{in}\,(F)(-1, \xi^b\eta_2 + X_2, \ldots, \xi^b\eta_n + X_n) + t^{d(F)+1} H(t, \xi^b\eta_2 + X_2, \ldots, \xi^b\eta_n + X_n).$$

Let now $\sigma : K[X_1, \ldots, X_n] \to K[X_1, \ldots, X_n]$ be the morphism such that

$$\sigma(X_1) = -X_1, \quad \sigma(X_i) = X_i \text{ for } i > 1$$

and let

$$\eta^* := (\xi^b\eta_2, \ldots, \xi^b\eta_n).$$

It is immediate that

$$\rho\psi_\eta = \psi_{\eta^*}\sigma.$$

Let $\pi : K[X_1, \ldots, X_n] \to K[X_2, \ldots, X_n]$ be the projection such that $\pi(X_1) = 1$, $\pi(X_i) = X_i$. Let $J = \mathrm{in}\,(I) = (\mathrm{in}\,(F_1), \ldots, \mathrm{in}\,(F_s))$, so that

$$\pi(J) = (\mathrm{in}\,(F_1)(1, X_2, \ldots, X_n), \ldots, \mathrm{in}\,(F_s)(1, X_2, \ldots, X_n)).$$

Let $\bar{\eta} := (\eta_2, \ldots, \eta_n)$ be a zero of $\pi(J)$ so that $\bar{\eta}_{2i} := (\eta_2\xi^{2i}, \ldots, \eta_n\xi^{2i})$ is a zero of $\pi(J)$ for each i and $(t^a, \eta_2 t^b, \ldots, \eta_n t^b)$ is the generic zero of an irreducible component of the variety defined by J. Then $(-t^a, \eta_2\xi^b t^b, \ldots, \eta_n\xi^b t^b)$ is the generic zero of an irreducible component of the variety defined by $\sigma(J)$, so that, for each i, $\bar{\eta}_{2i+1} := (\eta_2\xi^{2i+1}, \ldots, \eta_n\xi^{2i+1})$ is a zero of $\pi\sigma(J) = (\mathrm{in}\,(F_1)(1, X_2, \ldots, X_n), \ldots, \mathrm{in}\,(F_s)(1, X_2, \ldots, X_n))$.

Extending the results of Proposition 1, we have:

Proposition 4 *The zeroes $\bar{\eta}_{2i}$ of $\pi(J)$ and the zeroes $\bar{\eta}_{2i+1}$ of $\pi\sigma(J)$ are equivalent. Moreover the $\bar{\eta}_{2i}$-transformations of Γ and the $\bar{\eta}_{2i+1}$-transformations of $\sigma(\Gamma)$ are all isomorphic. In particular this holds for the $\bar{\eta}$-transformation of Γ and the η^*-transformation of $\sigma(\Gamma)$.*

We have then the following analogon of Theorem 1:

Proposition 5 *Let $F_1, \dots, F_s$ be a standard basis of I for the weights $(a, b, \dots b)$, let $\nu := b/a$, with a even. Let $G_i(X_2, \dots, X_n) := \mathrm{in}\,(F_i)(-1, X_2, \dots, X_n)$ and let d_i be the pseudodegree of $\mathrm{in}\,(F_i)$. Then:*

1) $G_1 = \ldots = G_s = 0$ has either no solutions or finitely many solutions only.
Let $\bar{\eta} := (\eta_2, \dots, \eta_n)$ be a zero of $G_1 = \ldots = G_s = 0$ and let $R_i(t, X_2, \dots, X_n) = \psi_{\bar{\eta}}\sigma(F_i)/t^{d_i}$. Then:

2) $R_1, \dots, R_s$ locally define an admissible curve.

3) $F_i(-t^a, \eta_2 t^b + t^b Q_2(t), \dots, \eta_n t^b + t^b Q_n(t)) = 0$ for all i, if and only if

$$R_i(t, Q_2(t), \dots, Q_n(t)) = 0$$

for all i and for $Q_i(t) \in \mathbf{K}[[t]]_{\mathrm{Puis}}$.

Corollary 1 *Given a standard basis $F_1, \dots, F_s$ of I for the weights $(a, b, \dots b)$, let $\nu := b/a$, let d_i be the pseudodegree of $\mathrm{in}\,(F_i)$, let $\eta := (\eta_2, \dots, \eta_n)$ be a real zero of*

$$\mathrm{in}\,(F_1)(1, X_2, \dots, X_n) = \ldots = \mathrm{in}\,(F_s)(1, X_2, \dots, X_n) = 0$$

and let

$$R_i(t, X_2, \dots, X_n) := \psi_\eta(F_i)/t^{d_i}.$$

Let $((-1)^j u^c, V_2(u), \dots, V_n(u))$, $V_i \in \mathbf{R}[[u]]_{\mathrm{Puis}}$ be a real parametrization of a real branch of $R_1 = \ldots = R_s = 0$, i.e.

$$R_1((-1)^j u^c, V_2(u), \dots, V_n(u)) = \ldots = R_s((-1)^j u^c, V_2(u), \dots, V_n(u)) = 0.$$

Then

$$((-1)^{aj} u^{ac}, (-1)^{bj} \eta_2 u^{bc} + (-1)^{bj} u^{bc} V_2(u), \dots, (-1)^{bj} \eta_n u^{bc} + (-1)^{bj} u^{bc} V_n(u))$$

is a real parametrization of a real branch of $F_1 = \ldots = F_s = 0$.

Let now a be even, let $\eta := (\eta_2, \dots, \eta_n)$ be a real zero of

$$\mathrm{in}\,(F_1)(-1, X_2, \dots, X_n) = \ldots = \mathrm{in}\,(F_s)(-1, X_2, \dots, X_n) = 0$$

and let $R_i(t, X_2, \dots, X_n) := \psi_\eta \sigma(F_i)/t^{d_i}$. Let $((-1)^j u^c, V_2(u), \dots, V_n(u))$, $V_i \in \mathbf{R}[[u]]_{\mathrm{Puis}}$ be a real parametrization of a real branch of $R_1 = \ldots = R_s = 0$. Then

$$(u^{ac}, (-1)^{bj} \eta_2 u^{bc} + (-1)^{bj} u^{bc} V_2(u), \dots, (-1)^{bj} \eta_n u^{bc} + (-1)^{bj} u^{bc} V_n(u))$$

is a real parametrization of a real branch of $F_1 = \ldots = F_s = 0$.

Moreover a real branch of $F_1 = \ldots = F_s = 0$ has a parametrization obtained as above.

We are now able to describe the modifications we need to perform to the algorithm sketched in section 5.8, to compute only the real analytic branches; the modifications apply only to Steps (7) and (8) which are to be modified as follows:

7) For each *real* root $(\eta_2, \dots, \eta_n)$ of J_1:

   (a) compute $R_i(t, X_2, \dots, X_n) = H_i(t^a, t^b(\eta_2 + X_2), \dots, t^b(\eta_n + X_n))/t^{d_i}$

   (b) Compute $S_2, \dots, S_n$ satisfying the conditions of Proposition 1

   (c) return $[(t^a, \eta_2 t^b, \dots, \eta_n t^b);\ t^b;\ (S_2, \dots, S_n)]$

8) If a is even, let ξ be a primitive $2a^{th}$-root of unity and let L_1 be the ideal whose roots, all simple, are $\{(\eta_2\xi^b,\ldots,\eta_n\xi^b) : (\eta_2,\ldots,\eta_n)$ a root of $J_1\}$. For each real root $(\eta_2,\ldots,\eta_n)$ of L_1:

   (a) compute $R_i(t,X_2,\ldots,X_n) = H_i(-t^a, t^b(\eta_2+X_2),\ldots,t^b(\eta_n+X_n))/t^{d_i}$.
   (b) Compute $S_2,\ldots,S_n$ satisfying the conditions of Proposition 1.
   (c) return $[(t^a,\eta_2 t^b,\ldots,\eta_n t^b);\ t^b;\ (S_2,\ldots,S_n)]$

9) For each real root $(\eta_2,\ldots,\eta_n)$ of J_i, $i>1$:

   (a) compute $R_i(t,X_2,\ldots,X_n) = H_i(t^a, t^b(\eta_2+X_2),\ldots,t^b(\eta_n+X_n))/t^{d_i}$
   (b) compute all real solutions of $R_1(t,X_2,\ldots,X_n) = \ldots = R_t(t,X_2,\ldots,X_n) = 0$, i.e. compute all

$$[((-1)^j u^c, T_2(u),\ldots,T_n(u));\ (-1)^k u^d;$$
$$(S_2(u,X_2,\ldots,X_n),\ldots,S_n(u,X_2,\ldots,X_n))]$$

   where T_i, S_i have real coefficients and are such that, denoting $Q_i(u)$ the unique formal power series such that, for all i, $S_i(u,Q_2(u),\ldots,Q_n(u)) = 0$ and $V_i(u) := T_i(u) + (-1)^k u^d Q_i(u)$, one has:

$$R_1((-1)^j u^c, V_2(u),\ldots,V_n(u)) = \ldots = R_t((-1)^j u^c, V_2(u),\ldots,V_n(u)) = 0.$$

   (c) For each real solution

$$[((-1)^j u^c, T_2(u),\ldots,Tn(u));\ (-1)^k u^d;$$
$$(S_2(u,X_2,\ldots,X_n),\ldots,S_n(u,X_2,\ldots,X_n))]$$

   return

$$[((-1)^{aj}u^{ac}, (-1)^{bj}\eta_2 u^{bc} + (-1)^{bj}u^{bc}T_2(u),\ldots,(-1)^{bj}\eta_n u^{bc} + (-1)^{bj}u^{bc}T_n(u));$$
$$(-1)^{bj+k}u^{bc+d};\ (S_2(u,X_2,\ldots,X_n),\ldots,S_n(u,X_2,\ldots,X_n))]$$

   so that, letting

$$U_i(u) := (-1)^{bj}\eta_i u^{bc} + (-1)^{bj}u^{bc}V_i(u) =$$
$$(-1)^{bj}\eta_i u^{bc} + (-1)^{bj}u^{bc}T_i(u) + (-1)^{bj+k}u^{bc+d}Q_i(u)$$

   one has:

$$F_1((-1)^{aj}u^{ac}, U_2(u),\ldots,U_n(u)) = \ldots = F_s((-1)^{aj}u^{ac}, U_2(u),\ldots,U_n(u)) = 0.$$

10) If a is even, let ξ be a primitive $2a^{th}$-root of unity and let L_i be the ideal whose roots, all simple, are $\{(\eta_2\xi^b,\ldots,\eta_n\xi^b) : (\eta_2,\ldots,\eta_n)$ is a root of $J_i\}$. For each real root $(\eta_2,\ldots,\eta_n)$ of L_i, $i>1$:

   (a) compute $R_i(t,X_2,\ldots,X_n) = H_i(-t^a, t^b(\eta_2+X_2),\ldots,t^b(\eta_n+X_n))/t^{d_i}$.
   (b) Compute all real solutions of $R_1(t,X_2,\ldots,X_n) = \ldots = R_t(t,X_2,\ldots,X_n) = 0$:

$$[((-1)^j u^c, T_2(u),\ldots,T_n(u));\ (-1)^k u^d;$$
$$(S_2(u,X_2,\ldots,X_n),\ldots,S_n(u,X_2,\ldots,X_n))].$$

(c) For each real solution

$$[((-1)^j u^c, T_2(u), \ldots, T_n(u));\ (-1)^k u^d;$$
$$(S_2(u, X_2, \ldots, X_n), \ldots, S_n(u, X_2, \ldots, X_n))]$$

return

$$[(-u^{ac}, (-1)^{bj}\eta_2 u^{bc} + (-1)^{bj}u^{bc}T_2(u), \ldots, (-1)^{bj}\eta_n u^{bc} + (-1)^{bj}u^{bc}T_n(u));$$
$$(-1)^{bj+k}u^{bc+d};\ (S_2(u, X_2, \ldots, X_n), \ldots, S_n(u, X_2, \ldots, X_n))].$$

5.11 The Algebraic Computations

Our algorithm requires extensive recourse to solving 0-dimensional systems of polynomial equations and to dealing with the arithmetics of algebraic numbers. Efficient techniques for both problems are therefore crucial for the performance of the algorithms we are discussing. However we have chosen here to avoid discussing how to deal with these questions and just to give the existence of algorithms for both problems as granted.

The reason is that not only the current theoretical state of the art of polynomial system solving is much advanced in respect of the currently available implementations (not only in general purpose symbolic computation systems, but even in fairly specialized ones), but that even in this respect, according to the major experts in the area, further advances will be available in the next future (see [10]). Implementations reflecting these theoretical advances are to be considered as forthcoming and will have then an impact on the practical performance of our proposals.

The general philosophy underlying all recent advances in polynomial system solving is that there is no need to actually compute zeroes $(\eta_2, \ldots, \eta_n)$ of a 0-dimensional ideal $I \subset K[X_2, \ldots, X_n]$ provided that one is able to perform arithmetical operations in $K(\eta_2, \ldots, \eta_n)$, so that the effort is in devising effective (and efficient) schemes to perform arithmetics in $K(\eta_2, \ldots, \eta_n)$ when only a 0-dimensional ideal I is given which has $(\eta_2, \ldots, \eta_n)$ as a zero.

Here we discuss briefly how to use for our algorithm some recent ideas in this direction which are contained in the forthcoming paper [15], together with the use of seminumerical techniques for real roots, advocated in [4] and [3].

The basic ideas of [15] are the following:

1. there is a surjection from $K[X_2, \ldots, X_n]/I$ to $K(\eta_2, \ldots, \eta_n)$;
2. $K[X_2, \ldots, X_n]/I$ is a K-vector space of finite dimension $h = \text{mult}(I)$;
3. if I is given by a Gröbner basis, then a subvector space of $V \subset K[X_2, \ldots, X_n]$ can be given which is isomorphic to $K[X_2, \ldots, X_n]/I$;
4. therefore each element of $K(\eta_2, \ldots, \eta_n)$ can be represented (not in a unique way) by an element of V.
5. linear algebra algorithms can be used to perform sums and products of elements of $K[X_2, \ldots, X_n]/I$ and so of the corresponding elements of $K(\eta_2, \ldots, \eta_n)$;
6. moreover if $g \in V$, linear algebra algorithms can be used to compute Gröbner bases (and so vectorial representations) of both $K[X_2, \ldots, X_n]/I_0$ and $K[X_2, \ldots, X_n]/I_1$, where $I_0 := I + (g)$ and $I_1 := I : g$.

While sums and products in $K(\eta_2,\ldots,\eta_n)$ can be computed because of 5), computing inverses is less trivial. Let in fact $g(X_2,\ldots,X_n)\in V$, $g\neq 0$. Not only (because of non-uniqueness of representation) it is possible that $g(\eta_2,\ldots,\eta_n)=0$ but, what is worse, $g(\eta_2,\ldots,\eta_n)$ could be 0 for some zeroes of I, while being different from 0 for other zeroes of I.

However it is possible to prove that denoting $I_0 := I+(g)$ and $I_1 := I : g^h$, the zeroes of I_0 are exactly those zeroes $(\eta_2,\ldots,\eta_n)$ of I such that $g(\eta_2,\ldots,\eta_n)=0$, while the zeroes of I_1 are exactly those zeroes $(\eta_2,\ldots,\eta_n)$ of I such that $g(\eta_2,\ldots,\eta_n)\neq 0$. Moreover by 6) it is possible to compute Gröbner bases of both I_0 and I_1 and (again by linear algebra) a polynomial h such that $gh = 1 \bmod I_1$ so that $h(\eta_2,\ldots,\eta_n)$ is the inverse of $g(\eta_2,\ldots,\eta_n)$ in $K(\eta_2,\ldots,\eta_n)$ for each zero $(\eta_2,\ldots,\eta_n)$ of I_1.

Particular care must be devoted to polynomial arithmetics over $K(\eta_2,\ldots,\eta_n)$ if performed in this model; in fact a polynomial in $K(\eta_2,\ldots,\eta_n)[Z]$ is represented by a polynomial in Z whose coefficients are polynomials in $V\subset K[X_2,\ldots,X_n]$; such coefficients could be zero for some value of $(\eta_2,\ldots,\eta_n)$ and not zero for others; since polynomial division requires taking the non-zero coefficient of highest degree in a polynomial, a polynomial division algorithm will perform differently for different zeroes of I. This means that, while performing polynomial division, several zero-checkings and inverse computations are required, each of them gives a splitting of I into I_0 and I_1; the polynomial division algorithm is then to be performed both over $K[X_2,\ldots,X_n]/I_0$ and $K[X_2,\ldots,X_n]/I_1$.

To make this point more clear, let us compute the squarefree polynomial SQFR(f) associated to $f = Z^3-bZ^2+2aZ$, where (a,b) is a root of $I=(X^2+X, XY-X, Y^2-Y)$, i.e. is either $(0,0)$, $(0,1)$ or $(-1,1)$. To do this we have to compute $h=\gcd(f,f')$, since SQFR$(f) = f/h$. Since $f' = 3Z^2-2bZ+2a$ and its leading coefficient is not zero in $\mathbf{Q}(a,b)$, we perform pseudodivision of f by f', obtaining $9f = q(Z)f' + g(Z)$, for a suitable $q(Z)$ and for $g(Z) = (12a-2b)Z+2a$.

One has:

$$I_0 := I + (12X-2Y) = (X,Y);$$

for (a,b) a root of I_0, one has $g=0$, $h:=\gcd(f,f') = Z^2$, SQFR$(f) = f/h = Z$,

$$I_1 := I : (12X-2Y) = (X^2+X, Y-1)$$

and

$$(12a-2b)(-1/2-3/7a) = -6a-36/7a^2+b+6/7ab = -6a+36/7a+1+6/7a = 1$$

for (a,b) a root of I_1.

Performing division of f' by g one obtains

$$f' = q'(Z)g + 81/49a$$

and then:

$$I_{10} := I_1 + (X) = (X, Y-1);$$
$$I_{11} := I_1 : X = (X+1, Y-1);$$

for (a,b) a root of I_{10}, one has: $81/49a = 0$, so $h := \gcd(f,f') = g(Z) = -2Z$, SQFR$(f) = X^2-X$; for (a,b) a root of I_{11}, one has: $81/49a \neq 0$, so $\gcd(f,f')=1$ and SQFR$(f) = f$.

In fact the only root of I_0 is $(0,0)$ for which $f = X^3$, SQFR$(f) = X$; the only root of I_{10} is $(0,1)$ for which $f = X^3-X^2$, SQFR$(f) = X^2-X$; the only root of I_{11} is $(-1,1)$ for which $f = X^3-X^2-2X$ is squarefree.

An advantage of this approach is that if, in our algorithm, we represent the roots of J by using a Gröbner basis of J (obtained in Step 4), according to the scheme described here, and we directly compute in this model

$$R_i(t, X_2, \dots, X_n) := H_i(t^a, t^b(\eta_2 + X_2), \dots, t^b(\eta_n + X_n))/t^{d_i}$$

for all i and then a standard basis of $(R_1, \dots, R_t)$ the ideal J will be naturally splitted into ideals J_i so that:

1. the union of the zeroes of J_i gives the zeroes of J;
2. for each J_i we will obtains polynomials $S_i(z_2, \dots, z_n, t, X_2, \dots, X_n)$, $i = 1 \dots t_i$ such that for each zero $\eta := (\eta_2, \dots, \eta_n)$ of J_i a standard basis of $R_1, \dots, R_t$ is given by
$$\{S_i(\eta_2, \dots, \eta_n, t, X_2, \dots, X_n\}_{i=1\dots t_i}$$
3. since $\{S_i(\eta_2, \dots, \eta_n, t, X_2, \dots, X_n\}$, $i = 1 \dots t_i$ is a standard basis of $\mathrm{in}(\phi_\eta(J_i))$ the zeroes of J_i are equivalent as required in Step 6).

This hopefully explains why we chose a so careful formulation of step 6) instead of the conceptually simpler, but computationally less efficient, formulation "Compute all roots of J".

We are still left to discuss how to deal with real roots. Our proposal is the following:

1) There are multivariate versions of Sturm Theorem to decide whether a 0-dimensional ideal J_i has real zeroes; we use such techniques and we apply Steps 7) and 9) only if J_i has real zeroes, and Steps 8) and 10) only if L_i has real zeroes.

2) To compute L_i from J_i we do the following: since J_i is known through a Gröbner basis, there are techniques to compute a pseudohomogeneous ideal I_i such that $J_i = \pi(I_i)$; then L_i is simply obtained by $L_i = \pi\sigma(I_i)$.

3) The recursive calls of the algorithm are still performed in the computational model for algebraic numbers described above; so they are done for all roots of J_i (resp. L_i) provided that J_i (resp. L_i) has at least one real root. However when performing Steps 7.c), 8.c), 9.c) and 10.c) in the outmost level of recursion, we use the seminumerical techniques advocated in [4, 3] to obtain floating point approximations of all real algebraic numbers appearing as coefficients.

4) Therefore in particular the poynomials S_i will have floating point coefficients and the computation of polynomial approximations of the formal power series solutions of the S_i will be performed numerically, so that our final output will be polynomial approximations of Puiseux series with floating point coefficients.

5.12 A Complete Example

We apply our algorithm now to compute the real branches of the fourth curve discussed in MacMillan's paper, which is generated by the following polynomials in $\mathbf{Q}[z, x, y]$:

$$F_1 := (x^9 + y^9) + (x^6 + y^6)z + xyz^2 + z^5$$

$$F_2 := y^{10} + x^4 z + y^2(x - y)z + z^3$$

We are looking for Puiseux expansions in $\mathbf{R}[[z]]_{\mathrm{Puis}}$.

The computations have been performed partially by hand and mainly by using the system CoCoA developed in Genova (see Section 5.13).

The multiplicity of the origin as a zero of

$$(F_1(0,x,y),\ F_2(0,x,y)) = (x^9+y^9,\ y^{10})$$

is 90 so we have altogether 90 solutions centered at the origin. MacMillan's exponents are $1/7,\ 2/7,\ 2/3,\ 3/2$.

A standard basis computation shows that (F_1,F_2) is a standard basis for the weights $(7,1,1)$. The ideal

$$J_0 = (x^9+y^9,\ y^{10}+xy^2-y^3)$$

has multiplicity 90, so all solutions have initial exponent $\nu \geq 1/7$. The ideal

$$J := (x^9+y^9,\ y^{10}+xy^2-y^3) : (x^{90},\ y^{90})$$

is equal to the ideal

$$(x+y^8-y,\ y^{63}-9y^{56}+36y^{49}-84y^{42}+126y^{35}-126y^{28}+84y^{21}-36y^{14}+9y^7-2).$$

Its roots are all simple and its multiplicity is 63; so there are 63 solutions with initial exponent $\nu \geq 1/7$ corresponding to 9 branches and 27 solutions with initial exponent $\nu > 1/7$. Only one of these 9 branches is real and has initial approximation $\eta := (-\sqrt[7]{2}, \sqrt[7]{2})$. Substitution gives the polynomials:

$$t^{11}+210b^6y^4+252b^5y^5+210b^4y^6+120b^3y^7+45b^2y^8+10by^9+y^{10}+b^4t-4b^3tx+6b^2tx^2-4btx^3+tx^4+b^2x+15b^2y+2bxy+86by^2+239y^3$$

$$t^{26}+2b^6t^4-6b^5t^4x+15b^4t^4x^2-20b^3t^4x^3+15b^2t^4x^4-6bt^4x^5+t^4x^6+6b^5t^4y+15b^4t^4y^2+20b^3t^4y^3+15b^2t^4y^4+6bt^4y^5+t^4y^6-b^2t^7+bt^7x+84b^6x^3-126b^5x^4+126b^4x^5-84b^3x^6+36b^2x^7-9bx^8+x^9-bt^7y+t^7xy+84b^6y^3+126b^5y^4+126b^4y^5+84b^3y^6+36b^2y^7+9by^8+y^9+18bx-72x^2+18by+72y^2,$$

where b denotes $\sqrt[7]{2}$, from which we obtain:

$$1/2b^5t^{11}+5b^6y^9+1/2b^5y^{10}-2b^6tx^3+1/2b^5tx^4+b^6xy+43b^6y^2+1/2b^5xy^2+239/2b^5y^3+210b^4y^4+252b^3y^5+210b^2y^6+120by^7+45y^8+b^2t-4btx+6tx^2+x+15y,$$

$$-1/504b^6t^{26}+1/28b^5t^{11}-1/504b^6t^4x^6-1/504b^6t^4y^6-1/504b^6x^9-1/504b^6t^7xy+179/504b^6y^9+1/28b^5y^{10}-1/7b^6tx^3+1/28b^5tx^4-1/126b^5t^4+1/42b^4t^4x-5/84b^3t^4x^2+5/63b^2t^4x^3-5/84bt^4x^4+1/42t^4x^5-1/42b^4t^4y-5/84b^3t^4y^2-5/63b^2t^4y^3-5/84bt^4y^4-1/42t^4y^5+1/252bt^7-1/252t^7x+1/7b^6x^2-1/3b^5x^3+1/2b^4x^4-1/2b^3x^5+1/3b^2x^6-1/7bx^7+1/28x^8+1/252t^7y+1/14b^6xy+41/14b^6y^2+1/28b^5xy^2+689/84b^5y^3+29/2b^4y^4+35/2b^3y^5+44/3b^2y^6+59/7by^7+89/28y^8+1/14b^2t-2/7btx+3/7tx^2+y$$

which allow to apply the Implicit Function Theorem.

We now investigate the next initial exponent $2/7$. A standard basis for the weights $(7,2,2)$ is given by:

$$F_1 = (x^9+y^9)+(x^6+y^6)z+xyz^2+z^5,$$

$$F_2 = y^{10}+x^4z+y^2(x-y)z+z^3,$$

$$F_3 = y^2z^6-x^8z^3-x^7yz^3-x^6y^2z^3-x^5y^3z^3-x^4y^4z^3-x^3y^5z^3-x^2y^6z^3-xy^7z^3-y^8z^3-x^8y^{10}-x^7y^{11}-x^6y^{12}-x^5y^{13}-x^4y^{14}-x^3y^{15}-x^2y^{16}-xy^{17}-y^{18}-x^{12}z-x^{11}yz-x^{10}y^2z-x^9y^3z-x^8y^4z-x^7y^5z-x^6y^6z-x^5y^7z-x^4y^8z+x^6y^2z^2+y^8z^2+2y^{11}z+xy^3z^3.$$

The ideal J_0 has multiplicity 27 (so that there are no solutions with initial exponent ν, $1/7 < \nu < 2/7$). The ideal J is $(x-y,\ y^7+1/2)$ and has multiplicity 7. There are therefore 20 solutions with initial exponent $\nu > 2/7$ and 7 solutions (corresponding to a single branch) with initial exponent $2/7$.

The ideal J has the single real root $\eta := (-1/\sqrt[7]{2}, -1/\sqrt[7]{2})$ which is simple. Step 7 gives then:

$2/7b^6t^{17} - 10b^6t^7y^9 - b^5t^7y^{10} + 2/7b^6x^9 + 45/2b^6t^7y^2 + 60b^5t^7y^3 + 105b^4t^7y^4 + 126b^3t^7y^5 + 105b^2t^7y^6 + 60bt^7y^7 + 45/2t^7y^8 + 2/7b^6y^9 - b^5t^8 + 2/7b^6tx^6 + 2/7b^6ty^6 - 4b^6t^2x^3 - b^5t^2x^4 - 1/4bt^7 - 36/7b^6x^2 - 12b^5x^3 - 18b^4x^4 - 18b^3x^5 - 12b^2x^6 - 36/7bx^7 - 9/7x^8 - 5/2t^7y - 12/7b^6xy - 22/7b^6y^2 - b^5xy^2 - 11b^5y^3 - 18b^4y^4 - 18b^3y^5 - 12b^2y^6 - 36/7by^7 - 9/7y^8 - 2/7b^5t - 6/7b^4tx - 15/7b^3tx^2 - 20/7b^2tx^3 - 15/7btx^4 - 6/7tx^5 - 6/7b^4ty - 15/7b^3ty^2 - 20/7b^2ty^3 - 15/7bty^4 - 6/7ty^5 + 1/2b^2t^2 + 2bt^2x + 3t^2x^2 + x$

$2/7b^6t^{17} + 10b^6t^7y^9 + b^5t^7y^{10} + 2/7b^6x^9 - 45/2b^6t^7y^2 - 60b^5t^7y^3 - 105b^4t^7y^4 - 126b^3t^7y^5 - 105b^2t^7y^6 - 60bt^7y^7 - 45/2t^7y^8 + 2/7b^6y^9 + b^5t^8 + 2/7b^6tx^6 + 2/7b^6ty^6 + 4b^6t^2x^3 + b^5t^2x^4 + 1/4bt^7 - 36/7b^6x^2 - 12b^5x^3 - 18b^4x^4 - 18b^3x^5 - 12b^2x^6 - 36/7bx^7 - 9/7x^8 + 5/2t^7y + 16/7b^6xy - 50/7b^6y^2 + b^5xy^2 - 13b^5y^3 - 18b^4y^4 - 18b^3y^5 - 12b^2y^6 - 36/7by^7 - 9/7y^8 - 2/7b^5t - 6/7b^4tx - 15/7b^3tx^2 - 20/7b^2tx^3 - 15/7btx^4 - 6/7tx^5 - 6/7b^4ty - 15/7b^3ty^2 - 20/7b^2ty^3 - 15/7bty^4 - 6/7ty^5 - 1/2b^2t^2 - 2bt^2x - 3t^2x^2 + y$

where $b = -1/\sqrt[7]{2}$.

We now investigate the initial exponent $2/3$. A standard basis for the weights $(3,2,2)$ is:

$F_1 = x^9 + y^9 + x^6z + y^6z + z^5 + xyz^2,$

$F_2 = -x^6y^{10} - y^{16} - y^{10}z^4 - xy^{11}z + x^{13} + x^4y^9 + x^{10}y^2 - x^9y^3 + xy^{11} - y^{12} + x^9z^2 + y^9z^2J,$

$F_3 = y^{10} + x^4z + xy^2z - y^3z + z^3,$

$F_4 = -y^{10}z + x^9y + y^{10} + x^6yz + y^7z + yz^5 - x^4z^2 + y^3z^2 - z^4,$

$F_5 = xy^{10}z - x^{10}y + x^9y^2 - xy^{10} + y^{11} - x^7yz + x^6y^2z - xy^7z + y^8z - xyz^5 + y^2z^5 + x^5z^2 + xz^4.$

The ideal J_0 has multiplicity 3 (so that there are 17 solutions with initial exponent ν, $2/7 < n < 2/3$). The ideal J is $(x,\ y^3 - 1)$ which has multiplicity 3 and has only simple roots, the only real one being $(0,1)$, which corresponds to a real analytic branch of index 3 and intial exponent $2/3$. Step 7 gives then:

$t^8x^9 + t^8y^9 + 9t^8y^8 + 36t^8y^7 + 84t^8y^6 + 126t^8y^5 + 126t^8y^4 + t^5x^6 + 84t^8y^3 + t^5y^6 + 36t^8y^2 + 6t^5y^5 + 9t^8y + 15t^5y^4 + t^8 + 20t^5y^3 + 15t^5y^2 + 6t^5y + 2t^5 + xy + x$
J

J $-1/3t^{11}y^{10} - 10/3t^{11}y^9 - 15t^{11}y^8 - 40t^{11}y^7 + 1/3t^8x^9 - 70t^{11}y^6 + 1/3t^8y^9 - 84t^{11}y^5 + 3t^8y^8 - 70t^{11}y^4 + 12t^8y^7 - 40t^{11}y^3 + 28t^8y^6 - 15t^{11}y^2 + 42t^8y^5 - 10/3t^{11}y + 42t^8y^4 - 1/3t^{11} + 1/3t^5x^6 + 28t^8y^3 + 1/3t^5y^6 + 12t^8y^2 + 2t^5y^5 + 3t^8y + 5t^5y^4 + 1/3t^8 + 20/3t^5y^3 + 5t^5y^2 - 1/3t^2x^4 + 2t^5y + 2/3t^5 - 1/3xy^2 + 1/3y^3 - 1/3xy + y^2 + y$

Since the origin is not a solution of J_0 there are no solution with initial exponent $\nu > 2/3$. We have yet to find the 17 solutions with initial exponent ν, $2/7 < \nu < 2/3$. The standard basis for the weights $(3,2,2)$ gives a potential MacMillan exponent $\nu = 1/2$, while the standard basis for the weights $(7,2,2)$ does not introduce new initial exponents in the range $(2/7, 2/3)$. Computing a standard basis for the weights $(1,1,2)$ gives:

$F_1 = x^9 + y^9 + x^6z + y^6z + z^5 + xyz^2,$

$F_2 = -x^6y^{10} - y^{16} - xy^{11}z + x^4z^5 + z^7 + x^{13} + x^4y^9 + x^9z^2 + y^9z^2 + xy^2z^5 - y^3z^5 + x^{10}y^2 - x^9y^3 + xy^{11} - y^{12},$

$F_3 = y^{10} + x^4z + xy^2z - y^3z + z^3,$

$F_4 = -y^{10}z + x^9y + y^{10} + x^6yz + y^7z + yz^5 - x^4z^2 + y^3z^2 - z^4\ ,$

$$F_5 = xy^{10}z - x^{10}y + x^9y^2 - xy^{10} + y^{11} - x^7yz + x^6y^2z - xy^7z + y^8z - xyz^5 + y^2z^5 + x^5z^2 + xz^4.$$

The ideal J_0 has multiplicity 7 and the ideal $J = (y,\ x^4 + 1)$ multiplicity 4. Therefore there are 13 solutions with initial exponent ν, $2/7 < \nu < 1/2$, 3 with initial exponent $\nu > 1/2$ (which are the ones already found with initial exponent $\nu = 2/3$) and 4 with initial exponent $1/2$. The sistem $y = x^4 + 1 = 0$ has no real solution. Homogeneization of J gives the ideal ${}^hJ = (y,\ x^4 + z^2)$, so $\sigma({}^h(J)) = {}^hJ$, and the ideal whose roots are $\{(\xi\eta_2, \xi\eta_3) : (\eta_2, \eta_3) \text{ root of } J\}$, where ξ is a 4th-root of 1, is again J. In fact the roots of J are $\{(\zeta^{2i+1}, 0) : i = 0\dots 3\}$ where ζ^8 is a primitive 8th-root of 1; we can choose $\xi = \zeta^2$ so that $(\xi\zeta^{2i+1}, 0) = (\zeta^{2i+3}, 0)$. The two branches with initial exponent $1/2$ are therefore both complex.

We have still to find the 13 solutions with initial exponent in the range $(7/2, 1/2)$. The standard basis for the weights $(2,1,1)$ gives a new candidate exponent $1/3$. Computing a standard basis for the weights $(3,1,1)$ we obtain that J_0 has the origin as its only root of multiplicity 7, so that the missing thirteen solutions have initial exponent in the range $(7/2, 1/3)$. The standard basis also introduces the new candidate exponent $4/13$. With respect to the weights $(13,4,4)$ we obtain that J_0 has multiplicity 20 and that $J = (y, x^{13} + 1)$ has multiplicity 13. The single branch of index 13 and initial exponent $4/13$ is real since J has the single real root $(-1, 0)$, to which Step 7 can be applied. In conclusion there are 4 real branches of indexes 7, 7, 3, 13 and initial exponents $1/7$, $2/7$, $2/3$, $4/13$. Moreover there are 8 complex branches of index 7 and initial exponent $1/7$ and 2 complex branches of index 2 and initial exponent $1/2$.

Since in the previous example, we never needed to perform our "main" transformation, since we never encountered multiple roots, we perform here a partial computation on the second MacMillan example, where multiple roots are met.

The ideal is in $\mathbf{Q}[u, x, y, z]$ and we look for Puiseux expansions in $\mathbf{K}[[u]]_{\text{Puis}}$. It is generated by

$$(x^3 + (x + y + z)u + u^2,\ y^4 + (x - y + z)u - u^3,\ z^5 + (x^2 + y^2 + zx)u + xu^2 + u^4)$$

We look for Puiseux expansions of initial exponent $1/3$. The ideal

$$J_0 = (x^3,\ y^4 + x - y + z,\ z^5 + x^2 + y^2 + xz)$$

has multiplicity 60, the origin as a root of multiplicity 6 and other 18 triple roots which are the roots of the ideal

$$J_1 = (x,\ y - z^{10} - z,\ z^{18} + 2z^9 + z^3 + 1).$$

We perform therefore the transformation: $x = xt$, $y = (b + y)t$, $z = (c + z)t$, $u = t^3$ where (b, c) is a root of $(b - c^{10} - c, c^{18} + 2c^9 + c^3 + 1)$ and we obtain the curve defined by the polynomials:

$$c^{10}t + 2ct + x^3 + xt + yt + zt + t^3,$$

$$-4c^{15}y + 4c^{10}y^3 - 4c^6y - 6c^5y^2 + 4cy^3 + x + y^4 - y + z - t^5,$$

$$2c^{10}y + 5c4z + 10c^3z^2 + 10c^2z^3 + cx + 2cy + 5cz^4 + x^2 + xz + xt^2 + y^2 + z^5 + t^7.$$

We choose the MacMillan exponent $1/3$ and we compute a standard basis for the weights $(3, 1, 1, 1)$. The ideal J is generated by:

$x^3 + c^{10} + 2c$,

$z - 14357/49781xc^{15} + 10250/149343xc^{12} - 4920/49781xc^9 - 84800/149343xc^6 +$
$22550/149343xc^3 - 25181/49781x$,

$y - 29651/149343xc^{15} + 10250/49781xc^{12} - 14760/49781xc^9 - 55276/149343xc^6 +$
$22550/49781xc^3 - 25762/49781x$

where c satisfies $c^{18} + 2c^9 + c^3 + 1$. For each of the 18 values of c there are exactly 3 simple roots of the system.

Remark that a single computation is all we need to perform Step 7 for all 4-tuples (c, x, y, z) satisfying the conditions above.

5.13 Some Experimentations in CoCoA

While the algorithm outlined here is not implemented, we have performed some experimentation in CoCoA (version 1.5.3), a system for symbolic computations in Commutative Algebra and Algebraic Geometry developed at the Mathematics Department of the University of Genoa by A. Giovini and G. Niesi [1, 7]. This system, written partly in Pascal and partly in C, runs on any computer of the Macintosh or MS-DOS family. It allows to compute Gröbner bases and standard bases of polynomial ideals over $\mathbf{Q}$ or $\mathbf{Z}_p$, to perform ideal operations and to compute invariants of ideals. In particular it computes multiplicities of ideals. It must be remarked that the algorithms to perform these operations are not (in the 0-dimensional case) the efficient linear algebra ones, but are founded on Gröbner basis computations; in particular the division $J : (X_2^h, \dots, X_n^h)$ requires n Gröbner basis computations and can be quite costly. CoCoA allows also to compute Gröbner bases for polynomial ideals over a field $\mathbf{Q}(\eta_2, \dots, \eta_n)$ where $(\eta_2, \dots, \eta_n)$ is a root of a zero-dimensional ideal I, at least in the case in which the surjection $\mathbf{Q}[X_2, \dots, X_n]/I \to \mathbf{Q}(\eta_2, \dots, \eta_n)$ is actually a bijection. Standard basis computations in this setting are not available in CoCoA 1.5.3, but are present in an experimental version. In both cases, the algorithms can be applied also in case the surjection is not a bijection, but then a careful interpretation of the output is needed. CoCoA has no facility to recognize real roots of systems not for multiplicity handling. Therefore it can be used for all steps of the algorithm outlined in section 7 except Step (6) and real-root recognition where ad hoc hand-driven computations are required.

Here we focus on Steps (1) to (5) which can be performed by the following instructions in CoCoA.

$h = \mathsf{TangentCone}(i)$; $h = h[z = 1]$
computes the standard basis $(H_1, \dots, H_t)$ of $(F_1, \dots, F_s)$ and returns J_0.

$\mathsf{Mult}(R/h)$
counts the number of roots with multiplicity.

$j = h : \mathsf{ideal}(x^m, y^m)$; $k = \mathsf{gbasis}(j)$; $\mathsf{Mult}(S/j)$
removes the null root, counts the number of non zero roots, with multiplicity, and gives the ideal J; m is the number of roots with multiplicity.

The tables below report an outline of the computation (with timings in secs.) for the example discussed in the previous paragraph and for another example in MacMillan, i.e. the curve with equations:

$$X^3 + (X^2 - Y^2)Z + Z^4, \quad Y^3 + (X^2 - Y^2)Z - Z^4$$

Both examples have been computed on a Macintosh SE, with 2MB RAM, which is among the slowest computers in the Macintosh family.

TABLE 5.1. Invariants & Timings

Initial exp. ν	1/7	2/7	2/3	1/2*	1/3*	4/13*
# roots, μ	90	27	3	7	7	20
# non-zero roots, μ_1	63	7	3	4	0	13
mult(0), μ_0	27	20	0	3	7	7
h=TangentCone(i)	0.86	2.59	2.15	1.98	5.50+	5.50+
$h = h[z = 1]$	0.16	0.20	0.25	0.13	0.21	0.20
Mult (R/h)	0.96	1.14	0.94	1.01	1.04	0.98
$j = h : \text{ideal}(x^m, y^m)$	5.91	4.38	2.36	2.95		2.56
$k = \text{gbasis}(j)$	1.79	0.73	0.75	0.66		0.56
Mult(S/j)	0.50	0.21	0.18	0.20		0.23

* the initial exponent has been derived by slopes derived from a standard basis computation.
\+ the standard basis computation has been truncated (cf. Appendix B).

TABLE 5.2. Invariants & Timings

Initial exp. ν	1	3/2	4/3*	9/7*	5/4*
# roots, μ	9	0	3	3	7
# non-zero roots, μ_1	2		0	0	4
mult(0), μ_0	7		0	3	3
h=TangentCone(i)	0.26	0.41	0.78	0.71	0.75
$h = h[z = 1]$	0.20	0.18	0.25	0.20	0.23
Mult (R/h)	0.30	0.26	0.31	0.36	0.43
$j = h : \text{ideal}(x^m, y^m)$	1.79		0.51		0.83
$k = \text{gbasis}(j)$	0.31		0.15		0.15
Mult(S/j)	0.18		0.20		0.18

* the initial exponent has been derived by slopes derived from a standard basis computation.

5.14 Appendix A. Algebraic Preprocessing

Throughout the paper we have assumed that our input is a set of polynomials $F_1, \ldots, F_s$ in $K[X_1, \ldots, X_n]$, where K is a finite algebraic extension of the rationals, which define a curve Γ, whose irreducible components are all simple. To test this, by symbolic computation methods, we have first to compute a Gröbner basis of $I = (F_1, \ldots, F_s)$, by which we can read the dimension of I; if the dimension of I is 1 then the components of Γ are only curves and points.

It is then possible to compute the top-radical of $I := (F_1, \ldots, F_s)$ which is the ideal defining a variety whose irreducible components are all simple and are the components of Γ of maximal dimension. If I coincides with its top-radical, then I satisfies the assumptions; otherwise the algorithm can be applied to the top-radical of I.

Assuming now that I defines a curve Γ with no multiple irreducible components, the singular points of Γ, which are finite in number are the roots of the 0-dimensional ideal generated by $(F_1, \ldots, F_s)$ and by the maximal minors of the Jacobian matrix $(\partial F_i/\partial X_j)_{ij}$.

If $\alpha := (\alpha_1, \ldots, \alpha_n)$ is a singular point of the curve, by the translation $X_i \to X_i - \alpha_i$,Jwe can assume that α is the origin. Our final assumption is then that none of the irreducible components of Γ passing through the origin is contained in the hyperplane $X_1 = 0$.

To check this, let $\pi : K[X_1, \ldots, X_n] \to K[X_2, \ldots, X_n]$ be the projection such that $\pi(X_1) = 0$, $\pi(X_i) = X_i$ for $i > 1$. The local dimension of $\pi(I)$ is 1 if and only if there are components of Γ passing through the origin and contained in the hyperplane $X_1 = 0$. Then the ideal $J := I : \pi(I)$ locally defines a curve whose components are the components of I not contained in the hyperplane $X_1 = 0$. We can then apply our algorithm separately to J and to $\pi(I)$.

5.15 Appendix B. Truncated Standard Basis Computations

In the example presented in Section 5.12, the standard basis computation for weights $(3,1,1)$ and $(13,4,4)$ has been truncated after five elements have been computed; there are at least two more elements in it, but a complete computation failed after approx. 20 mins. for memory overflow. In both cases however the missing elements would not change the ideal J_0. In fact the only possible effects of truncating a standard basis computation are that the resulting ideal J_0 has higher multiplicity than the correct output, i.e.:

1) the resulting ideal has some more zeroes than the correct output; this will be detected when applying recursively the algorithm to the transformed ideal: a complete standard basis computation of the R_i's will produce in this case an ideal with no zeroes;

2) the resulting ideal has some zeroes in common with the correct output, but with higher multiplicity; again a complete standard basis computation of the R_i's will allow to find the correct multiplicity;

3) the multiplicity of the origin in the resulting ideal is higher than μ_0; this can be tested by comparing the resulting μ_0 with the value of μ for the next higher initial exponent; if the two values are equal, then the obtained value for μ_0 is the correct one; otherwise the standard basis computation must be resumed.

In our example, for the weights $(3,1,1)$ we obtain $\mu = 7$; since this is the maximal possible value for for the multiplicity of J_0, we must have obtained the correct output. For the weights $(4,4,13)$ we obtain $\mu_1 = 13$; it could be that the correct value is less (so that there are solutions with initial exponent ν, $2/7 < \nu < 13/4$). Since each root however returns a solution (as it is checked at the next recursion level), this is not the case.

It is to be remarked that a common feature of standard (and Gröbner) basis computations is that the most expensive part of the algorithm often occurs after a standard basis has been obtained and is only necessary to certify the result. Moreover standard basis elements entering late in the output have higher chances to modify the standard basis only but not the ideal J_0. It would be worthwhile to investigate how much it is possible to make recourse in an automatic way to truncated standard basis computations, while recovering missing informations on the number of solutions by backtracking through the recursion levels.

5.16 References

[1] E. Armando, A. Giovini and G. Niesi. *CoCoA User's Manual, (v. 1.5)*. Dipartimento di Matematica, Università di Genova, 1991.

[2] A. D. Bruno and A. Soleev. The local uniformization of branches of a curve. Preprint IHES 34, 1990.

[3] P. Cellini, P. Gianni and C. Traverso. Shape of Curves and Surfaces: the Combinatorics In *Computer Graphics and Mathematics*, EurographicSeminar Series, Springer-Verlag, eds. B. Falcidieno, I. Herman and C. Pienovi, 1992 (present book).

[4] P. Cellini, P. Gianni and C. Traverso. Algorithms for the shape of semialgebraic sets. A new approach. In *Proc. AAECC 9*, number 539 in Lect. Notes in Comp. Sci., pages 1–18. Springer-Verlag, 1991.

[5] C.L. Bajaj, C.M. Hoffmann, R.E. Lynch and J.E. Hopcroft. Tracing surfaces intersections. *Computer Aided Geometric Design*, 5:285–307, 1988.

[6] F. Cucker, L. Pardo, T. Recio, M.F. Roy and M. Raimondo. On the computation of the local and global branches of real algebraic curves. In *Proc. AAECC 6*, number 356 in Lect. Notes in Comp. Sci., pages 161–181. Springer-Verlag, 1989.

[7] A. Giovini and G. Niesi. CoCoA: A User-Friendly System for Commutative Algebra. In *Proc. DISCO 90*, number 429 in Lect. Notes in Comp. Sci., pages 20–29. Springer-Verlag, 1990.

[8] W. Groebner. *Algebraische Geometrie*. Bibliographisches Institut Mannheim, 1968-70.

[9] C.M. Hoffmann. Algebraic curves. In *Mathematical aspects of Scientific Software*, number 14 in IMA Volumes in Mathematics and Applications, pages 101–122. Springer-Verlag, 1988.

[10] G. Cardanus jr. POSSO: a report on the expected state of the art in POlynomial System SOlving. in preparation.

[11] Y. Lakshman. *On the complexity of computing Groebner bases for zero dimensional polynomial ideals*. PhD thesis, Rensselaer Polytechnique Institute, 1991.

[12] W.D. MacMillan. A method for determining the solutions of a system of analytic functions in the neighborhood of a branch point. *Math.Ann.*, 72:180–202, 1912.

[13] J. Maurer. Puiseux expansions for space curves. *Manuscripta Math.*, 32:91–100, 1980.

[14] T. Mora, G. Pfister, and C. Traverso. An introduction to the tangent cone algorithm. In *Issuses in robotics and non-linear geometry*. JAI Press, 1991.

[15] T. Mora and C. Traverso. Representing algebraic numbers by Groebner bases. Proc. MEGA 92. (submitted).

[16] C. Traverso. Communication at the workshop Computing in Algebraic Geometry. Cortona, 1991.

6 Shape of Curves and Surfaces: the Combinatorics

Paola Cellini
Patrizia Gianni
Carlo Traverso

ABSTRACT This paper develops some points that were only sketched in the companion paper [6], in particular explaining the combinatorial part of the algorithms for the computation of the topology of real plane curves and space surfaces. The origin of these algorithms is [11].
In particular, for the curves, we compute the shape through a computation of the intersection of the curve with the vertical tangents. From the number and placement of the roots of the intersection we can derive an explicit description of the shape.
For the surfaces, we derive the information on the topology considering the change of the shape of a plane section, through a computation of the rank of the homology groups of the connected components.

6.1 The shape of curves

We consider pairs (X, Y) of a topological set X and a subset $Y \subseteq X$; two pairs (X, Y), (X', Y') have the same *shape* if an homeomorphism $\phi : X \to X'$ exists such that $\phi(Y) = Y'$.

For some classes of pairs a classification of shapes can be given, We are interested in particular when X is either

1. $\mathbf{PR}^2$, the projective 2-space, or
2. $\mathbf{R}^2$, or
3. a closed rectangle, product of two closed intervals in $\mathbf{R}$,

and Y is a non-singular real algebraic curve.

In these cases, Y is a union of a finite number of connected components, being homeomorphic either:

- to the circle S^1, or
- to an open segment (only in case 2), or
- to a closed segment (only in case 3).

Any such component S disconnects the complementary $X - S$ with one exception: when the pair (X, S) has the shape of $(\mathbf{P}^2, \mathbf{P}^1)$, a projective line in a projective plane. Otherwise, if S is a circle, then $X - S$ has two components, which are identified as the interior (being homeomorphic to a disk) and the exterior, being homeomorphic either to:

- a Möbius band in case 1,
- $\{(x, y) \in \mathbf{R}^2 \mid x^2 + y^2 > 1\}$ in case 2,
- $\{(x, y) \in \mathbf{R}^2 \mid 2 \geq x^2 + y^2 > 1\}$ in case 3.

In the projective case, only one component does not disconnect the complementary (and only if the curve has odd degree), and is called the *odd branch*; the other components are called *ovals*. The shape is completely defined giving the "inclusion" between ovals (an oval is included in another if it is contained in its interior).

An alternative way to define the shape, always in the projective case, is the following: give the set of connected components of the curve C, the set of the connected components of the complementary $\mathbf{P}^2 - C$ (called *regions*), and the incidence relation between them, that is true if a component of the curve is in the boundary of a component of the complementary. If the degree is odd this is sufficient (the unique odd branch is identified since it is the unique component of the curve that bounds only one component of the complementary), otherwise one needs the further information that identifies the unique region that is non-orientable. From these data the shape, as inclusion between the ovals, can be recovered as follows.

The region bounded by the odd branch (in the odd degree case) or the non-orientable region are the outermost regions; any component can be connected uniquely to any region by a chain without repetitions (component, region, component, region,...) where two subsequent terms are incident; a component C_1 is in the interior of a component C_2 if and only if the unique chain connecting C_1 to the outermost region contains C_2.

In the case of the rectangle X, the shape is determined by the following data:

1. the set of components of C in the interior of X (components tangent to the boundary are hence split in two) (*components*)
2. the set of connected components of $\mathbf{P}^2 - C$ (*regions*)
3. the set P of points of C that lie on the boundary of X (*boundary points*)
4. the set of components of the boundary minus P (*boundary components*)
5. the incidence relation between boundary points and components or boundary components
6. the incidence relation between components or boundary components and regions.

This description is of course also valid for any region homeomorphic to X, for example a disk.

There are some conditions on these data in order to be the data corresponding to a curve:

- there are as many boundary components as boundary points;
- any boundary point is adjacent to two boundary components and to one component (in special cases, when the curve is tangent to the boundary of the rectangle, to zero or two components);
- any component is adjacent to two regions;
- the graph having the regions as nodes, the components as sides, and the adjacency as incidence, is acyclic;
- any boundary component is adjacent to one region;
- any component is adjacent to zero, one or two boundary points;

- any boundary component is adjacent to two boundary points (with a unique exception when there are no boundary points or one point and there is then one boundary component);
- the graph consisting of the boundary components and the boundary points consists of a unique cycle (with the same exception as above);
- if C, C' are components and P, Q (resp. P', Q') are their boundary points, then P, P', Q, Q' cannot appear in this order inside the cycle described in the preceding point.

If these conditions are satisfied, then there is a curve in a square whose shape is given by these data.

In the case of curves in $\mathbf{R}^2$, the situation is strictly analogous to the previous case, but instead of the boundary points we have to consider the asymptotes of C, (considering one boundary point for every branch) as a limit case of increasing size of rectangles in $\mathbf{R}^2$.

6.2 An algorithm for the shape of plane curves

To our knowledge, the first algorithm for the determination of the shape of plane algebraic curves is in [11]; the algorithm was independently found and generalized, but this remained the unique account that gives an intrinsic description of the shape. Here we generalize the algorithm with a better and more general description.

The algorithm for the determination of the shape of a curve starts from the rectangle case; from this we derive the shape in $\mathbf{R}^2$ as a limit case, and in $\mathbf{P}^2$ with a supplementary discussion.

We assume hence that we have a rectangle X, and a curve $C \subseteq X$. The "top", "bottom", "left" and "right" side of a rectangle $[r, s] \times [a, b]$ are precisely what you imagine: for example, $\{r\} \times [a, b]$ is the left side.

A first intermediate step is to compute an integer n_0 and three finite sequences of integers, $n_1, \ldots, n_d$, $m_1, \ldots, m_d$, $k_1, \ldots, k_d$. In the "generic" case these are defined as follows.

Consider the set $P_1, \ldots, P_d$ composed of the vertical tangency points of C contained in the interior of X, and of the intersections of C with the boundary of X. We assume that no vertical line contains more than one such point (if this is false, we may deform — topologically — the curve to obtain this "generic position", see [11]). We assume that the $P_1, \ldots, P_d$ are ordered "from left to right", i.e. if $P_i = (x_i, y_i)$, $P_{i+1} = (x_{i+1}, y_{i+1})$ then $x_i < x_{i+1}$.

Then:

- n_0 is the number of points on the left side of X
- n_i is the number of points of C in the vertical segment passing through P_i
- m_i is the number of points of C on the part of the vertical segment passing through P_i and below P_i
- k_i is a number in a finite range that classifies the different type of P_i: either it is an internal vertical tangency point, and may have two local half-branches at the left, or two local half-branches at the right, or one branch at each side; or it is a boundary point, and it is on the upper side, or on the lower side, and may have an half-branch on the left or not, and one on the right or not.

These data are not independent: n_i, k_i and k_{i+1} determine $n_i + 1$.

We show how these data allow to reconstruct the shape; in [6] we have discussed how to find these m_i, n_i, k_i.

The algorithm proceeds scanning X from left to right; more precisely, let X_t be $\{(x,y) \in X | x \leq t\}$; we show how the shape varies when t increases passing one of the X_i. We keep track of a little more than the shape of X_t: we have to know on which side are the boundary points, and in which sequence (counterclockwise) they are on the sides. This implies that we know which boundary components appear on each side, and in which sequence. In particular, the first boundary component of the right side will be the bottom one, bounded by point number 1 (if it exists), the i-th will be bounded by the $i-1$-th and the i-th point, (if they exist), and so on. It may happen that the top and the bottom component are the same, if the other sides do not have boundary points; or that only one boundary component appears on the right side, if it has no boundary points.

The shape of X_t varies corresponding to the variation of its right side R_t, and this changes only for values for which an X_i exists.

To simplify the description we assume that the rectangle is $[0,1] \times [0,1]$, that the boundary points on the left or right side have only one branch, and that there are no boundary points on the corners of X; in that case, the shape of X_ϵ, for small ϵ is the following: we have $2n_0$ boundary points, n_0 components, n_0+1 regions, $2n_0-2$ boundary components (0 if $n_0 = 0$), and the boundary relations between them are easy (but long) to specify. The extra information too is easy to describe: n_0 of the boundary points are on the left side, and n_0 on the right.

Now we describe how these data change when we pass through one of the P_i; to simplify the description, we only show the passage from $X_{t-\epsilon}$ to $X_{t+\epsilon}$, when t corresponds to P_i. The description that follows has some obvious exceptions when there are 0 or 1 boundary points.

There are two typical transformations: a new component appears, or two component endpoints are identified. These operations have a series of side effects.

In the first case, the appearance is inside one of the boundary components: two endpoints appear inside this boundary component, (the endpoints of the new component), and split it in three boundary components; a new region appears, that is incident to the new component and to the middle new boundary component.

In the second case, the identification of the two endpoints causes the two components to become one, and a series of identifications happens. Here we give the precise details.

A) If P_i is internal to X, then:

1. If P_i has one branch on each side, nothing happens.
2. If P_i has two branches on the right,
   (a) a new component appears;
   (b) a new region appears;
   (c) two new boundary points appear in position m_i+1 and m_i+2 in the right side (i.e. inside the m_i+1-th boundary component in the right side);
   (d) the $m+1$ boundary component is split in 3;
   (e) the three boundary components arising from the splitting and the two new boundary points are inserted in the cycle graph in an obvious way (obvious exception when there was only one boundary component);
   (f) of the three boundary components arising from the splitting, the top and bottom ones inherit the incidence to the region, and the middle one is incident to the new region;

(g) the new component is incident to the new region and to the region that was incident to the boundary region that was split.

3. If P_i has two branches on the left,
   (a) two boundary points, in position $m_i + 1$ and $m_i + 2$, disappear;
   (b) the boundary component between these points disappears, and the other two boundary components bounded by them are identified;
   (c) the two boundary component identified in the previous point are each incident to one region; these two regions are identified;
   (d) the components incident to the points that disappear are identified and inherit the incidence with the regions (no further identification is needed);
   (e) all the other incidence relations are inherited.

B) If P_i is on the upper boundary

1. if it has one branch on the right
   (a) a new component appears;
   (b) a new region appears;
   (c) two new boundary points appear, one on the top side and another in the right side in top position;
   (d) the top boundary component of the right side is split in three;
   (e) the incidence relations are settled as in case A2.
2. if it has one branch on the left, the top boundary point of the right side migrates to the top side;
3. if it has one branch on the left and one on the right, this is equivalent to adding first a point with a branch on the left, then a point with a branch on the right, then identifying the two points deleting the boundary component between them;
4. if it has no branches, then a new boundary point appears, a boundary component is split, and the boundary relations are inherited.

(remark that cases 3. and 4. only appear when C is tangent to the upper side, from below in case 3., from above in case 4.).

C) If P_i is on the lower boundary, the discussion is the same as in B, with some obvious modifications.

This settles the computation of the shape in a rectangle. For the shape in $\mathbf{R}^2$, we can first consider the case of a vertical strip as a limit of the rectangle when the height goes to infinity, then the case of $\mathbf{R}^2$ as the limit of the case of the strip when the width goes to infinity. Hence, in the case of $\mathbf{R}^2$ the vertical asymptotes appear as limit of boundary points on the top and bottom side, and the other asymptotes (points at infinity) as limit of boundary points on the right and left side.

To compute the shape in the case of the projective space, one has to make identifications of "minus infinity" and "plus infinity"; more precisely, one begins remarking that we can have only one vertical asymptote (otherwise one has a multiple point at infinity, at the point corresponding to the vertical direction); then we identify the two boundary points corresponding to them; moreover we identify the boundary points at the left (at minus infinity) and at the right (plus infinity) reversing the order (at left from top to bottom, at the right from bottom to top), identifying moreover the components and the components at the infinity that bound identified points and the regions bounded by identified boundary regions; in the case of curves of even degree, there is one and only one region that is bounded by two boundary components that are identified: this region is non-orientable in the projective space, hence it is identified as the outermost region.

Discarding the boundary points and components, we have a set of components and a set of regions; in the odd degree case there is a unique component that bounds only one region, in the even degree case an outermost region has been identified; this allows to find whether a component is in the interior of another one.

6.3 The topology of surfaces

There is no easy complete description in finite terms of the shape of a surface in space. For example, a toroidal surface can be embedded in the space with a knot; two toroidal surfaces can be linked in infinitely many ways; and these knots and links are not completely classified. The only information that can be easily decribed is the topology of the surface.

The topology of a non-singular surface with boundary S is uniquely characterized by the number of connected components and, for each of them, by the genus, the number of boundary components, and the orientability (a surface is either orientable or non-orientable). Instead of the genus we can give the dimension of the first homology group H_1.

We consider here only the simplest case: the topology of compact surfaces in $\mathbf{R}^2$. We find it, in analogy to the case of the curves, considering the variation of the topology of the intersection of the surface with an halfspace.

We assume hence that the surface S is contained in the interior of the cube $[0,1] \times [0,1] \times [0,1]$, and in "generic position" (we explain this later), and we study how the intersection S_t of S with $X_t = [0,1] \times [0,1] \times [0,t]$ changes.

Consider the function $\zeta(x,y,z) = z$ on the surface S. The topology of S_t changes only for values t that correspond to critical points of the function ζ on S. The genericity hypothesis on S consists on the fact that these critical points are non-degenerate (the hessian is non-degenerate) and different critical points correspond to different values of the function ζ (almost all linear changes of coordinates put S in generic position,).

To study the change of the topology of S_t we use Morse theory, see [14]. The passage through a critical point corresponds up to homotopy to the addition of a cell of dimension i, where i is the index of the critical point, i.e. the number of negative eigenvalues of the Hessian at the critical point. This addition changes the component and in particular their homology. We describe the possible types of these changes, depending on the geometry of the situation; this analysis will allow to understand the change in the topology from geometric data.

Passing through a critical points the curve section C_t acquires a singular point, and changes with the following pattern, as can be proved by Meyer-Vietoris sequence, as detailed in [11].

1. If the critical point has index 0, the curve acquires an isolated point that then becomes a small oval. S_t changes acquiring a new component with trivial homology and one oval as boundary.

2. If the critical point has index 2, the curve shrinks an oval to a point that then disappears. The homology can change in two different ways:

- if the connected component of S_t corresponding to the critical point had only one boundary component, then the corresponding homology group H_2 passes from 0 to $\mathbf{Z}$;
- otherwise the rank of H_1 decreases by 1.

3. If the critical point has index 1, C_t has a node; there are three possibilities:

   (a) one oval acquires a double point, and then is split into two ovals

   (b) two ovals approach, acquiring a double point, and then are combined in one; the two ovals were part of the boundary of the same connected component of S_t

   (c) the same as above, but the two ovals were part of the boundary of different components.

   In the first two cases the rank of H_1 increases by 1, in the second case the two components are joined, and the H_1 of the resulting component is the direct sum of the two H_1 of the two original components.

If we are able to discriminate the different subcases, in particular if we are able to associate ovals of C_t to components of S_t, we can inductively describe the topology of S_t. Components can be identified by their critical points (we can keep track of them); hence the problem is solved if we are able to find a continuous curve lying on S and connecting the oval to a critical point. Alternatively, we can find a continuous curve connecting the oval and the critical point from the inside, and never meeting S outside the endpoints (we know that it is possible to retract this curve to the surface). This can be found through a roadmap (see [5] for a general definition of roadmaps, or [11] for a less general but more focused approach).

Many details have to be filled to complete the algorithm: they are too complex to fit here. In a following work these details will be given, together with a treatement of the case of general surfaces (not in generic position, not compact in $\mathbf{R}^3$, in $\mathbf{P}^3$, singular surfaces, etc.).

6.4 Curves in higher dimensional spaces

The question of the shape of curves in $\mathbf{R}^3$ or in $\mathbf{P}^3$ cannot be easily settled since again we fall into a classification of knots; however, finding a description of the knot defined by a space curve is possible; we give an outline only.

We can project the curve to a plane, and if the projection is generic only a finite number of nodes appears. We have to describe an algorithm similar to the one for plane curves, but allowing the presence of knots. When we arrive at a knot, the two branches cross, and the corresponding boundary points are exchanged. Moreover we have to know which branch is "above the other". This is not difficult to find.

The problem is slightly harder when the projection is not generic; the problem is one of infinitesimal deformations of singular maps, and can be solved with a local analysis. The details of this question too are postponed to a further work.

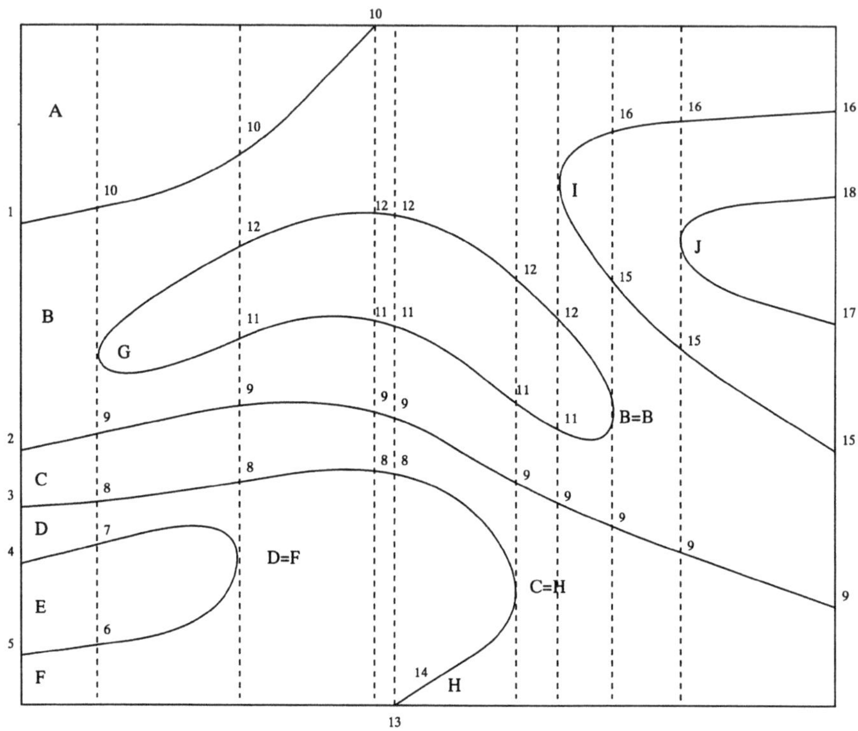

FIGURE 6.1. An example of a curve in a rectangle

6.5 An example

We compute one example of a curve in full. We start from a drawing (and not from an equation) since we are not interested here in the root finding part; moreover, any differentiable curve in a bounded region can be approximated by an algebraic curve.

Our example is a curve in a rectangle, but we will also compute the shape from it in the projective plane, assuming that the shape in the rectangle is equal to the limit of shapes in rectangles of increasing size.

The sequences associated to the curve are respectively: $n_i = (5,6,6,5,5,4,4,4,4)$, $m_i = (5,1,5,1,1,4,2,2)$, and $k_i = (1,2,3,4,2,1,2,1)$ (encoding $1 =$ two branches on right, $2 =$ two branches on left, $3 =$ top side, one branch on left, $4 =$ bottom side, one branch on right).

A component (or a boundary component) is uniquely determined by the regions and the boundary points by which it is adjacent; moreover, we distinguish a boundary component since it is adjacent to one region (components are adjacent to two). Hence we identify a shape giving the set of regions and points adjacent to the components and boundary components.

In the following table, we show the running of the computation. The columns correspond to the state of the shape between vertical lines, assuming that an empty item repeats the content of the preceding column, and a % means that no item appears. The first part of the

(1,2,3,4,5) (6,7,8,9,10)	(6,7,8,9,11,12,10)	(8,9,11,12,10)	(8,9,11,12) 10	13 (14,8,9,11,12)	(9,11,12)	(9,11,12,15,16)	(9,15,16)	(9,15,17,18,16)
(A,1,10) (B,1,2) (C,2,3) (D,3,4) (E,4,5) (F,5,6) (E,6,7) (D,7,8) (C,8,9) (B,9,10)	(B,9,11) (G,11,12) (B,10,12)	(D,5,8) % %		(D,5,13) (H,13,14) (D,14,8)	(C,13,9) % %	(B,15,12) (I,15,16) (B,10,16)	(B,9,15) % %	(I,16,18) (J,18,17) (I,17,15)
(A,B,1,10) (B,C,2,9) (C,D,3,8) (D,E,4,7) (E,F,5,6)	(B,G,11,12)	(D,E,4,5) %		(D,H,13,14)	(C,D,3,13)	(B,I,15,16)	(B,G)	(I,J,17,18)
		D=F			C=H		B=B	

TABLE 6.1. The shape computation

table shows the points in the four sides, the second part shows the boundary components, the third part shows the components, and the fourth part shows the region identifications.

At the end the shape consists of 12 boundary points $(1,2,3,4,5,13,9,15,17,18,16,10)$, 8 regions (A,B,C,D,E,G,I,J), 12 boundary components, $(A,1,10)$, $(B,1,2)$, $(C,2,3)$, $(D,3,4)$, $(E,4,5)$, $(D,5,13)$, $(C,13,9)$, $(B,9,15)$, $(I,15,17)$, $(J,17,18)$, $(I,18,16)$, $(B,16,10)$, 7 components, $(A,B,1,10)$, $(B,C,2,9)$, $(C,D,3,13)$, $(D,E,4,5)$, (B,G), $(B,I,15,16)$, $(I,J,17,18)$.

Now we assume that the situation is the limit situation for big rectangles, and we want to compute the shape for the projective curve: we have to identify 1 with 9, 2 with 15, 3 with 17, 4 with 18, 5 with 16 and 13 with 10; hence $(A,1,10)$ with $(C,13,9)$, $(B,1,2)$ with $(B,9,15)$, $(C,2,3)$ with $(I,15,17)$, $(D,3,4)$ with $(J,17,18)$, $(E,4,5)$ with $(I,18,16)$ and $(D,5,13)$ with $(B,16,10)$; hence A with C, B with B (showing that B becomes non-orientable), C with I, D with J, E with I and D with B. At the end we have three regions: A (identified with C, I, E), B (identified with D) and G.

Moreover, we have identifications of components, following the endpoint identifications: $(A,B,1,10)$, $(B,C,2,9)$, $(B,I,15,16)$, $(D,E,4,5)$, $(I,J,17,18)$ and $(C,D,3,13)$ are one component (with the region identifications it becomes (A,B)); the other component is (B,G). Since B is non-orientable, we have two mutually exclusive ovals.

6.6 References

[1] M. E. Alonso and M. Raimondo. The computation of the topology of a planar semi-algebraic set. *Rend. Sem. Mat. Univ. Politecnico Torino*, 1989.

[2] S. Arnborg and H.Fen. Algebraic decomposition of regular curves. *J. of Symb. Comp.*, 5, 1988.

[3] D. Arnon, G. Collins, and S. MacCallum. Cylindrical algebraic decomposition, I and II. *SIAM J. of Computing*, 13, 1984.

[4] D. Arnon and S. MacCallum. A polynomial-time algorithm for the topological type of a real algebraic curve. *J. of Symb. Comp.*, 5, 1988.

[5] J. F. Canny. *The complexity of robot motion planning.* The MIT Press, 1987. Ph.D. thesis.

[6] P. Cellini, P. Gianni, and C. Traverso. Algorithms for the shape of semialgebraic sets: a new approach. In *AAECC-9.* Springer Verlag, 1991.

[7] G. E. Collins. Quantifier elimination for real closed fields by cylindrical algebraic decomposition. In *Proc. II GI Conf. on Automata and Formal Languages.* Springer Verlag, 1975.

[8] F. Cucker, L. Gonzalez Vega, and F. Rossello. On algorithms for real algebraic plane curves. In *Effective methods in algebraic geometry.* Birkhauser, 1991.

[9] F. Cucker, L. M. Pardo, M. Raimondo, T. Recio, and M. F. Roy. Computation of the local and global analytic structure of a real curve. In *AAECC-5.* Springer Verlag, 1988.

[10] P. Gianni and C. Traverso. La forma delle curve algebriche reali piane. Realizzazione di un algoritmo. In *Curve algebriche. Atti del convegno di Geometria Algebrica, Firenze.* Consiglio Nazionale delle Ricerche, IAGA, 1981.

[11] P. Gianni and C. Traverso. Shape determination of real curves and surfaces. *Ann. Univ. Ferrara, Sez. VII, Sec. Math.*, XXIX, 1983.

[12] A. Gonzalez Corbalan. *Reconocimiento de formas en curvas planas.* PhD thesis, Universitad de Cantabria, 1990.

[13] L. Gonzalez Vega. Working with algebraic plane curves in Reduce: the GCUR package. In *ISSAC-91.* ACM, 1991.

[14] J. Milnor. *Morse Theory*, volume 51 of *Annals of Math. Studies.* Cambridge Univ. Press, 1963.

7 Two Surfaces Suffice

Hang K. Du
Ronald N. Goldman

ABSTRACT Classical elimination theory is the study of conditions that guarantee the existence of solutions to systems of polynomial equations. The main goal is to construct a set of polynomial expressions in the coefficients of the original polynomial equations, which are satisfied exactly when the original polynomials have a common root. For n polynomials in $n-1$ unknowns, there is a single unique condition called the *resultant*. More generally, for an arbitrary number of equations and unknowns many conditions may be required; these conditions are called *resolvents*. Unfortunately, classical resolvents are not practical because they yield far too many conditions which are very expensive to compute.

In this paper, we develop a new resolvent which is computationally efficient, and we apply it to solve the space curve/curve intersection problem in Computer Aided Geometric Design (CAGD).

One way to solve the curve/curve intersection problem is to solve the implicitization and inversion problems. By applying our new resolvent, we find that generally two surfaces of relatively low degree suffice to implicitize any $3D$, faithfully parametrized, polynomial curve, and the inversion equation is also a rational expression of relatively low degree. Thus we provide a very efficient curve/curve intersection algorithm for low degree, $3D$, polynomial curves.

7.1 Introduction

Classical elimination theory is the study of conditions for the existence of solutions to systems of polynomial equations. The main idea is to construct a polynomial function or functions of the coefficients of the original polynomials whose vanishing signals that there is a common root. For n polynomials in $n-1$ unknowns, we have the resultant.

Definition 7.1.1 : *A* resultant *is a irreducible polynomial expression in the coefficients of a set of n polynomials in $n-1$ variables such that the vanishing of the resultant is a necessary and sufficient condition for these polynomials to have a common root.*

There are many different resultant expressions [4], but resultants are unique up to constant multiples [6]. More generally, for an arbitrary number of unknowns and equations, we have the resolvent.

Definition 7.1.2 : *A* resolvent *is a set of polynomial expressions in the coefficients of a set of n polynomials in k variables, $n > k+1$, such that the vanishing of the resolvent is a necessary and sufficient condition for these polynomials to have a common root.*

Unlike resultants, resolvents are not unique. Different methods yield different sets of conditions (see below).

One application of resultants and resolvents in CAGD (Computer Aided Geometric Design) is to finding the intersection points of rational curves [4, 5, 7, 8, 11]. Given two rational curves $P(t) = (x_1 = f_1(t)/w(t), \ldots, x_r = f_r(t)/w(t))$ and $Q(s) = (x_1 = \hat{f}_1(s)/\hat{w}(s), \ldots, x_r = \hat{f}_r(s)/\hat{w}(s))$ for $r \geq 2$, we want to find the intersection points in the parameters s, t, and the coordinates $(x_1, \ldots, x_r)$. In CAGD, the intersection points in the parameters s and t are usually restricted to bounded intervals $[s_0, s_1]$ and $[t_0, t_1]$; in practice, often $s_0 = t_0 = 0$ and $s_1 = t_1 = 1$. The intersection points in the coordinates $(x_1, \ldots, x_r)$ are in $\Re^r$. One way to tackle this problem is to solve the implicitization problem – find a set of implicit hypersurfaces whose intersection is a given parametric curve – and then substitute the parametric equations of one curve, say $Q(s)$, into the implicit equations of the other curve, $P(t)$. This reduces the intersection problem to solving a system of univariate polynomial equations in s. Substituting the intersection points in s into the parametric equations of $Q(s)$, we obtain the intersection points in $(x_1, \ldots, x_r)$. However, to compute the intersection points in t efficiently, we have to solve the inversion problem for $P(t)$: given the coordinates $(x_1, \ldots, x_r)$ of a point on the curve $P(t)$, find the corresponding parameter t.

When $r = 2$, $P(t) = (x = f(t)/w(t), y = g(t)/w(t))$ and $Q(s) = (x = \hat{f}(s)/\hat{w}(s), y = \hat{g}(s)/\hat{w}(s))$ are planar curves, and the implicitization problem can be solved using resultants. Indeed a point (x, y) lies on the curve $P(t)$ if and only if

$$f_x(t) = xw(t) - f(t) = 0, \quad g_y(t) = yw(t) - g(t) = 0. \tag{7.1}$$

Thus if $\mathcal{R}_e$ is a resultant, then eliminating t from (7.1) we find that (x, y) lies on $P(t)$ if and only if $s(x, y) = \mathcal{R}_e(f_x(t), g_y(t)) = 0$.

We could try to implicitize a $3D$, rational curve $P(t) = (x = f(t)/w(t), y = g(t)/w(t), z = h(t)/w(t))$ using resultants by projecting $P(t)$ on to the xy, xz, yz planes. Implicitizing these planar curves using resultants generates three cylinders. But the degrees of these cylinders are unnecessarily high and the intersection of these cylinders only contains $P(t)$; it is not necessarily equal to $P(t)$ [6].

A better approach is to apply resolvents. Again, a point (x, y, z) lies on the curve $P(t)$ if and only if

$$f_x(t) = xw(t) - f(t) = 0, \quad g_y(t) = yw(t) - g(t) = 0, \quad h_z(t) = zw(t) - h(t) = 0. \tag{7.2}$$

Thus if $\{\mathcal{R}_i\}_{i=1,\ldots,l}$ is a resolvent, then eliminating t from (7.2) we find that (x, y, z) lies on $P(t)$ if and only if

$$s_i(x, y, z) = \mathcal{R}_i(f_x(t), g_y(t), h_z(t)) = 0 \quad i = 1, 2, \ldots, l.$$

The surfaces $s_i(x, y, z) = 0$ are implicit equations for the parametric curve $P(t)$. Note that for planar rational curves, the implicit equation $s(x, y) = 0$ is unique because the resultant is unique, but for non-planar curves, the surfaces $s_i(x, y, z) = 0$ are not unique because the resolvent is not unique. Thus, while this method works well in theory, in order to make this technique practical, we must construct our resolvents carefully so that we generate only a small number of surfaces all of which have relatively low degree.

We begin by reviewing the disadvantages of existing resolvent techniques in Section 7.2. In Section 7.3, we develop our new Sylvester resolvent for m linearly independent univariate polynomials of degree m. Then, in Section 7.4, we show how to obtain two resolvent expressions for 3 polynomials of degree $n \geq 3$ by constructing a list, $\mathcal{S}$, of m linearly independent polynomials of degree m from 3 polynomials of degree $n \geq 3$, and applying our Sylvester resolvent to the polynomials in $\mathcal{S}$. In Section 7.5, we use the Sylvester resolvent for 3 polynomials to solve the implicitization problem for $3D$, rational curves. One of our main results is that generally, in real affine space, two surfaces of relatively low degree, $\lceil \frac{n}{2} \rceil$ and $\lfloor \frac{n}{2} \rfloor + 1$, suffice to implicitize any faithfully parametrized, 3–dimensional, polynomial curve of degree n, and we give a simple constructive algorithm based on resolvents for finding the equations of these two surfaces. We then use these surfaces to construct a general intersection algorithm for two $3D$ polynomial curves. Although we get two surfaces of degrees $\lceil \frac{n}{2} \rceil$ and $\lfloor \frac{n}{2} \rfloor + 1$ that intersect in a curve of degree n in real affine space, these surfaces, when embedded in complex projective space, intersect in a curve of degree $\lceil \frac{n}{2} \rceil (\lfloor \frac{n}{2} \rfloor + 1)$. Thus our result does not contradict Bezout's theorem. We briefly summarize our results and discuss some future research in Section 7.6.

7.2 Weaknesses of Known Methods

7.2.1 Kronecker's Resolvent

This method is due to Kronecker [12]. Given r univariate polynomials $f_1(t), \ldots, f_r(t)$, form

$$\begin{aligned} g_u(t) &= u_1 f_1(t) + u_2 f_2(t) + \cdots + u_r f_r(t), \\ g_v(t) &= v_1 f_1(t) + v_2 f_2(t) + \cdots + v_r f_r(t), \end{aligned}$$

where the u_i's and v_i's are indeterminates, and compute

$$\text{Resultant}(g_u(t), g_v(t)) = \sum_{i_1,\ldots,i_r,j_1,\ldots,j_r} u_1^{i_1} \ldots u_r^{i_r} v_1^{j_1} \ldots v_r^{j_r} r_{i_1,\ldots,i_r,j_1,\ldots,j_r}$$

where $r_{i_1,\ldots,i_r,j_1,\ldots,j_r}$ is a polynomial in the coefficients of the polynomials $f_1(t), \ldots, f_r(t)$. Then, by the properties of resultants, the following three statements are equivalent:

- $f_1(t), \ldots, f_r(t)$ have a common root.
- $g_u(t), g_v(t)$ have a common root for all u, v.
- $r_{i_1,\ldots,i_r,j_1,\ldots,j_r} = 0$ for all $i_1, \ldots, i_r, j_1, \ldots, j_r$.

The set of all $r_{i_1,\ldots,i_r,j_1,\ldots,j_r}$ is called the Kronecker resolvent of $f_1(t), \ldots, f_r(t)$.

The main drawback of this technique is that there are far too many functions $r_{i_1,\ldots,i_r,j_1,\ldots,j_r}$. Although we could use this method to implicitize an arbitrary non-planar, parametric, rational curve, it yields too many hypersurfaces with too high degree. For example, using this method we would obtain a $3D$, parametric, quartic curve as the intersection of 495 quartic surfaces.

7.2.2 Kakié's Resolvent

Given $f_i(t) = a_{i,n_i} t^{n_i} + a_{i,n_i-1} t^{n_i-1} + \cdots + a_{i,1} t + a_{i,0}$ for $i = 1, 2, \ldots, r$, we let

$$\hat{n} = \max(n_1, \ldots, n_r) \quad \text{and} \quad \bar{n} = \min(n_1, \ldots, n_r).$$

Then, for each integer $l \geq \hat{n}$, we construct the matrix

$$\boldsymbol{R}_l = \begin{bmatrix} a_{1,0} & a_{1,1} & \cdots & a_{1,n_1} & & \\ & \cdots & \cdots & \cdots & \cdots & \\ & & a_{1,0} & a_{1,1} & \cdots & a_{1,n_1} \\ \cdots & \cdots & \cdots & \cdots & \cdots & \cdots \\ a_{r,0} & a_{r,1} & \cdots & a_{r,n_r} & & \\ & \cdots & \cdots & \cdots & \cdots & \\ & & a_{r,0} & a_{r,1} & \cdots & a_{r,n_r} \end{bmatrix},$$

with $l - n_i + 1$ rows and $l + 1$ columns of the coefficients $a_{i,j}, j = 0, \ldots, n_i$ for each $i = 1, \ldots, r$, the blank spaces being filled with zeros. Kakié [9] proved the following theorem:

Theorem 7.2.1 *In order that the polynomials $f_1(t), f_2(t), \ldots, f_r(t)$ have a common factor of degree greater than or equal to k, it is necessary and sufficient that* $\operatorname{rank}(\boldsymbol{R}_{\hat{n}+\bar{n}-k})$ *is less than or equal to* $\hat{n}+\bar{n}-2k+1$.

Here the necessary and sufficient condition for 3 polynomials $f_1(t), f_2(t), f_3(t)$ to have a common root is that $\operatorname{rank}(\boldsymbol{R}_{2n-1}) < 2n$. Using this condition to implicitize a $3D$, rational curve, we would have to compute all $\binom{3n}{2n}$ of the $2n$ x $2n$ subdeterminants of $\boldsymbol{R}_{2n-1}$.

Again the main drawback to this technique is that there are way too many $2n$ x $2n$ subdeterminants of $\boldsymbol{R}_{2n-1}$. Using Kakié's resolvent to implicitize a non-planar, parametric, rational curve, we get some low degree surfaces, but there are still too many. For instance, this technique produces 120 quadric, cubic, and quartic surfaces for a $3D$, parametric, quartic curve.

7.3 Sylvester's Resolvent for m Linearly Independent Polynomials of Degree m

In this section, we construct necessary and sufficient conditions for m linearly independent polynomials $f_1(t), \ldots, f_m(t)$ of degree m to have a common root. The main idea here is to extend the Sylvester technique for computing the resultant of two univariate quadratic polynomials.

For two linearly independent polynomials $f_1(t), f_2(t)$ of degree two, Sylvester's method for finding the resultant is to compute

$$\mathcal{R}_s = \begin{vmatrix} 0 & a_{1,2} & a_{1,1} & a_{1,0} \\ 0 & a_{2,2} & a_{2,1} & a_{2,0} \\ a_{1,2} & a_{1,1} & a_{1,0} & 0 \\ a_{2,2} & a_{2,1} & a_{2,0} & 0 \end{vmatrix}$$

where $\mathcal{R}_s$ is the determinant of the coefficient matrix of $f_1(t), f_2(t), tf_1(t), tf_2(t)$. It is not hard to show that $f_1(t), f_2(t)$ have a common root if and only if $\mathcal{R}_s = 0$. That is, $f_1(t), f_2(t)$ have a common root if and only if the polynomials $f_1(t), f_2(t), tf_1(t), tf_2(t)$ are linearly dependent. The value $\mathcal{R}_s$ is called the *Sylvester resultant* of $f_1(t), f_2(t)$. We now extend Sylvester's method to find resolvents for m linearly independent univariate polynomials of degree m.

7.3.1 Linearly Independent

Consider m linearly independent polynomials

$$\begin{aligned} f_1(t) &= a_{1,m}t^m + a_{1,m-1}t^{m-1} + \cdots + a_{1,1}t + a_{1,0}, \\ &\cdots\cdots\cdots \\ f_m(t) &= a_{m,m}t^m + a_{m,m-1}t^{m-1} + \cdots + a_{m,1}t + a_{m,0} \end{aligned}$$

where $a_{i,j} \in \mathcal{F}$, a field of characteristic zero, and $a_{m,m} \neq 0$. Then the m x $m+1$ matrix

$$\boldsymbol{U} = \begin{bmatrix} a_{1,m} & a_{1,m-1} & \cdots & a_{1,1} & a_{1,0} \\ a_{2,m} & a_{2,m-1} & \cdots & a_{2,1} & a_{2,0} \\ \vdots & \vdots & \cdots & \vdots & \vdots \\ a_{m-1,m} & a_{m-1,m-1} & \cdots & a_{m-1,1} & a_{m-1,0} \\ a_{m,m} & a_{m,m-1} & \cdots & a_{m,1} & a_{m,0} \end{bmatrix} \tag{7.3}$$

has rank m. But the matrix $\boldsymbol{U}$ has rank m if and only if $|\boldsymbol{U}\boldsymbol{U}^T| \neq 0$. Let $\boldsymbol{U}_i$ be the m x m matrix obtained by removing the i–th column of $\boldsymbol{U}$. It is shown in [1] that

$$|\boldsymbol{U}\boldsymbol{U}^T| = \sum_{i=1}^{m+1} |\boldsymbol{U}_i|^2. \tag{7.4}$$

Thus the matrix $\boldsymbol{U}$ has rank m if and only if

$$|\boldsymbol{U}\boldsymbol{U}^T| = \sum_{i=1}^{m+1} |\boldsymbol{U}_i|^2 > 0.$$

Therefore the polynomials $f_1(t), f_2(t), \ldots, f_{m-1}(t), f_m(t)$ are linearly independent if and only if $\sum_{i=1}^{m+1} |\boldsymbol{U}_i|^2 > 0$.

7.3.2 The Sylvester Resolvent $\mathcal{R}_s$

Let $\deg(f_i(t))$ denote the degree of the polynomial $f_i(t)$ in the variable t. Without loss of generality we can assume that $\deg(f_m(t)) = m$. In addition, let

$$\mathbf{p}(t) = \begin{bmatrix} f_1(t) \\ \vdots \\ f_m(t) \\ tf_m(t) \end{bmatrix} = \boldsymbol{C}\mathbf{t} = \begin{bmatrix} 0 & a_{1,m} & \cdots & a_{1,1} & a_{1,0} \\ \vdots & \vdots & & \vdots & \vdots \\ 0 & a_{m,m} & \cdots & a_{m,1} & a_{m,0} \\ a_{m,m} & \cdots & a_{m,1} & a_{m,0} & 0 \end{bmatrix} \begin{bmatrix} t^{m+1} \\ \vdots \\ t \\ 1 \end{bmatrix}.$$

Then $\boldsymbol{C}$ is an $m+1$ x $m+2$ matrix. Performing Gaussian elimination on $\boldsymbol{C}$, we get

$$\hat{\mathbf{p}}(t) = \begin{bmatrix} \hat{f}_1(t) \\ \vdots \\ \hat{f}_m(t) \\ t\hat{f}_m(t) \end{bmatrix} = \hat{\boldsymbol{C}}\mathbf{t} = \begin{bmatrix} 0 & \cdots & 0 & \hat{a}_{1,1} & \hat{a}_{1,0} \\ \vdots & \vdots & & \vdots & \vdots \\ 0 & \hat{a}_{m,m} & \cdots & \hat{a}_{m,1} & \hat{a}_{m,0} \\ \hat{a}_{m,m} & \cdots & \hat{a}_{m,1} & \hat{a}_{m,0} & 0 \end{bmatrix} \begin{bmatrix} t^{m+1} \\ \vdots \\ t \\ 1 \end{bmatrix},$$

where $\hat{f}_m(t) = f_m(t)$, $0 \leq \deg(\hat{f}_1(t)) \leq \cdots \leq \deg(\hat{f}_m(t)) = m$, and $\deg(\hat{f}_i(t)) = \deg(\hat{f}_j(t))$ for $j \neq i$ if and only if $\hat{f}_i(t) = \hat{f}_j(t) \equiv 0$.

Lemma 7.3.1 :

(1) *$f_1(t), \ldots, f_m(t)$ and $\hat{f}_1(t), \ldots, \hat{f}_m(t)$ span the same subspace of polynomials of degree m.*

(2) *$f_1(t), \ldots, f_m(t)$ are linearly independent polynomials if and only if $\hat{f}_1(t), \ldots, \hat{f}_m(t)$ are linearly independent polynomials.*

(3) *$f_1(t), \ldots, f_m(t)$ have a common root if and only if $\hat{f}_1(t), \ldots, \hat{f}_m(t)$ have a common root.*

Proof. **(1)**, **(2)**, and **(3)** follow directly from standard properties of linear algebra. □

Now define a column vector with $m+2$ polynomials of degree $m+1$ by setting

$$\mathbf{p}_k(t) = \begin{bmatrix} f_1(t) \\ \vdots \\ f_m(t) \\ tf_m(t) \\ tf_k(t) \end{bmatrix} = \boldsymbol{C}_k\mathbf{t} = \begin{bmatrix} 0 & a_{1,m} & \cdots & a_{1,0} \\ \vdots & \vdots & & \vdots \\ 0 & a_{m,m} & \cdots & a_{m,0} \\ a_{m,m} & \cdots & a_{m,0} & 0 \\ a_{k,m} & \cdots & a_{k,0} & 0 \end{bmatrix} \begin{bmatrix} t^{m+1} \\ \vdots \\ t^2 \\ t \\ 1 \end{bmatrix}$$

for $k = 1, \ldots, m-1$. Then clearly the polynomials in $\boldsymbol{p}_k(t)$ are linearly dependent if and only if the matrix $\boldsymbol{C}_k$ is singular. Moreover, it is easy to see that if $f_1(t), \ldots, f_m(t)$ have a common root, then the polynomials in $\boldsymbol{p}_k(t)$ must be linearly dependent, so $|\boldsymbol{C}_k| \equiv 0$ for $k = 1, \ldots, m-1$. We shall show that the converse is also true.

Define

$$\bar{\boldsymbol{p}}_q(t) = \begin{bmatrix} \hat{f}_1(t) \\ \vdots \\ \hat{f}_m(t) \\ t\hat{f}_m(t) \\ tf_q(t) \end{bmatrix} = \bar{\boldsymbol{C}}_q \boldsymbol{t}, \text{ and } \hat{\boldsymbol{p}}_k(t) = \begin{bmatrix} \hat{f}_1(t) \\ \vdots \\ \hat{f}_m(t) \\ t\hat{f}_m(t) \\ t\hat{f}_k(t) \end{bmatrix} = \hat{\boldsymbol{C}}_k \boldsymbol{t},$$

where $\bar{\boldsymbol{C}}_q$ and $\hat{\boldsymbol{C}}_k$ are the coefficient matrices of $\bar{\boldsymbol{p}}_q(t)$ and $\hat{\boldsymbol{p}}_k(t)$, respectively. Notice that $|\bar{\boldsymbol{C}}_q| = \pm|\boldsymbol{C}_q|$ because $\bar{\boldsymbol{C}}_q$ is obtained by applying Gaussian elimination to $\boldsymbol{C}_q$. Now we have the following lemma:

Lemma 7.3.2 : *Suppose that $f_1(t), f_2(t), \ldots, f_m(t)$ are linearly independent polynomials. Then $|\hat{\boldsymbol{C}}_k| \equiv 0$ for $k = 1, \ldots, m-1$ if and only if $|\bar{\boldsymbol{C}}_q| = \pm|\boldsymbol{C}_q| \equiv 0$ for $q = 1, \ldots, m-1$.*

Proof. By the construction of $\hat{f}_k(t)$, we have

$$\begin{bmatrix} \hat{f}_1(t) \\ \vdots \\ \hat{f}_m(t) \end{bmatrix} = \boldsymbol{D} \begin{bmatrix} f_1(t) \\ \vdots \\ f_m(t) \end{bmatrix} \tag{7.5}$$

where $\boldsymbol{D}$ is an m x m matrix. Conversely, by Lemma 7.3.1 part **(1)**, we have

$$\begin{bmatrix} f_1(t) \\ \vdots \\ f_m(t) \end{bmatrix} = \hat{\boldsymbol{D}} \begin{bmatrix} \hat{f}_1(t) \\ \vdots \\ \hat{f}_m(t) \end{bmatrix}$$

where, again, $\hat{\boldsymbol{D}}$ is an m x m matrix.

Suppose that $|\bar{\boldsymbol{C}}_q| = \pm|\boldsymbol{C}_q| \equiv 0$ for $q = 1, \ldots, m-1$. Then

$$\begin{bmatrix} tf_1(t) \\ \vdots \\ tf_m(t) \end{bmatrix} = \boldsymbol{T} \begin{bmatrix} \hat{f}_1(t) \\ \vdots \\ \hat{f}_m(t) \end{bmatrix} + t\hat{f}_m(t)\, \boldsymbol{v}$$

where $\boldsymbol{T}$ is an m x m matrix whose last row is $\boldsymbol{0}$ and $\boldsymbol{v}$ is an m-element column vector whose last element is 1 because $t\hat{f}_m(t) = tf_m(t)$. Therefore, by Equation (7.5), we have

$$\begin{bmatrix} t\hat{f}_1(t) \\ \vdots \\ t\hat{f}_m(t) \end{bmatrix} = \boldsymbol{D} \begin{bmatrix} tf_1(t) \\ \vdots \\ tf_m(t) \end{bmatrix} = (\boldsymbol{DT}) \begin{bmatrix} \hat{f}_1(t) \\ \vdots \\ \hat{f}_m(t) \end{bmatrix} + (\boldsymbol{Dv})t\hat{f}_m(t).$$

Thus $t\hat{f}_k(t)$ can be written as a linear combination of $\hat{f}_1(t), \ldots, \hat{f}_m(t), t\hat{f}_m(t)$. Therefore $|\hat{\boldsymbol{C}}_k| \equiv 0$ for $k = 1, \ldots, m-1$.

The converse works similarly. □

Lemma 7.3.3 : *Suppose that $\hat{f}_1(t), \hat{f}_2(t), \ldots, \hat{f}_m(t)$ are linearly independent polynomials. If $\hat{f}_1(t) \equiv \hat{a}_{1,0} \neq 0$, then $|\hat{\boldsymbol{C}}_k| \neq 0$ for at least one $k \in \{1, \ldots, m-1\}$.*

Proof. Suppose that $\hat{f}_1(t) \equiv \hat{a}_{1,0} \neq 0$. If $\deg(\hat{f}_i(t)) = i$ for $i = 2, \ldots, m-1$, then clearly

$$|\hat{C}_1| = \pm \begin{vmatrix} \hat{a}_{m,m} & \cdots & \hat{a}_{m,2} & \hat{a}_{m,1} & \hat{a}_{m,0} & 0 \\ & \hat{a}_{m,m} & \cdots & \hat{a}_{m,2} & \hat{a}_{m,1} & \hat{a}_{m,0} \\ & & \ddots & \vdots & \vdots & \vdots \\ & & & \hat{a}_{2,2} & \hat{a}_{2,1} & \hat{a}_{2,0} \\ & & & & \hat{a}_{1,0} & 0 \\ & & & & & \hat{a}_{1,0} \end{vmatrix} \neq 0.$$

On the other hand, if $\deg(\hat{f}_i(t)) = i$ for $i = j+1, \ldots, m-1$ and $\deg(\hat{f}_j(t)) \leq j-1$, then $\deg(\hat{f}_q(t)) \leq q-1$ for $q = 2, \ldots, j-1$. That is, if one polynomial $\hat{f}_j(t)$ does not have degree j, then the polynomials $\hat{f}_{j-1}(t), \ldots, \hat{f}_2(t)$ do not have full degree. Consider the column vector

$$[t\hat{f}_m(t), \hat{f}_m(t), \ldots, \hat{f}_{j+1}(t), t\hat{f}_j(t), \hat{f}_j(t), \ldots, \hat{f}_1(t)]^T = \boldsymbol{Dt}$$

where

$$\boldsymbol{Dt} = \begin{bmatrix} \hat{a}_{m,m} & \cdots & \hat{a}_{m,5} & \hat{a}_{m,4} & \hat{a}_{m,3} & \hat{a}_{m,2} & \hat{a}_{m,1} & \hat{a}_{m,0} & 0 \\ & \hat{a}_{m,m} & \cdots & \hat{a}_{m,5} & \hat{a}_{m,4} & \hat{a}_{m,3} & \hat{a}_{m,2} & \hat{a}_{m,1} & \hat{a}_{m,0} \\ & & \ddots & & & \vdots & \vdots & \vdots & \vdots \\ & & & \hat{a}_{j+1,j+1} & \cdots & \hat{a}_{j+1,3} & \hat{a}_{j+1,2} & \hat{a}_{j+1,1} & \hat{a}_{j+1,0} \\ & & & & \hat{a}_{j,j-1} & \cdots & \hat{a}_{j,1} & \hat{a}_{j,0} & 0 \\ & & & & & \hat{a}_{j,j-1} & \cdots & \hat{a}_{j,1} & \hat{a}_{j,0} \\ & & & & & & \ddots & \vdots & \vdots \\ & & & & & & & \hat{a}_{2,1} & \hat{a}_{2,0} \\ & & & & & & & & \hat{a}_{1,0} \end{bmatrix} \begin{bmatrix} t^{m+1} \\ t^m \\ \vdots \\ t^{j+1} \\ t^j \\ t^{j-1} \\ \vdots \\ 1 \end{bmatrix}$$

and $|\hat{C}_j| = \pm|\boldsymbol{D}|$. The j x j submatrix

$$\boldsymbol{E} = \begin{bmatrix} \hat{a}_{j,j-1} & \cdots & \hat{a}_{j,1} & \hat{a}_{j,0} \\ & \ddots & \vdots & \vdots \\ & & \hat{a}_{2,1} & \hat{a}_{2,0} \\ & & & \hat{a}_{1,0} \end{bmatrix}$$

is nonsingular since it is the coefficient matrix of the linearly independent polynomials $\hat{f}_j(t), \ldots, \hat{f}_1(t)$. Thus $|\hat{C}_j| = \pm|\boldsymbol{D}| \neq 0$ when $\deg(\hat{f}_j(t)) \leq j-1$. Note that the case $\deg(\hat{f}_j(t)) \leq j-1$ and $\deg(\hat{f}_q(t)) \leq q-2$ for $q = 2, \ldots, j-1$ does not happen because in order for $\hat{f}_j(t), \ldots, \hat{f}_1(t)$ to be linearly independent, the diagonal entries of $\boldsymbol{C}$ must be nonzero. □

Let $\boldsymbol{a}_k = [a_{1,k}, \ldots, a_{m,k}]^T$ for $k = 0, \ldots, m$. Now we are in a position to prove the following important theorem:

Theorem 7.3.4 : *Suppose that $f_1(t), f_2(t), \ldots, f_m(t)$ are linearly independent polynomials, and*

$\deg(f_m(t)) = m$. Then these polynomials have a common root at $t = t_0$ if and only if $|\boldsymbol{C}_j| = 0$ for $j = 1, \ldots, m-1$. Moreover,

$$t_0 = -\frac{|\boldsymbol{a}_m \ \ \boldsymbol{a}_{m-1} \ \cdots \ \boldsymbol{a}_2 \ \ \boldsymbol{a}_0|}{|\boldsymbol{a}_m \ \ \boldsymbol{a}_{m-1} \ \cdots \ \boldsymbol{a}_2 \ \ \boldsymbol{a}_1|}.$$

Proof. If $f_1(t), \ldots, f_m(t)$ have a common root, then $\boldsymbol{p}_k(t)$ contains $m+2$ linearly dependent polynomials since these polynomials have degree at most $m+1$, and they have a common root. Therefore $|\boldsymbol{C}_j| = 0$ for $j = 1, \ldots, m-1$.

Conversely, suppose that $|\boldsymbol{C}_j| = 0$ for $j = 1, \ldots, m-1$. Then, by Lemma 7.3.2, $|\hat{\boldsymbol{C}}_k| = 0$ for $k = 1, \ldots, m-1$. First, we consider what can happen to the polynomial $\hat{f}_1(t)$. If $\hat{f}_1(t) \equiv 0$, then the polynomials $\hat{f}_1(t), \ldots, \hat{f}_m(t)$ are linearly dependent. This means, by Lemma 7.3.1, that $f_1(t), \ldots, f_m(t)$ are linearly dependent polynomials. Hence $\hat{f}_1(t) \not\equiv 0$. In addition, if $\hat{f}_1(t) \equiv \hat{a}_{1,0} \neq 0$, then by Lemma 7.3.3, at least one of the $|\hat{\boldsymbol{C}}_k| \neq 0$. But by Lemma 7.3.2 this implies that at least one of the $|\boldsymbol{C}_j| \neq 0$ which contradicts the assumption that all the $|\boldsymbol{C}_j|$ vanish. Again, this case is not possible. Therefore the only case left is that $\deg(\hat{f}_1(t)) = 1$. As a consequence, $\deg(\hat{f}_i(t)) = i$ for $i = 2, \ldots, m$.

Second, since $|\hat{\boldsymbol{C}}_k| = 0$, the polynomials in $\hat{\boldsymbol{p}}_k(t)$ are linearly dependent for $k = 1, \ldots, m-1$. That is, we have $m-1$ homogeneous equations, where the k-th equation is

$$(\tau_{k,m}t + \varepsilon_{k,m})\hat{f}_m(t) + \sum_{i=k+1}^{m-1} \varepsilon_{k,i}\hat{f}_i(t) + (\tau_{k,k}t + \varepsilon_{k,k})\hat{f}_k(t) + \sum_{i=1}^{k-1} \varepsilon_{k,i}\hat{f}_i(t) = 0. \tag{7.6}$$

Because $\deg(\hat{f}_m(t)) > \ldots > \deg(\hat{f}_{k+1}(t))$, the constants $\varepsilon_{k,k+2}, \ldots, \varepsilon_{k,m}, \tau_{k,m}$ must equal zero. Thus Equation (7.6) becomes

$$\varepsilon_{k,k+1}\hat{f}_{k+1}(t) + (\tau_{k,k}t + \varepsilon_{k,k})\hat{f}_k(t) + \sum_{i=1}^{k-1} \varepsilon_{k,i}\hat{f}_i(t) = 0,$$

where not all $\varepsilon_{k,1}, \ldots, \varepsilon_{k,k+1}, \tau_{k,k}$ equal zero. In matrix form, we have the following system of $m-1$ homogeneous equations:

$$\begin{bmatrix} h_{1,1}^1 & h_{1,2}^0 & 0 & \cdots & 0 & 0 \\ h_{2,1}^0 & h_{2,2}^1 & h_{2,3}^0 & \cdots & 0 & 0 \\ \vdots & \vdots & \vdots & \ddots & & \vdots \\ h_{m-2,1}^0 & h_{m-2,2}^0 & \cdots & h_{m-2,m-2}^1 & h_{m-2,m-1}^0 & 0 \\ h_{m-1,1}^0 & h_{m-1,2}^0 & \cdots & h_{m-1,m-2}^0 & h_{m-1,m-1}^1 & h_{m-1,m}^0 \end{bmatrix} \begin{bmatrix} \hat{f}_1(t) \\ \hat{f}_2(t) \\ \vdots \\ \hat{f}_{m-1}(t) \\ \hat{f}_m(t) \end{bmatrix} = \boldsymbol{0}, \tag{7.7}$$

where $h_{i,j}^0 = \varepsilon_{i,j}$ and $h_{i,i}^1 = \tau_{i,i}t + \varepsilon_{i,i}$, and at least one of the $\varepsilon_{i,j}$ or $\tau_{i,i} \neq 0$ for $i = 1, \ldots, m-1$ and $j = 1, \ldots, i+1$. Notice that $h_{i,i+1}^0 = \varepsilon_{i,i+1} \neq 0$, because if they are zero, then $h_{i,i}^1 = 0$. As a consequence, all the $\varepsilon_{i,j} = 0$ for $j = 1, \ldots, i-1$, contrary to assumption.

Consider the first equation of System (7.7),

$$h_{1,1}^1\hat{f}_1(t) + h_{1,2}^0\hat{f}_2(t) = 0.$$

Because $\deg(\hat{f}_1(t)) = 1$ and $\deg(\hat{f}_2(t)) = 2$, clearly $\hat{f}_1(t)$ and $\hat{f}_2(t)$ have a common root. Let $q(t)$ be the common factor of $\hat{f}_1(t)$ and $\hat{f}_2(t)$. Then we can write the second equation of System (7.7) as

$$q(t)g_2(t) + h_{2,3}^0\hat{f}_3(t) = 0, \quad \text{where } q(t)g_2(t) = h_{2,1}^0\hat{f}_1(t) + h_{2,2}^1\hat{f}_2(t),$$

and $\deg(g_2(t)) = 2$. Thus $q(t)$ divides $\hat{f}_3(t)$ because $h_{2,3}^0 \neq 0$. In general, if $\hat{f}_1(t), \ldots, \hat{f}_i(t)$ have a common factor $q(t)$, then we can rewrite the i-th equation as

$$q(t)g_i(t) + h_{i,i+1}^0\hat{f}_{i+1}(t) = 0, \quad \text{where } q(t)g_i(t) = h_{i,i}^1\hat{f}_i(t) + \sum_{j=1}^{i-1} h_{i,j}^0\hat{f}_j(t),$$

with $h^0_{i,i+1} \neq 0$ and $\deg(g_i(t)) = i$. Hence $\hat{f}_1(t), \ldots, \hat{f}_{i+1}(t)$ have a common factor $q(t)$ for $i = 1, \ldots, m-1$. Therefore from System (7.7) we can conclude that if all the $|\hat{C}_k|$ vanish, then $\hat{f}_1(t), \ldots, \hat{f}_m(t)$ have a common root. Thus, by Lemma 7.3.1, $f_1(t), \ldots, f_m(t)$ have a common root.

To find the common root t_0 of $f_1(t), \ldots, f_m(t)$ is easy. Recall that we have the system of homogeneous equations

$$C_k t_o = \begin{bmatrix} 0 & a_m & \cdots & a_1 & a_0 \\ a_{m,m} & \cdots & a_{m,1} & a_{m,0} & 0 \\ a_{k,m} & \cdots & a_{k,1} & a_{k,0} & 0 \end{bmatrix} \begin{bmatrix} t_0^{m+1} \\ \vdots \\ 1 \end{bmatrix} = \begin{bmatrix} f_1(t_0) \\ \vdots \\ tf_k(t_0) \end{bmatrix} = \mathbf{0}. \tag{7.8}$$

Since $|C_k| = 0$, we can write System (7.8) as

$$\begin{bmatrix} 0 & a_m & a_{m-1} & \cdots & a_1 \\ a_{m,m} & a_{m,m-1} & \cdots & a_{m,1} & a_{m,0} \end{bmatrix} \begin{bmatrix} t_0^{m+1} \\ \vdots \\ t_0 \end{bmatrix} = A \begin{bmatrix} t_0^{m+1} \\ \vdots \\ t_0 \end{bmatrix} = - \begin{bmatrix} a_0 \\ 0 \end{bmatrix}.$$

Moreover, $|D| = \begin{vmatrix} a_m & a_{m-1} & \cdots & a_2 & a_1 \end{vmatrix} \neq 0$. We can see it in the following manner: If $|D| = 0$, then the polynomials $(f_1(t) - a_{1,0}), \ldots, (f_m(t) - a_{m,0})$ are linearly dependent. Hence there exist constants $\epsilon_1, \ldots, \epsilon_m$ such that

$$\sum_{k=1}^{m} \epsilon_k (f_k(t) - a_{k,0}) = 0. \tag{7.9}$$

Substituting the root $t = t_0$ in Equation 7.9, we get $\sum_{k=1}^{m} \epsilon_k a_{k,0} = 0$. Hence

$$0 = \sum_{k=1}^{m} \epsilon_k (f_k(t) - a_{k,0}) = \sum_{k=1}^{m} \epsilon_k f_k(t) - \sum_{k=1}^{m} \epsilon_k a_{k,0} = \sum_{k=1}^{m} \epsilon_k f_k(t)$$

contradicting the linear independence of the polynomials $f_1(t), \ldots, f_m(t)$. Hence $|D| \neq 0$. Thus $|A| \neq 0$ because $a_{m,m} \neq 0$. Using Crammer's rule

$$\begin{aligned} t_0 &= -\frac{\begin{vmatrix} 0 & a_m & a_{m-1} & \cdots & a_2 & a_0 \\ a_{m,m} & a_{m,m-1} & \cdots & a_{m,2} & a_{m,1} & 0 \end{vmatrix}}{\begin{vmatrix} 0 & a_m & a_{m-1} & \cdots & a_2 & a_1 \\ a_{m,m} & a_{m,m-1} & \cdots & a_{m,2} & a_{m,1} & a_{m,0} \end{vmatrix}} \\ &= -\frac{(-1)^{m+2} a_{m,m} \begin{vmatrix} a_m & a_{m-1} & \cdots & a_2 & a_0 \end{vmatrix}}{(-1)^{m+2} a_{m,m} \begin{vmatrix} a_m & a_{m-1} & \cdots & a_2 & a_1 \end{vmatrix}} \\ &= -\frac{\begin{vmatrix} a_m & a_{m-1} & \cdots & a_2 & a_0 \end{vmatrix}}{\begin{vmatrix} a_m & a_{m-1} & \cdots & a_2 & a_1 \end{vmatrix}}. \quad \square \end{aligned}$$

Let $\hat{\mathcal{R}}_s = \{|\hat{C}_k| \mid k = 1, \ldots, m-1\}$ and $\mathcal{R}_s = \{|\bar{C}_q| = \pm|C_q| \mid q = 1, \ldots, m-1\}$. Then $f_1(t), \ldots, f_m(t)$ have a common root if and only if $\hat{\mathcal{R}}_s = \mathcal{R}_s = \{0\}$. $\hat{\mathcal{R}}_s$ and $\mathcal{R}_s$ are called the *Sylvester resolvents* of $f_1(t), \ldots, f_m(t)$.

Notice that if $f_1(t), \ldots, f_m(t)$ do not have a common root, then $\hat{\mathcal{R}}_s$ is not necessarily equal to $\mathcal{R}_s$. For example, consider the polynomials

$$\begin{aligned} f_1(t) &= -8t^2 + 3t + 3, \\ f_2(t) &= 14t^3 - 9t^2 + 7t + 3 \\ f_3(t) &= t^3 - 9t^2 + 22t + 7. \end{aligned}$$

Reducing the leading coefficients of $f_1(t)$ and $f_2(t)$, we obtain the following polynomials:

$$\begin{aligned}\hat{f}_1(t) &= 2057t + 409,\\ \hat{f}_2(t) &= -8t^2 + 3t + 3,\\ \hat{f}_3(t) &= t^3 - 9t^2 + 22t + 7.\end{aligned}$$

The Sylvester resolvents are

$$\hat{\mathcal{R}}_s = \{|\hat{C}_1| = -883156, |\hat{C}_2| = 5288\}, \quad \mathcal{R}_s = \{|\bar{C}_1| = 5288, |\bar{C}_2| = 1026608\}.$$

Thus using either $\hat{\mathcal{R}}_s$ or $\mathcal{R}_s$ we conclude that $f_1(t), f_2(t), f_3(t)$ do not have a common root.

7.4 Sylvester's Resolvent for Three Polynomials

Given three linearly independent polynomials $\hat{f}(t), \hat{g}(t), \hat{h}(t)$ of degrees $n_1 \geq n_2 \geq n_3 \geq 1$ with $n_1 \geq 3$, we want to find necessary and sufficient conditions for these polynomials to have a common root. To begin, we construct a list

$$\mathcal{S} = [\hat{f}(t), t\hat{f}(t), \ldots, t^{p_1}\hat{f}(t), \hat{g}(t), t\hat{g}(t), \ldots, t^{p_2}\hat{g}(t), \hat{h}(t), t\hat{h}(t), \ldots, t^{p_3}\hat{h}(t)],$$

where $0 \leq p_1 \leq p_2 \leq p_3$, such that

$$m = p_1 + p_2 + p_3 + 3 = \max(n_1 + p_1, n_2 + p_2, n_3 + p_3).$$

The condition $0 \leq p_1 \leq p_2 \leq p_3$ is imposed to keep m as small as possible. The list $\mathcal{S}$ contains m polynomials of degree m, and these polynomials have a common root if and only if $\hat{f}(t), \hat{g}(t), \hat{h}(t)$ have a common root. In general, the m polynomials in $\mathcal{S}$ are linearly independent, but degenerate cases where these polynomials are linearly dependent do occur. For example, $\hat{f}(t), \hat{g}(t), \hat{h}(t)$ may have more than one common root or they may be composite functions. We will deal with these cases in a future paper; see also [6]. ¿From here on we shall assume that the polynomials in $\mathcal{S}$ are linearly independent, and we will apply the Sylvester resolvent to the polynomials in $\mathcal{S}$ to get two resolvent expressions for $\hat{f}(t), \hat{g}(t), \hat{h}(t)$.

Our first problem is to find minimal values for p_3 and m. In Section 7.5, we will use these resolvents to implicitize $3D$, parametric, polynomial curves, and we shall show that the degrees of the resulting surfaces depend only on p_3. Thus our first priority is to minimize p_3. To minimize both p_3 and m, we could use integer programming. But this method is known to be NP-complete [10]. Therefore, in this section, we derive formulas for p_1, p_2, p_3 with minimal p_3. Moreover, if $n_2 = n_1 - k, k \geq 1$, then we also get the minimal m.

In CAGD, usually $\hat{f}(t), \hat{g}(t), \hat{h}(t)$ have the same degree. We can reduce the degrees of $\hat{f}(t), \hat{g}(t), \hat{h}(t)$ by eliminating the leading coefficients of $\hat{g}(t), \hat{h}(t)$ to obtain polynomials $f(t) = \hat{f}(t)$ and $g(t), h(t)$ such that $\deg(g(t)) = n_2 \leq n_1 - k$ and $\deg(h(t)) = n_3 \leq n_1 - 2k$, where $k = \min(n_1 - n_2, n_2 - n_3)$. Explicitly,

$$\begin{aligned}f(t) &= a_{n_1}t^{n_1} + a_{n_1-1}t^{n_1-1} + \cdots + a_1t + a_0,\\ g(t) &= b_{n_2}t^{n_2} + b_{n_2-1}t^{n_2-1} + \cdots + b_1t + b_0,\\ h(t) &= c_{n_3}t^{n_3} + c_{n_3-1}t^{n_3-1} + \cdots + c_1t + c_0\end{aligned}$$

where $a_{n_1}, b_{n_2}, c_{n_3} \neq 0$. Moreover, $f(t), g(t), h(t)$ have a common root if and only if $\hat{f}(t), \hat{g}(t), \hat{h}(t)$ have a common root.

Since minimizing p_3 is our first priority, when we construct the list S we have to increase p_1 and p_2 until either the size m of S is equal to the highest degree d in S or $m+1=d$. Now, if $m=d$, then we are done. In this case $p_3=p_2$. Otherwise, $m+1=d$, and we have to add the polynomial $t^{p_2+1}h(t)$ to S. Hence $p_3=p_2+1$. Thus $p_3=p_2+r$ where $r=0$ or 1. The following lemma tells us how to compute these integers.

Lemma 7.4.1 : *Given three integers $n_1, n_2 \le n_1-k, n_3 \le n_1-2k$ where $n_1 \ge 3$ and $k=\min(n_1-n_2, n_2-n_3) \ge 1$, then*

$$p_1 = \max(0, \lceil \frac{n_1}{2} \rceil - (k+2)),\ p_2 = \lceil \frac{n_1}{2} \rceil - 2,\ p_3 = \lfloor \frac{n_1}{2} \rfloor - 1$$

satisfy the following:

(a) $p_1 \le p_2 \le p_3$.

(b) $m = p_1+p_2+p_3+3 = \max(n_1+p_1, n_2+p_2, n_3+p_3) = n_1+p_1$.

(c) *p_3 is minimal with respect to* **(a)** *and* **(b)**.

(d) *when p_3 is minimal, p_2 is also minimal.*

(e) *when p_3 and p_2 are minimal and $p_1+n_1=p_2+n_2$, m is also minimal.*

Proof. **(a)** is obvious from the definition of p_1, p_2, p_3.

(b) To prove that p_1, p_2, p_3 satisfy condition **(b)** is easy because

$$\begin{aligned} m = p_1+p_2+p_3+3 &= p_1 + \lceil \frac{n_1}{2} \rceil - 2 + \lfloor \frac{n_1}{2} \rfloor - 1 + 3 = n_1 + p_1 \\ &\ge n_1 - k + p_2 \ge n_2 + p_2 \\ &\ge n_1 + \lfloor \frac{n_1}{2} \rfloor - 2k - 1 \ge n_1 - 2k + p_3 \ge n_3 + p_3. \end{aligned}$$

(c) We suppose that $p_3 = \lfloor \frac{n_1}{2} \rfloor - 1$ is not minimal, i.e. there exist integers

$$0 \le \hat{p}_1 \le \hat{p}_2 = \hat{p}_3 - r \le \hat{p}_3 = p_3 - q, q \ge 1$$

that satisfy the condition

$$\hat{m} = \hat{p}_1 + \hat{p}_2 + \hat{p}_3 + 3 = \max(n_1+\hat{p}_1, n_2+\hat{p}_2, n_3+\hat{p}_3),$$

and prove that no such $\hat{p}_3$ exists. Notice that $\hat{p}_2 = \hat{p}_3 - r = p_3 - q - r$ for $r=0$ or 1. Now we have three cases to consider.

Case 1 : $\hat{m} = n_1 + \hat{p}_1$. This implies that

$$n_1 = \hat{p}_2 + \hat{p}_3 + 3 = 2\lfloor \frac{n_1}{2} \rfloor - 2q - r + 1. \tag{7.10}$$

But the left hand side is not equal to the right hand side of Equation (7.10) for $q \ge 1$ and $r=0$ or 1. Hence this case is not possible.

Case 2 : $\hat{m} = n_2 + \hat{p}_2 > n_1 + \hat{p}_1$. The Equation $\hat{m} = n_2 + \hat{p}_2$ implies that

$$\hat{p}_1 = n_2 - 3 - \hat{p}_3 = n_2 - \lfloor \frac{n_1}{2} \rfloor - 2 + q. \tag{7.11}$$

On the other hand, $n_1 + \hat{p}_1 < n_2 + \hat{p}_2$, so

$$\hat{p}_1 < n_2 - n_1 + \hat{p}_2 = n_2 - \lceil \frac{n_1}{2} \rceil - 1 - q - r, \tag{7.12}$$

since $\hat{p}_2 = \hat{p}_3 - r$. ¿From Equation (7.11) and Relation (7.12) we get the relation

$$\lfloor \frac{n_1}{2} \rfloor - \lceil \frac{n_1}{2} \rceil + 1 - 2q - r > 0, \tag{7.13}$$

which is impossible because $q \geq 1$ and $r = 0$ or 1. Therefore *Case 2* is not possible.

Case 3 : $\hat{m} = n_3 + \hat{p}_3 > n_1 + \hat{p}_1, n_2 + \hat{p}_2$. ¿From $\hat{m} = n_3 + \hat{p}_3$ and $n_3 + \hat{p}_3 > n_1 + \hat{p}_1$, we get

$$\hat{p}_1 = n_3 - 3 - \hat{p}_2 = n_3 - \lfloor \frac{n_1}{2} \rfloor - 2 + q + r, \tag{7.14}$$

$$\hat{p}_1 < n_3 - n_1 + \hat{p}_3 = n_3 - \lceil \frac{n_1}{2} \rceil - 1 - q. \tag{7.15}$$

But again we obtain the impossible Relation (7.13) from Equation (7.14) and Relation (7.15). Hence this case is also impossible. Therefore $p_3 = \lfloor \frac{n_1}{2} \rfloor - 1$ is minimal.

(d) Given $p_3 = \lfloor \frac{n_1}{2} \rfloor - 1$, we suppose that $p_2 = \lceil \frac{n_1}{2} \rceil - 2$ is not minimal, i.e. there exist integers $0 \leq \check{p}_1 \leq \check{p}_2 = p_2 - q \leq p_3, q \geq 1$ that satisfy the condition

$$\check{m} = \check{p}_1 + \check{p}_2 + p_3 + 3 = \max(n_1 + \check{p}_1, n_2 + \check{p}_2, n_3 + p_3),$$

and prove that no such $\check{p}_2$ exists. Since $\check{p}_2 + p_3 = n_1 - 3 - q$, we know that $\check{m} > n_1 + \check{p}_1$. Thus we have only two cases to consider here.

Case 4 : $\check{m} = n_2 + \check{p}_2 > n_1 + \check{p}_1$. The Equation $\check{m} = n_2 + \check{p}_2$ implies that

$$\check{p}_1 = n_2 - 3 - p_3 = n_2 - \lfloor \frac{n_1}{2} \rfloor - 2. \tag{7.16}$$

On the other hand, $n_1 + \check{p}_1 < n_2 + \check{p}_2$, so

$$\check{p}_1 < n_2 - n_1 + \check{p}_2 = n_2 - \lfloor \frac{n_1}{2} \rfloor - 2 - q. \tag{7.17}$$

Clearly Equation (7.16) and Relation (7.17) imply $q < 0$ contradicting the assumption that $q \geq 1$. Therefore *Case 4* is not possible.

Case 5 : $\check{m} = n_3 + p_3 > n_1 + \check{p}_1, n_2 + \check{p}_2$. ¿From $\check{m} = n_3 + p_3$ and $n_3 + p_3 > n_1 + \check{p}_1$, we get

$$\check{p}_1 = n_3 - 3 - \check{p}_2 = n_3 - \lceil \frac{n_1}{2} \rceil - 1 + q, \tag{7.18}$$

$$\check{p}_1 < n_3 - n_1 + p_3 = n_3 - \lceil \frac{n_1}{2} \rceil - 1. \tag{7.19}$$

But again we obtain the impossible Inequality $q < 0$ from Equation (7.18) and Relation (7.19). Hence this case is also impossible. Therefore $p_2 = \lceil \frac{n_1}{2} \rceil - 2$ is minimal.

(e) Given $p_1 + n_1 = p_2 + n_2$, then clearly, $n_2 = n_1 - k$, and

$$p_3 + n_3 = n_3 + \lfloor \frac{n_1}{2} \rfloor - 1 \leq p_2 + n_2 = n_2 + \lceil \frac{n_1}{2} \rceil - 2$$

In order to prove that m is also minimal, we must show that $p_1 = \lceil \frac{n_1}{2} \rceil - (2 + k)$ is also minimal. Suppose that there exists an integer $q \geq 1$ such that

$$0 \leq \bar{p}_1 = p_1 - q \leq p_2 \leq p_3$$

and

$$\bar{m} = \bar{p}_1 + p_2 + p_3 + 3 = \max(n_1 + \bar{p}_1, n_2 + p_2, n_3 + p_3).$$

Notice that since $p_2 + p_3 = n_1 - 3$, $\bar{m} = \bar{p}_1 + n_1$. But $\bar{m} = n_1 + \bar{p}_1 = n_1 + p_1 - q < n_2 + p_2$, because $p_1 + n_1 = p_2 + n_2$. Thus $q = 0$. □

Example 4.1 : (a) If $n_1 = 8, n_2 = 7, n_3 = 6$, then $k = 1, p_1 = 1, p_2 = 2, p_3 = 3$. Thus

$$\mathcal{S}_1 = [f(t), tf(t), g(t), tg(t), t^2g(t), h(t), th(t), t^2h(t), t^3h(t)]$$

has 9 polynomials of degree 9. Since $n_1 + p_1 = n_2 + p_2$, $m = 9$ is minimal.

(b) If $n_1 = 7, n_2 = 4, n_3 = 3$, then $k = 1$. Using our formulas, we get $p_1 = 1, p_2 = p_3 = 2$. Therefore

$$\mathcal{S}_2 = [f(t), tf(t), g(t), tg(t), t^2g(t), h(t), th(t), t^2h(t)].$$

Notice that $n_1 + p_1 = 8 > n_2 + p_2 = 6$. Thus $m = 8$ is not minimal. We can have

$$\hat{\mathcal{S}}_2 = [f(t), g(t), tg(t), t^2g(t), h(t), th(t), t^2h(t)]$$

which contains only 7 polynomials of degree 7.

Now recall that

$$\mathcal{S} = [f(t), tf(t), \ldots, t^{p_1}f(t), g(t), tg(t), \ldots, t^{p_2}g(t), h(t), th(t), \ldots, t^{p_3}h(t)]$$

where $m = p_1 + n_1$. By the Sylvester resolvent (Theorem 7.3.4), the polynomials in $\mathcal{S}$ have a common root at $t = t_0$ if and only if the following vectors contain linearly dependent polynomials:

$$\boldsymbol{p}_i(t) = [\mathcal{S}, t^{p_1+1}f(t), tq_i(t)]^T = \boldsymbol{C}_i[t^{m+1}, \ldots, 1]^T,$$

where $q_i(t)$ is the i-th element in $\mathcal{S}$ not equal to $t^{p_1}f(t)$. But the only polynomials $tq_i(t)$ which are not one of the $m-1$ polynomials in the list $\mathcal{S}$ are $t^{p_2+1}g(t)$ and $t^{p_3+1}h(t)$. Hence we only have to check the two determinants corresponding to the two vectors

$$\begin{aligned} \boldsymbol{p}_1(t) &= [\mathcal{S}, t^{p_1+1}f(t), t^{p_2+1}g(t)]^T = \boldsymbol{C}_1[t^{m+1}, \ldots, 1]^T, \\ \boldsymbol{p}_2(t) &= [\mathcal{S}, t^{p_1+1}f(t), t^{p_3+1}h(t)]^T = \boldsymbol{C}_2[t^{m+1}, \ldots, 1]^T. \end{aligned}$$

(Similarly using the Sylvester resolvent for two polynomials $f(t), g(t)$ of degree n_1, n_2, we find that $f(t), g(t)$ have a common root if and only if the coefficient matrix of the column vector

$$[f(t), tf(t), \ldots, t^{n_2-1}f(t), g(t), tg(t), \ldots, t^{n_1-1}g(t)]^T$$

is singular. This determinant condition is the Sylvester resultant of two polynomials.)

Let

$$v_j = \begin{cases} a_j & 0 \le j \le n_1, \\ 0 & \text{otherwise.} \end{cases} ; \quad u_j = \begin{cases} b_j & 0 \le j \le n_2, \\ 0 & \text{otherwise.} \end{cases} ; \quad w_j = \begin{cases} c_j & 0 \le j \le n_3, \\ 0 & \text{otherwise.} \end{cases} ;$$

$$\boldsymbol{v}_j = \begin{bmatrix} v_{j-p_1} & \cdots & v_j & u_{j-p_2} & \cdots & u_j & w_{j-p_3} & \cdots & w_j \end{bmatrix}^T. \tag{7.20}$$

Then we have proved the following theorem:

Theorem 7.4.2 : *Suppose that the polynomials in the list $\mathcal{S}$ are linearly independent. Then the polynomials $\hat{f}(t), \hat{g}(t), \hat{h}(t)$ have a common root at $t = t_0$ if and only if*

$$|\boldsymbol{C}_1| = \begin{vmatrix} \boldsymbol{0} & \boldsymbol{v}_m & \boldsymbol{v}_{m-1} & \cdots & \cdots & \boldsymbol{v}_1 & \boldsymbol{v}_0 \\ u_{n_1} & \cdots & u_0 & 0 & \cdots & \cdots & 0 \\ \boldsymbol{0}_{l_1} & v_{n_2} & \cdots & v_0 & 0 & \cdots & 0 \end{vmatrix} = 0,$$

$$|\boldsymbol{C}_2| = \begin{vmatrix} \boldsymbol{0} & \boldsymbol{v}_m & \boldsymbol{v}_{m-1} & \cdots & \cdots & \boldsymbol{v}_1 & \boldsymbol{v}_0 \\ u_{n_1} & \cdots & u_0 & 0 & \cdots & \cdots & 0 \\ \boldsymbol{0}_{l_2} & w_{n_3} & \cdots & w_0 & 0 & \cdots & 0 \end{vmatrix} = 0$$

where $\mathbf{0}_{l_1}$ *and* $\mathbf{0}_{l_2}$ *contain* $l_1 = n_1+p_1-n_2-p_2$ *and* $l_2 = n_1+p_1-n_3-p_3$ *zeroes,* p_1, p_2, p_3 *are given in Lemma* 7.4.1, *and* v_i, u_i, w_i, *and* $\boldsymbol{v}_i$ *are given in Equation* (7.20). *Moreover,*

$$t_0 = -\frac{\begin{vmatrix} \boldsymbol{v}_m & \boldsymbol{v}_{m-1} & \cdots & \boldsymbol{v}_2 & \boldsymbol{v}_0 \end{vmatrix}}{\begin{vmatrix} \boldsymbol{v}_m & \boldsymbol{v}_{m-1} & \cdots & \boldsymbol{v}_2 & \boldsymbol{v}_1 \end{vmatrix}}.$$

Notice that p_2 and p_3 depend only on n_1 and do not depend on any relationship between n_1, n_2, and n_3. Moreover, if m is not minimal, then $p_1 + n_1 > p_2 + n_2 \geq p_3 + n_3$. Thus $|C_1| = a_{n_1}^{q_1}|\hat{C}_1|$ and $|C_2| = a_{n_1}^{q_2}|\hat{C}_2|$ where $q_1 = p_1 + n_1 - p_2 - n_2, q_2 = p_1 + n_1 - p_3 - n_3$, and $\hat{C}_1, \hat{C}_2$ are $m + 2 - q_1$ x $m + 2 - q_1$ and $m + 2 - q_2$ x $m + 2 - q_2$ matrices. But, since $a_{n_1} \neq 0$, the polynomials $\hat{f}(t), \hat{g}(t), \hat{h}(t)$ have a common root at $t = t_0$ if and only if $|\hat{C}_1| = |\hat{C}_2| = 0$.

7.5 Intersection of $3D$ Polynomial Curves

In this section we show how to apply our new resolvent to implicitize and invert faithfully parametrized $3D$, polynomial curves. Then, using this tool, we give a general curve/curve intersection algorithm.

Let a non–planar, $3D$, parametric, polynomial curve $P(t)$ of degree n_1 be defined by

$$x = f(t) = \sum_{i=0}^{n_1} \hat{a}_i t^i, \quad y = g(t) = \sum_{i=0}^{n_1} \hat{b}_i t^i, \quad z = h(t) = \sum_{i=0}^{n_1} \hat{c}_i t^i$$

where at least one of the $\hat{a}_{n_1}, \hat{b}_{n_1}, \hat{c}_{n_1} \neq 0$. Let $f_x(t) = f(t) - x,\ g_y(t) = g(t) - y,\ h_z(t) = h(t) - z$. Then, without loss of generality, we assume that $\hat{a}_{n_1} \neq 0$, and reduce the degrees of $g_y(t)$ and $h_z(t)$ to obtain

$$\begin{aligned} f_x(t) &= \left(\sum_{i=1}^{n_1} \hat{a}_i t^i\right) + \hat{a}_0 - x = \left(\sum_{i=1}^{n_1} a_i t^i\right) + a_0(x), \\ g_{xy}(t) &= g_y(t) - \frac{\hat{b}_{n_1}}{a_{n_1}} f_x(t) = \left(\sum_{i=1}^{n_2} b_i t^i\right) + b_0(x,y), \\ h_{xz}(t) &= h_z(t) - \frac{\hat{c}_{n_1}}{a_{n_1}} h_z(t) = \left(\sum_{i=1}^{\hat{n}_3} \bar{c}_i t^i\right) + \bar{c}_0(x,z). \end{aligned}$$

Again, without loss of generality, assume that $n_2 \geq \hat{n}_3$. Now if $n_2 = \hat{n}_3$, then eliminating the leading coefficient of $h_{xz}(t)$, we have

$$h_{xyz}(t) = h_{xz}(t) - \frac{\bar{c}_{n_2}}{b_{n_2}} g_{xy}(t) = \left(\sum_{i=1}^{n_3} c_i t^i\right) + c_0(x,y,z).$$

Notice that $f_x(t),\ g_y(t),\ h_z(t)$ have a common root if and only if $f_x(t),\ g_{xy}(t),\ h_{xyz}(t)$ have a common root.

7.5.1 Implicitization and Inversion: Two Surfaces Suffice

In order to apply the Sylvester resolvent, the polynomials in

$$\mathcal{S} = [f_x(t), tf_x(t), \ldots, t^{p_1} f_x(t), g_{xy}(t), tg_{xy}(t), \ldots, t^{p_2} g_{xy}(t), h_{xyz}(t), th_{xyz}(t), \ldots, t^{p_3} h_{xyz}(t)]$$

(see Section 7.4) have to be linearly independent for almost all points $(\hat{x}, \hat{y}, \hat{z})$ on the curve $P(t)$ (except for a finite number of singular points). But the polynomials in $\mathcal{S}$ are linearly independent if and only if the m x $m+1$ coefficient matrix $\boldsymbol{U}$ has rank m. Now recall from Section 7.3.1 that $\boldsymbol{U}$ has rank m if and only if

$$|\boldsymbol{U}\boldsymbol{U}^T| = \sum_{i=1}^{m+1} |\boldsymbol{U}_i|^2 \neq 0,$$

where $\boldsymbol{U}_i$ is the m x m matrix obtained by removing the i–th column of the matrix $\boldsymbol{U}$. Thus the polynomials in S are linearly independent if and only if

$$s_0(\hat{x}, \hat{y}, \hat{z}) = \sum_{i=1}^{m+1} (|\boldsymbol{U}_i|(\hat{x}, \hat{y}, \hat{z}))^2 \neq 0 \tag{7.21}$$

for almost all points $(\hat{x}, \hat{y}, \hat{z})$ on the curve $P(t)$ – that is, if and only if the polynomial $s_0(f(t), g(t), h(t)) \not\equiv 0$. Here we shall assume that the polynomials in S are linearly independent for almost all points on the curve $P(t)$. We will deal with the case of linear dependence in the future paper. The following theorem gives us the degrees and explicit formulas for the implicit surfaces as well as the inversion equation.

Theorem 7.5.1 : *Suppose that $P(t) = (x = f(t), y = g(t), z = h(t))$ is a non-planar, 3D, polynomial curve of degree n_1, $f_x(t)$, $g_{xy}(t)$, $h_{xyz}(t)$ have degrees n_1, n_2, n_3, $k = \min(n_1 - n_2, n_2 - n_3) \geq 1$, $p_1 = \max(0, \lceil \frac{n_1}{2} \rceil - k - 2)$, $p_2 = \lceil \frac{n_1}{2} \rceil - 2$, $p_3 = \lfloor \frac{n_1}{2} \rfloor - 1$, and $s_0(f(t), g(t), h(t)) \not\equiv 0$, where $s_0(x, y, z)$ is given in Equation (7.21). Then $P(t)$ is the real affine intersection of two surfaces*

$$s_1(x,y,z) = |\boldsymbol{C}_1|(x,y,z) = \begin{vmatrix} \boldsymbol{0} & \boldsymbol{v}_m & \boldsymbol{v}_{m-1} & \cdots & \cdots & \boldsymbol{v}_1 & \boldsymbol{v}_0 \\ u_{n_1} & \cdots & u_0 & 0 & \cdots & \cdots & 0 \\ \boldsymbol{0}_{l_1} & v_{n_2} & \cdots & v_0 & 0 & \cdots & 0 \end{vmatrix} = 0,$$

$$s_2(x,y,z) = |\boldsymbol{C}_2|(x,y,z) = \begin{vmatrix} \boldsymbol{0} & \boldsymbol{v}_m & \boldsymbol{v}_{m-1} & \cdots & \cdots & \boldsymbol{v}_1 & \boldsymbol{v}_0 \\ u_{n_1} & \cdots & u_0 & 0 & \cdots & \cdots & 0 \\ \boldsymbol{0}_{l_2} & w_{n_3} & \cdots & w_0 & 0 & \cdots & 0 \end{vmatrix} = 0$$

where $\boldsymbol{0}_{l_1}$ and $\boldsymbol{0}_{l_2}$ contain $l_1 = n_1 + p_1 - n_2 - p_2$ and $l_2 = n_1 + p_1 - n_3 - p_3$ zeroes, and v_i, u_i, w_i, and $\boldsymbol{v}_i$ are given in Equation (7.20). The degree of $s_1(x, y, z)$ is at most $\lceil \frac{n_1}{2} \rceil$ and the degree of $s_2(x, y, z)$ is at most $\lfloor \frac{n_1}{2} \rfloor + 1$. Moreover, the inversion equation is given by

$$t_0(x,y,z) = -\frac{|\boldsymbol{V}_1|(x,y,z)}{|\boldsymbol{V}_2|(x,y,z)} = -\frac{\begin{vmatrix} \boldsymbol{v}_m & \boldsymbol{v}_{m-1} & \cdots & \boldsymbol{v}_2 & \boldsymbol{v}_0 \end{vmatrix}}{\begin{vmatrix} \boldsymbol{v}_m & \boldsymbol{v}_{m-1} & \cdots & \boldsymbol{v}_2 & \boldsymbol{v}_1 \end{vmatrix}}$$

which is is a linear polynomial if $n_1 = 3$ and a rational polynomial of degree at most $\lfloor \frac{n_1}{2} \rfloor - 1$ if $n_1 > 3$.

Proof. We know that a point (x, y, z) is on the curve $P(t)$ if and only if the polynomials $f_x(t), g_y(t), h_z(t)$ have a common root. But these polynomials have a common root at $t = t_0$ if and only if the polynomials $f_x(t), g_{xy}(t), h_{xyz}(t)$ have common root at $t = t_0$. Since $s_0(f(t), g(t), h(t)) \not\equiv 0$, the polynomials in the list S are linearly independent for almost all points $(\hat{x}, \hat{y}, \hat{z})$ on the curve $P(t)$. By Theorem 7.4.2, the polynomials $f_x(t), g_{xy}(t), h_{xyz}(t)$ have common root at $t = t_0$ if and only if the coefficient matrices $\boldsymbol{C}_1$ and $\boldsymbol{C}_2$ of the column vectors

$$\begin{aligned} \boldsymbol{p}_1(t) &= [f_x(t), \ldots, t^{p_1+1} f_x(t), g_{xy}(t), \ldots, t^{p_2+1} g_{xy}(t), h_{xyz}(t), \ldots, t^{p_3} h_{xyz}(t)]^T, \\ \boldsymbol{p}_2(t) &= [f_x(t), \ldots, t^{p_1+1} f_x(t), g_{xy}(t), \ldots, t^{p_2} g_{xy}(t), h_{xyz}(t), \ldots, t^{p_3+1} h_{xyz}(t)]^T, \end{aligned}$$

are singular. Since $p_2 + 1 \geq p_3$, only the last $p_2 + 2$ columns of $\boldsymbol{C}_1$ contain the linear polynomials $a_0(x), b_0(x,y)$, and $c_0(x,y,z)$; the other columns contain only constants. Thus $\deg(|\boldsymbol{C}_1|(x,y,z)) \leq p_2 + 2 = \lceil \frac{n_1}{2} \rceil$. Similarly, $\deg(|\boldsymbol{C}_2|(x,y,z)) \leq p_3 + 2 = \lfloor \frac{n_1}{2} \rfloor + 1$.

By Theorem 7.4.2, the inversion equation is

$$t_0(x,y,z) = -\frac{|\boldsymbol{V}_1|(x,y,z)}{|\boldsymbol{V}_2|(x,y,z)} = -\frac{\begin{vmatrix} \boldsymbol{v}_m & \boldsymbol{v}_{m-1} & \cdots & \boldsymbol{v}_2 & \boldsymbol{v}_0 \end{vmatrix}}{\begin{vmatrix} \boldsymbol{v}_m & \boldsymbol{v}_{m-1} & \cdots & \boldsymbol{v}_2 & \boldsymbol{v}_1 \end{vmatrix}},$$

where $\boldsymbol{v}_j$ is given in Equation (7.20). When $n_1 = 3$,

$$|\boldsymbol{V}_1|(x,y,z) = \begin{vmatrix} a_3 & a_2 & a_0(x) \\ 0 & b_2 & b_0(x,y) \\ 0 & 0 & c_0(x,y,z) \end{vmatrix}; \quad |\boldsymbol{V}_2|(x,y,z) = \begin{vmatrix} a_3 & a_2 & a_1 \\ 0 & b_2 & b_1 \\ 0 & 0 & c_1 \end{vmatrix}.$$

Hence $t_0(x,y,z) = -qc_0(x,y,z)$ where q is a constant. When $n_1 > 3$, only the last p_3 columns of $\boldsymbol{V}_1$ and $\boldsymbol{V}_2$ contain $a_0(x), b_0(x,y)$, and $c_0(x,y,z)$. Thus $|\boldsymbol{V}_1|(x,y,z)$ and $|\boldsymbol{V}_2|(x,y,z)$ have degree at most $p_3 = \lfloor \frac{n_1}{2} \rfloor - 1$. □

Implicitizing faithfully parametrized, $3D$, rational curves of degree n_1 using the Sylvester resolvent still generates two surfaces of degrees less than or equal to $\lceil \frac{n_1}{2} \rceil$ and $\lfloor \frac{n_1}{2} \rfloor + 1$, but it yields two additional linear inequalities; the inversion equation is still a rational polynomial of degree $\lfloor \frac{n_1}{2} \rfloor$. Unfortunately, the proof is somewhat more complicated, so we defer it to a future paper. In the meantime, the proof for rational curves can be found in [6].

Theorem 7.5.2 : *The surface $s_1(x,y,z)$ is a lowest degree surface on which $P(t)$ lies for* $3 \leq n_1 \leq 12$.

Proof. We use the method of *undetermined coefficients* to prove this theorem. Let

$$s(x,y,z) = \sum_{\substack{0 \leq i,j,k \leq r \\ 0 \leq i+j+k \leq r}} d_{i,j,k} x^i y^j z^k \tag{7.22}$$

be a surface of degree r. Note that we have $\binom{r+3}{3}$ undetermined coefficients $d_{i,j,k}$. Substituting $x = f(t), y = g(t), z = h(t)$ in to Equation (7.22), we obtain

$$p(t) = s(f(t), g(t), h(t)) = \sum_{q=0}^{rn_1} l_q(d_{i,j,k}) t^q,$$

where $l_q(d_{i,j,k})$ are linear polynomials. The curve $P(t)$ lies on the surface $s(x,y,z)$ if and only if $p(t) \equiv 0$. This is equivalent to saying that the system of $rn_1 + 1$ homogeneous linear equations, $l_q(d_{i,j,k}) = 0$, in $\binom{r+3}{3}$ unknowns $d_{i,j,k}$ has nontrivial solutions. But, the condition for this system of homogeneous linear equations to have nontrivial solutions is that the number of unknowns must be greater than the number of equations, i.e.

$$\binom{r+3}{3} - rn_1 - 1 = \frac{(r+3)(r+2)(r+1) - 6rn_1 - 6}{6} > 0.$$

Using Maple [3] to solve $q(r) = (r+3)(r+2)(r+1) - 6rn_1 - 6 = 0$ for r in terms of n_1, we get the solutions $r = \{0, \sqrt{6n_1 - 2} - 3, -(\sqrt{6n_1 - 2} + 3)\}$. Since we are only interested in positive integer values of r, we restrict our attention to the solution $\hat{r} = \lceil \sqrt{6n_1 - 2} - 3 \rceil$. To show that $q(\hat{r}) > 0$, we have to show that $q'(\hat{r}) > 0$ for $n_1 \geq 3$. Now,

$$q'(\hat{r}) = 3\hat{r}^2 + 12\hat{r} - 6n_1 + 11 \geq w(n_1) = 12n_1 - 6\sqrt{6n_1 - 2} - 4,$$

and

$$w'(n_1) = \frac{12\sqrt{3n_1 - 1} - 9\sqrt{2}}{\sqrt{3n_1 - 1}} > 0$$

for $n_1 \geq 3$. Thus $q'(\hat{r}) > 0$. Computing $\lceil \sqrt{6n_1 - 2} - 3 \rceil$ and $\lceil \frac{n_1}{2} \rceil$ for $3 \leq n_1 \leq 12$, we see that these two values are equal. □

Example 7.5.1 : Let $P(t)$ be the polynomial quartic curve defined by

$$\begin{aligned} x = f(t) &= (t-1)(t^3 + t^2 + t + 1), \\ y = g(t) &= (t-1)(t^3 + 3t^2 + 3t + 1), \\ z = h(t) &= (t-1)(2t^3 - t^2 + 3t + 1). \end{aligned}$$

Let $f_x(t) = f(t) - x$, $g_y(t) = g(t) - y$, and $h_z(t) = h(t) - z$. Then reducing the degrees of $g_y(t)$ and $h_z(t)$, we obtain

$$\begin{aligned} f_x(t) &= t^4 - (x+1), \\ g_{xy}(t) &= 2t^3 - 2t + (x - y), \\ h_{xyz}(t) &= 4t^2 - 5t + (3.5x - 1.5y - z + 1). \end{aligned}$$

Since $s_0(f(t), g(t), h(t)) = 4(t^8 + t^6 + t^4 + t^2 + 1)(16t^2 - 20t + 9)^2 \not\equiv 0$, using our implicitization technique, we get two implicit surfaces

$$\begin{aligned} s_1(x,y,z) &= -43x^2 + (24y + 17z - 25)x - 5y^2 - (z-5)y - 2z^2 + 20z, \\ s_2(x,y,z) &= 85.75x^3 - (110.25y + 73.5z + 80.5)x^2 + \\ &\quad (47.25y^2 + (63z + 101)y + 21z^2 - 26z - 85)x - \\ &\quad 6.75y^3 - (13.5z + 28.5)y^2 - (9z^2 - 2z - 5)y - 2z^3 + 14z^2 + 60z, \end{aligned}$$

and the inversion equation

$$t_0(x,y,z) = -\frac{19x + y - 10z + 10}{28x - 12y - 8z - 10}.$$

Note that $s_1(0,0,0) = s_2(0,0,0) = 0$, and $t_0(0,0,0) = 1$ because the polynomials $f(t), g(t), h(t)$ have a common root at $t_0 = 1$. Since the polynomial $s_0(f(t), g(t), h(t))$ does not have real roots, the intersection of $s_1(x,y,z)$ and $s_2(x,y,z)$ is exactly the curve $P(t)$ in real affine space.

7.5.2 Curve/Curve Intersection

Applying our implicitization and inversion equations, we have the following curve/curve intersection algorithm for two $3D$, polynomial curves $P(t) = (x = f(t), y = g(t), z = h(t))$, $t \in [\hat{t}_0, \hat{t}_1]$, and $Q(s) = (x = \hat{f}(s), y = \hat{g}(s), z = \hat{h}(s))$, $s \in [\hat{s}_0, \hat{s}_1]$:

1. Implicitize the curve $P(t)$, to get two implicit surfaces $s_1(x,y,z) = 0, s_2(x,y,z) = 0$.

2. Substitute the parametric equations $x = \hat{f}(s)$, $y = \hat{g}(s)$, $z = \hat{h}(s)$ of $Q(s)$ into the implicit equations $s_1(x,y,z) = 0$, $s_2(x,y,z) = 0$ of $P(t)$ to obtain two univariate polynomial equations $p_1(s) = 0$, $p_2(s) = 0$.

3. Find the list of common roots $\mathcal{R}_S = [s_1, \ldots, s_n]$ of $p_1(s) = 0$, $p_2(s) = 0$ where $s_i \in [\hat{s}_0, \hat{s}_1]$ $i = 1, \ldots, n$; these common roots are the possible s parameters of the intersection points.

4. Use the parametric equations of the curve $Q(s)$ to find the list $\mathcal{R}_{\mathcal{X}} = [\boldsymbol{x}_1, \ldots, \boldsymbol{x}_n]$ of possible (x, y, z)–coordinates of the intersection points.

5. Apply the inversion equation for the curve $P(t)$ to get the list $\mathcal{R}_{\mathcal{T}} = [t_1, \ldots, t_n]$ of the t parameters of each possible intersection point. Discard t_i and the corresponding $s_i, \boldsymbol{x}_i$ from the lists $\mathcal{R}_{\mathcal{T}}$ and $\mathcal{R}_{\mathcal{S}}, \mathcal{R}_{\mathcal{X}}$ if $t_i \notin [\hat{t}_0, \hat{t}_1]$.

6. Compute the polynomial $c_0(t) = s_0(f(t), g(t), h(t))$ where $s_0(x, y, z)$ is given in Equation (7.21). For each triple $(\boldsymbol{x}_i, s_i, t_i)$, if $c_0(t_i) = 0$ and either $f(t_i) \neq \hat{f}(s_i)$ or $g(t_i) \neq \hat{g}(s_i)$ or $h(t_i) \neq \hat{h}(s_i)$, then discard $s_i, \boldsymbol{x}_i, t_i$ from the lists $\mathcal{R}_{\mathcal{S}}, \mathcal{R}_{\mathcal{X}}, \mathcal{R}_{\mathcal{T}}$.

Notice that the most difficult step is step 1, and that the speed of this algorithm is controlled by the degrees of $p_1(s)$ and $p_2(s)$. Since our resolvents produce the optimal number of surfaces (two), with lowest possible degrees, for $3 \leq n_1 \leq 12$, we get a very efficient curve/curve intersection algorithm for low degree curves.

Example 7.5.2 : Find the intersection of the quartic curve $P(t)$, given in Example 7.5.1, and the quintic curve $Q(s) = (x = \hat{f}(s), y = \hat{g}(s), z = \hat{h}(s))$ where

$$\begin{aligned}
\hat{f}(s) &= -0.1395823661s^5 + 6.572154354s^4 + 7.048138640s^3 - \\
&\quad 1.783248943s^2 - 1.961290638s - 0.5445512233, \\
\hat{g}(s) &= 10.27811571s^5 + 13.90506016s^4 + 7.509569572s^3 - \\
&\quad 2.929542615s^2 - 4.323420917s - 0.5747012102, \\
\hat{h}(s) &= -45.95668839s^5 - 31.55432938s^4 + 51.95137996s^3 + \\
&\quad 8.71905529s^2 - 16.49118079s - 0.4026163216.
\end{aligned}$$

Substituting $x = \hat{f}(s), y = \hat{g}(s), z = \hat{h}(s)$ into the two surfaces $s_1(x, y, z)$ and $s_2(x, y, z)$ in Example 7.5.1, we obtain two polynomials in s :

$$\begin{aligned}
p_1(s) &= -8412.206319s^{10} - 19345.14628s^9 - 1902.313745s^8 + 21832.97613s^7 + \\
&\quad 6334.301597s^6 - 11489.53007s^5 - 2528.459202s^4 + 3267.695736s^3 + \\
&\quad 285.1914713s^2 - 375.2488831s + 8.826980095, \\
p_2(s) &= 54275.74118s^{15} + 182595.0683s^{14} - 100.3440s^{13} - 503334.0103s^{12} - \\
&\quad 157367.4536s^{11} + 663808.5615s^{10} + 205367.5091s^9 - 500129.0891s^8 - \\
&\quad 98803.5776s^7 + 211694.9743s^6 + 9123.93613s^5 - 45104.20685s^4 + \\
&\quad 5129.978001s^3 + 3362.175489s^2 - 812.2471806s + 14.08497243.
\end{aligned}$$

Using the Padé approximate method given in [2], we find that the floating point greatest common divisor of $p_1(s), p_2(s)$ is

$$g(s) = 5.184620386s^4 + 2.628525304s^3 - 4.277842759s^2 - 1.166120573s + 1.0.$$

Solving for the roots of this polynomial, we find

$$\mathcal{R}_{\mathcal{S}} = [-0.8572566864, -0.7498306384, 0.5000948900, 0.6000073310].$$

Hence the intersection points in the (x, y, z) coordinates are

$$\begin{aligned}
\mathcal{R}_{\mathcal{X}} = \; &(-0.9372803643, -0.1865624072, \;\; 1.5040035880), \\
&(-0.9999655793, -1.0011344800, -1.0024328120), \\
&(-0.6831396275, -1.3390326640, -0.8817637736), \\
&(\;\; 0.0000667047, \;\; 0.0001388948, \;\; 0.0000369274).
\end{aligned}$$

Substituting these values into the inversion equation for t, we get

$$\mathcal{R}_T = [-0.500355369, 0.001327378557, 0.7491501982, 1.000094258].$$

Since the intersection of the surfaces $s_1(x, y, z)$ and $s_2(x, y, z)$ is exactly the curve $P(t)$ in real affine space (see Example 7.5.1), all the intersection points are valid. Using rational arithmetic instead of floating point, the real intersection points were found to be

$$\begin{array}{l}
[\ -0.8571428571, -0.7500000000,\ \ 0.5000000000,\ \ 0.6000000000],\\
[\ -0.5000000000,\ \ 0.0000000000,\ \ 0.7500000000,\ \ 1.0000000000],\\
[(-0.9375000000, -0.1875000000,\ \ 1.500000000)\\
\ (-1.0000000000, -1.0000000000, -1.000000000)\\
\ (-0.6835937500, -1.3398437500, -0.882812500)\\
\ (\ \ 0.0000000000,\ \ 0.0000000000,\ \ 0.000000000)],
\end{array}$$

in s, t, and (x, y, z), respectively.

7.6 Conclusions and Future Work

We began by constructing a new resolvent for m linearly independent polynomials of degree m. We then showed how to build a list of m polynomials of degree m from three linearly independent polynomials of degree greater than or equal to 3. Thus we provided a mechanism for determining if a system of univariate polynomials has exactly one common root. This technique allowed us to generate two surfaces of low degree whose affine intersection is a given $3D$, polynomial curve. Thus this method leads to much better implicitization procedures and inversion equations for $3D$, polynomial curves. Therefore we get a much better algebraic procedure for intersecting two such curves.

There are many interesting problems in CAGD to which resolvents can be applied [6]. We are going to address some of them in future papers. In particular we plan to :

1. Discuss necessary and sufficient conditions in terms of the original polynomials $f_1(t), \ldots, f_m(t)$ for the polynomials in $\mathcal{S}$ to be linearly independent, and use these conditions to detect unfaithful parametrizations and recover the implicit equations without additional computation.
2. Provide a rigorous proof that our resolvent techniques apply to non–planar rational curves.
3. Extend our resolvent methods to r univariate polynomials and to rational curves in r–space.
4. Extend our resolvent methods to multivariate polynomials and multivariate surfaces.

7.7 References

[1] A. C. Aitken. *Determinants and Matrices.* Oliver and Boyd, London, England, 1956.

[2] S. Cabay and R. Meleshko. A Weekly Stable Algorithm for Padé Approximants and the Inversion of Hankel Matrices. Submitted to the *SIAM Journal on Matrix Analysis and Applications*, 1991.

[3] B. W. Char, K. O. Geddes, G. H. Gonnet, M. B. Monagan, and S. M. Watt. *Maple Reference Manual.* Watcom Publications, 1988.

[4] E. Chionh. *Base Points, Resultants, and the Implicit Representation of Rational Surfaces.* PhD thesis, Department of Computer Science, University of Waterloo, 1990.

[5] H. K. Du. Intersection of Curves and Surfaces Using Elimination Theory. Master's thesis, University of Alberta, Department of Computing Science, 1989.

[6] H. K. Du. *New Resolvent Methods with Applications to Curves and Surfaces in Geometric Modeling.* PhD thesis, Department of Computer Science, University of Waterloo, 1991.

[7] R. N. Goldman. The Method of Resolvents: A Technique for the Implicitization, Inversion, and Intersection of Non-Planar, Parametric, Rational Cubic Curves. *Computer Aided Geometric Design*, 2(4):237–255, December 1985.

[8] R. N. Goldman and T. W. Sederberg. An Analytic Approach to the Intersection of All Piecewise Parametric Rational Cubic Curves. *Computer-Aided Design*, 19(6):282–292, 1987.

[9] K. Kakié. The Resultant of Several Homogeneous Polynomials in Two Indeterminates. *Proc. of the American Mathematical Society*, 54:1–7, Jan 1976.

[10] A. Schrijver. *Theory of Integer Programming.* John Wiley & Sons Ltd., New York, 1986.

[11] T. W. Sederberg. *Implicit and Parametric Curves and Surfaces.* PhD thesis, Department of Mechanical Engineering, Purdue University, August 1983.

[12] B. L. van der Waerden. *Modern Algebra.* Frederick Ungar Publishing Co., New York, 2nd edition, 1950.

8 Does a Trigonometric Curve Cross an Algebraic Surface?

Daniel Richardson

ABSTRACT A symbolic method is given to decide whether or not a curve in three dimensional space given parametrically by three trigonometric polynomials

$$x = p_1(t, sin(t), cos(t))$$
$$y = p_2(t, sin(t), cos(t))$$
$$z = p_3(t, sin(t), cos(t))$$

crosses an algebraic surface, given as the zero set of a polynomial in three variables. The technique uses false derivative and local Sturm sequences.

8.1 Introduction

Let an algebriac surface be given in three dimensional real space by a polynomial equation $A(x, y, z) = 0$. Suppose also a curve is given parametically by three trigonometric polynomials:

$$x = p_1(t, sin(t), cos(t))$$
$$y = p_2(t, sin(t), cos(t))$$
$$z = p_3(t, sin(t), cos(t))$$

Assume that all these polynomials have rational coefficients. The problem solved in this article is to decide if the curve intersects the surface, and if it does to find how many intersections there are and to approximate the solutions as closely as desired.

At least in a bounded interval, this problem can also be approximately solved by numerical methods. The technique given below, however, gives an exact answer as to the number of roots on the whole real line, and uses symbolic methods. A typical application would be to describe the intersection of a helix with a hyperboloid.

Section two below gives a theorem which is useful in dealing with the implicitly defined constants which arise in the later computation. It is shown that two polynomials in t and $sin(t)$ with rational coefficients have a common root which is non zero only if they have a common factor.

In section three the main techniques used are introduced, false derivatives and local Sturm sequences. A false derivative is just a continuous function whose sign agrees with the sign of the true derivative wherever the original function is zero. Such false derivatives separate roots of the original function, as true derivatives do, but they may be much simpler than true derivatives. False derivatives may be understood as simplifications of true derivatives using the vanishing of the original function. A local Sturm sequence is just the same a Sturm sequence except that it is based on false derivatives rather than true ones. With these ideas many techniques of root finding and construction of cylindrical decompositions can be extended to non algebraic functions.

In section four, false derivatives are found for polynomials in t and $sin(t)$. This means that all the roots can be found by local Sturm methods. The construction of a false derivative given here is a special form of a more general method which applies to polynomial combinations of functions satisfying first order algebraic differential equations. See [10, 11].

The algorithm is summarized in section five.

In the conclusion it is remarked that graphics systems may use some transcendental functions directly, rather than by approximation.

8.2 Constants

The Sturm technique discussed below finds the number of roots of a function in an interval by counting the number of sign variations in a sequence of functions at the endpoints of the interval. So in order to apply this we need to be able to calculate the sign of a function in the sequence. The endpoints themselves will be implicitly defined. We can always approximate these endpoints, and thus we can approximate the values of the functions at the endpoints. However, what if one of the functions in the sequence was zero at one of the endpoints? We could never recognize zero-ness by approximation. We need therefore to have another way to decide if a function evaluated at an implicitly defined constant is zero. For our purposes this is done below.

8.2.1 Common Root Theorem

Theorem. Let $p(t, sin(t))$ and $q(t, sin(t))$ be polynomials in t and $sin(t)$ with rational coefficients. If $p(t, sin(t))$ and $q(t, sin(t))$ have a common root which is not zero, then they have a common factor.

Proof. Suppose not. We may, for the sake of a contradiction, assume that $p(t, sin(t))$ and $q(t, sin(t))$ are both irreducible. Let

$$R(t) = Res_s(p(t, s), q(t, s))$$

where Res_s means the resultant of the two polynomials with respect to s. If $p(t, s)$ and $q(t, s)$ don't have a common factor then $R(t)$ is not identically zero. However, $R(t)$ is zero at the common root. Thus this common root, say α, is algebraic, but we assume that it is not zero. We know in this case that $sin(\alpha)$ is transcendental. If we write $p(t, s)$ and $q(t, s)$ as polynomials in s with coefficients which are polynomials in t, we see that all these coefficient polynomials in t must be zero at α. Thus all the coefficient polynomials in t have a common factor with degree greater than zero, and this factor must be a factor of both $p(t, s)$ and of $q(t, s)$. Contradiction.

Corollary 1.

Let C be the set of real algebraic numbers union the set of numbers of the form $k\pi/2$, where k is an integer. Then given a number t in this set C and given any polynomial $p(t, sin(t))$ with rational coefficients, we can decide whether or not $p(t, sin(t)) = 0$. It is assumed here that if we are given t we know whether or not it is algebraic, and we can approximate it, and if it is algebraic we can find an irreducible defining polynomial and an interval in which it is the only root of the defining polynomial.

Proof. If t is algebraic, decide if it is zero. If not, decide if the defining polynomial of t and $p(t, sin(t))$ have any common factors.

Corollary 2.

If $p(t, sin(t))$ is square free and $p(t, sin(t)) = 0$ and t is not zero, then $p'(t, sin(t))$ is not zero. Thus if p is square free, it's non zero roots are all simple.

8.3 False derivatives and Local Sturm sequences

A continuous function $h(t)$ is defined to be a false derivative for a function $f(t)$ over an interval I if the sign of $h(t)$ is the same as the sign of the derivative of $f(t)$ whenever $f(t) = 0$, for t in the interval I.

For example, suppose $f(t)$ is $e^y - p(t)$, where $p(t)$ is a polynomial. Then $p(t) - p'(t)$ is a false derivative of $f(t)$ over any interval.

At a double root of a function $f(t)$ a false derivative is zero. Also, the roots of a function $f(t)$ on an interval I are separated by the roots of a false derivative of $f(t)$ on I.

Some of the other properties of derivatives carry over to false derivatives, as indicated very briefly below.

Define a Thom sequence for a function $f(t)$ on an interval I to be a sequence of continuous functions $f_0(t), f_1(t), ..., f_n(t)$, where $f_0(t)$ is identically equal to $f(t)$ and $f_n(t)$ is a non zero constant, and each $f_{i+1}(t)$ is a false derivative of $f_i(t)$ on the interval I.

The length of a Thom sequence for $f(t)$ on interval I is an upper bound for the number of distinct roots of $f(t)$ on I.

Define a sign condition for a Thom sequence $f_0(t), ..., f_n(t)$ to be the conjunction, for $i = 0$ up to n, of the statements $f_i(t)c_i0$, where each c_i is one of $<, \leq, =, \geq, >$.

Now, by the usual proof, we have:

8.3.1 Thom's Lemma

Lemma. For any Thom sequence, the set of real numbers satisfying a sign condition is either empty, a single point, or a single interval.

8.3.2 Local Sturm sequences

In this subsection we consider only polynomials in t and $sin(t)$ with rational coefficients.

Assume that over the interval I we have a false derivative for such a polynomial, and that the false derivative is another such polynomial.

Define a local Sturm sequence for $p(t, sin(t))$ over the interval I to be a sequence of polynomials in t and $sin(t)$

$f_0(t, sin(t)), f_1(t, sin(t)), ..., f_{n-1}(t, sin(t)), f_n(t)$, where f_0 is the same as $p(t, sin(t))$, f_1 is a false derivative of f with respect to t valid over I and for each $i > 0$, there exists some quotients Q_i which are polynomials in t and $sin(t)$, and some always non negative (but not identically zero) multipliers A_i, which are polynomials in t alone, so that for each i,

$$f_{i+1}(t) + Q_i f_i(t) + A_i f_{i-1}(t) = 0$$

over I, and f_{i+1} has lower degree as a polynomial in $sin(t)$ than f_i.

It is required that $f_n(t)$, the last function, and all the A_i multipliers be polynomials in t alone.

Local Sturm sequences may be constructed as follows. Think of $f(t, sin(t))$ as a polynomial in $sin(t)$, whose coefficients are polynomials in t. Find a false derivative, with respect to t, for $f(t, sin(t))$ over I which is also a polynomial in t and $sin(t)$. If this can be done then a local Sturm sequence can be constructed by taking a negative polynomial remainder sequence, obtained by Euclidean division of each function into the one before.

We work here in the field of rational functions in t , and the multipliers may be obtained by clearing denominators by squared factors.

As soon as two adjoining terms of a local Sturm sequence are zero, all the subsequent terms are also zero.

The requirement that the multipliers be non negative at all points is imposed so that whenever one of the interior terms of a local Sturm sequence is zero, the terms on either side have opposite signs unless all subsequent terms are zero.

If the last term in a local Sturm sequence is identically 0, the term before last is a common factor of the first two functions, considering these as polynomials in $sin(t)$.

In general, the last term in a local Sturm sequence is 0 if and only if the first two terms have a common factor, which is true if and only if the resultant of the first two terms with respect to $sin(t)$ is 0.

Let $F(t)$ be any analytic function. If $F(r) = 0$, define the order of r, as a root of F, to be the least number d so that the d th derivative of F is non zero at r. In case $F(r)$ is not zero the order of r, as a root of F, is defined to be 0. As previously shown, for square free polynomials in t and $sin(t)$ the order of a non zero root can not exceed 1.

Define $sign(F(r+))$, where F is analytic at r, to be the sign of the d th derivative of F at r, where d is the order of r as a root of F.

Define $sign(F(r-))$ to be $(-1)^d sign(F(r+))$, where d is the order of r as a root of F, and F is analytic at r.

The sign of $F(r+)$ is supposed to be the sign of $F(r+\epsilon)$ where ϵ is infinitesimal. Also $F(r-)$ has sign agreeing with $F(r-\epsilon)$ where ϵ is infinitesimal.

If $S = (f_0(t), ..., f_n(t))$ is a local Sturm sequence, define $Var(S, r+)$ and $Var(S, r-)$ to be the number of sign variations in the sequence S evaluated at $r+$ and $r-$ respectively.

Sturm Theorem. Let $f(t)$ be an analytic function of one variable t over open interval (a, b). Let $S = (f_0(t), ..., f_n(t))$ be a local Sturm sequence for $f(t)$ valid over interval (a, b) and suppose the last function $f_n(t)$ has no root in open interval (a, b). Then the number of distinct roots of $f(t)$ in (a, b) is

$$Var(S, a+) - Var(S, b-)$$

provided that the limiting signs at $a+$ and $b-$ are defined for all of S.

This means that in intervals which do not contain roots of the last function, a local Sturm sequence behaves like a Sturm sequence.

Following [3], we can also define a local Sturm sequence for $f(t)$ and $h(t)$ to be a sequence as above $f_0, f_1, ..., f_n$, except that f_1 is now $f*(t)h(t)$, where $f*(t)$ is a false (or true) derivative of $f(t)$. In this case the number of sign variations in an interval counts the number of roots of $f(t)$ at which $h(t)$ is positive less the number at which $h(t)$ is negative. If we also know the number of roots of $f(t)$ and the number of roots of $h(t)$ in the interval we can find the number of roots of $f(t)$ at which $h(t)$ is positive.

Note that if the second function in the sequence was not a false derivative but the negation of a false derivative the sequence would gain sign variations when going through roots rather than losing them. It would still be true that the difference in the number of sign variations would be the number of roots.

8.4 False derivatives for $p(t, sin(t))$

We have not yet found any false derivatives for polynomials in t and $sin(t)$. Let $p_s(t, sin(t))$ be the result of differentiating $p(t, sin(t))$ with respect to $sin(t)$. (In other words, substitute s for $sin(t)$, differentiate with respect to s and then substitute $sin(t)$ back.) It is shown below that if the interval is restricted sufficiently, either $p_s(t, sin(t))$ or its negation can be used as a false derivative for $p(t, sin(t))$.

Assume $p(t, sin(t))$ is square free, and $sin(t) - 1$ is not a factor.

Form:

1) $p(t, 1)$

2) $R(t) = Res_s(p(t, s), \partial p / \partial s)$

3) $R_Q(t) = Res_s(p(t, s), (s^2 + (p_t/p_s)^2 - 1)p_s^2))$

The first of these polynomials in t is not identically 0 since we assumed $sin(t) - 1$ was not a factor of $p(t, sin(t))$. The second, $R(t)$ is not identically 0 because of the assumption of square freeness.

It is claimed that $R_Q(t)$ is also not identically 0. Suppose, for the sake of contradiction, that it was. It would then follow that $p(t, s)$ and $(s^2 + (p_t/p_s)^2 - 1))p_s^2$ would have a common factor. Suppose the common factor is $F(t, s)$. (We know that $p(t, s)$ and p_s don't have a common factor, since $R(t)$ is not identically zero. So $F(t, s)$ is not a factor of p_s.) Pick any t with $R(t)$ not zero. Find s, complex if necessary, so that $F(t, s) = 0$. Near t, $s(t)$ is defined implicitly by $F(t, s) = 0$. But the derivative of $s(t)$, satisfies $s'(t) = -(p_t/p_s)$. Thus $s(t)$ satisfies the same algebraic differential equation as $sin(t)$, namely $s^2 + (s')^2 - 1 = 0$. This is not possible since $s(t)$ is algebraic.

We may assume then that none of the three polynomials above are identically zero.

By standard methods we can find all the real roots of these polynomials. Let CR be the finite set consisting of the union of the roots of $p(t, 1)$, $R(t)$, and $R_Q(t)$. Let S be the set of numbers of the form $k\pi/2$, where k is an integer and there exists x and y in CR so that $x < k\pi/2 < y$. Finially, let

$$\alpha_1 < \alpha_2 < \ldots < \alpha_n$$

be the result of ordering the union of CR and S.

Proposition. In $[\alpha_i, \alpha_{i+1}]$ one or the other of the two functions, $p_s(t, sin(t))$ or $-p_s(t, sin(t))$, is a false derivative for $p(t, sin(t))$.

Proof. Restrict t to the interval $[\alpha_i, \alpha_{i+1}]$ and s to the interval $[-1, 1]$. $R(t)$ is not zero in the interior of this rectangle, so, in the rectangle, $p(t, s) = 0$ defines s implicitly as one of, possibly, several functions of t. These functions of t do not intersect. In the regions between these implicitly defined curves, the signs of p alternate.

Over the graph of the zero set of $p(t, s)$, which may be visualized as a stack of regions and curves, superimpose the graph of $s = sin(t)$. The roots of $p(t, sin(t))$ occur inside the rectangle at the points of intesection of the trigonometric and the algebraic curves. At each such interior root, the sign of $p'(t, sin(t))$ is the same as the sign of the region to which the graph of $s = sin(t)$ is going as t increases. It is these signs which we want to capture in a false derivative. The sign is $p_s(t, s)$ if the graph of $s = sin(t)$ goes up through the curve and it is $-p_s(t, s)$ if the graph of $s = sin(t)$ goes down through the curve. (To say that $s = sin(t)$ goes up through the curve means that $sin'(t) > s'(t)$ at the point of intersection. On the other hand, if $sin'(t) < s'(t)$ at the point of intersection we will say that $sin(t)$ goes down through the algebraic curve. It can't happen that the derivatives are equal at the point of intersection, because this would constitute a root of $R_Q(t)$.)

It remains to show that the graph of $s = sin(t)$ cuts through all the algebraic curves in the same direction inside the rectangle. Suppose not, for the sake of a contradiction. Then there would have to be two adjoining intersections with different sense, a positive cut followed by a negative one, or vica versa. In either case $s = sin(t)$ has to cut twice through the same branch of $s(t)$. Let the two adjoining intersections be $(u, sin(u))$ and $(v, sin(v))$. Define $z(t)$ to be the square root of $1 - s^2(t)$, with sign so that $z(u) = cos(u)$, where as before $s(t)$ is defined by $p(t, s(t)) = 0$. Both the $s(t)$ graph and the $sin(t)$ graph lie strictly inside the rectangle, so s is not equal to 1 on either graph. Thus $z(v) = cos(v)$.

Since $z(t) - s'(t)$ changes sign in $[u, v]$, and both are continuous, there must be a point between u and v at which $z(t) = s'(t)$. At this point $R_Q(t) = 0$. But all the roots of $R_Q(t)$ were found, so the interval can't contain such a point. Contradiction.

8.5 The algorithm

Assume given $Q(t, sin(t), cos(t))$, a polynomial with rational coefficients in t, $sin(t)$, and $cos(t)$.

1) Express Q in the form

$$(A(t, sin(t)) + B(t, sin(t))cos(t))F_1(t, sin(t))$$

where A and B have no common factors. Set

$$F_2(t, sin(t)) = (A + Bcos(t))(A - Bcos(t))$$

So $F_2 = (A^2 - B^2(1 - sin^2(t)))$.

Now let $F(t, sin(t)) = F_1(t, sin(t))$.

2) Replace $F(t, sin(t))$ by its square free part. Also remove any factor of $sin(t) - 1$. Form the possible local Sturm sequence S of functions:

$$F, F_s, F_1, F_2, ..., F_n$$

where each F_{i+1} is the negative remainder from division of F_i into F_{i-1}, after clearing of denominators as explained earlier. The degree of $sin(t)$ decreases in this sequence, and F_n is a function of t alone.

Form the three polynomials $F(t, 1), R(t), R_Q(t)$ as explained in the previous section. Find the set CR of roots of these polynomials, and the set S of multiples of $\pi/2$ which have roots above and below them. Finially, order the union of S and CR to obtain an ordered finite set X.

3) For each interval between points of X, find the number of sign variations in the sequence of functions S at the endpoints. The absolute value of the difference in the number of variations gives the number of roots inside the interval. These roots can all be isolated and approximated by bisection. Also check to see if the points of X are roots of $F(t, sin(t))$.

4) In the infinite strips to the left and right of the set X, with s between -1 and 1, find the number of curves $s(t)$ so that $F(t, s(t)) = 0$. If there are any such curves, each one gives rise to infinitely many roots of $F(t, sin(t))$. There will be one such root between each intersection of $s = sin(t)$ with the lines $s = 1$, $s = -1$, so all these roots can be approximated.

5) Having done F_1, set $F = F_2$ and repeat steps 2,3,4.

6) We now have all the roots of $F_1(t, sin(t))$ and $F_2(t, sin(t))$. All the roots of F_1 are roots of Q. All the roots of Q are roots of either F_1 or F_2. But some of the roots of F_2 might not be roots of Q. So we have to check each root of F_2 to see if it is also a root of Q.

Suppose we know that $F_2(\alpha, sin(\alpha)) = 0$. So we know that

$$(A + Bcos(\alpha))(A - Bcos(\alpha)) = 0$$

Assume α is not zero.

First test if $cos(\alpha) = 0$. If so, test if α is a root of A. If this happens, then $Q(\alpha) = 0$.

It can't happen that α is a common root of A and B, since A and B have no common factors.

So if $cos(\alpha) = 0$, one of the two factors of F_2 is non zero. We can decide which by approximating. So in all cases we can decide whether or not $Q(\alpha) = 0$.

8.6 Conclusions

Graphics systems originally represented three dimensional objects as volumes bounded by planes, which were themselves bounded by lines. In some applications it would seem better not to use an approximation of this linear kind but rather to represent surfaces and volumes by algebraic conditions. This may give a much more succinct representation, in which small scale precision is not lost.

Not only is the world not linear, it is not polynomial. Of course it is possible to approximate a trigonometric curve such as a helix by a series of straight lines or by a series of algebraic curves. In some cases, however, it may be better to use the transcendental functions directly. It is true that there are some unsolvable problems involving transcendental functions. However, in many cases of practical interest the mathematical and computational difficulties can be overcome. For example, this paper has shown that we can find, in a bounded number of steps, the intersections of a trigonometric curve of a certain kind with an algebraic surface.

In the refererences, a selection of recent papers is given, which show progress being made in solution of geometric and topological problems involving algebraic and non algebraic functions.

8.7 References

[1] R. Bianconi. Some Results in the Model Theory of Analytic Functions. Ph.D. thesis, Oxford, 1990.

[2] J. Bochnak, M. Coste and M.F. Roy. *Geométrie Algébrique Réelle.* Ergebnisse vol 12, Springer-Verlag, 1987.

[3] M. Coste and M.F. Roy. Thom's Lemma, the Coding of Real Algebraic Numbers and the Computation of the Topology of Semi-algebraic Sets, *J. of Symbolic Computation*, 5:121-129, 1988.

[4] J.H. Davenport. Computer Algebra for Cylindric Algebraic Decomposition. TRITA-NA-8511, NADA, KTH, Stockholm 1985.

[5] L. Gonzalez, H. Lombardi, T. Recio and M.F. Roy. Sturm-Habicht Sequence. ISSAC, 1989.

[6] P. Milne. On the Algorithms and Implementation of a Geometric Algebra System. Ph.D. Thesis, University of Bath, 1990.

[7] P. Pedersen. Generalizing Sturm's theorem to N dimensions. New York University, Dept of Computer Science, Courant Institute, April 1990.

[8] D. Richardson. Finding Roots of Equations Involving Solutions of First Order Algebraic Differential Equations. In T. Mora and C. Traverso, editors, *Effective Methods in Algebraic Geometry*, Birkhauser, 427-440, 1991.

[9] D. Richardson. Wu's Method and the Khovanskii Finiteness Theorem. *Journal of Symbolic Computation* 12:127-141, 1991.

[10] D. Richardson. Computing (In a Bounded Part of the Plane) the Topology of a Real Curve Defined by Solutions of Algebraic Differential Equations. Submitted for publication, 1991.

[11] D. Richardson. Towards Non Algebraic Cylindrical Decomposition, ISSAC, 247-255, 1991.

[12] M.F. Roy. Computation of the Topology of a Real Curve. *Proceedings of the Conference on Computational Geometry and Topology*, Sevilla, 1987.

[13] M.F. Roy, M.F and A. Szpirglas. Complexity of Computations of Cylindrical Decomposition and Topology of Real Algebraic Curves Using Thom's Lemma. *Proceedings of the Conference on Real Algebraic Geometry*, Trento, 1988.

[14] J. Shackell Growth Estimates for Exp-log Functions. to appear in *Journal of Symbolic Computation.*

[15] L. van den Dries. On the Elementary Theory of Restricted Elementary Functions. *Journal of Symbolic Logic*, 53:796-808, 1988.

[16] L. van den Dries and J. Denef. P-adic and Real Subanalytic Sets. *Annals of Mathematics*, 128:79-138, 1988.

[17] N.N. Vorobjov. Deciding Consistency of Systems of Polynomial in Exponent Inequalities in Subexponential Time. in T. Mora and C. Traverso, editors, *Effective Methods in Algebraic Geometry*, Birkhauser, 491-500, 1991.

[18] A.J. Wilkie. Some Model Completeness Results for Expansions of the Ordered Field of Real Numbers by Pfaffian Functions. preprint, Mathematics Institute, Oxford, 1991.

9 Incidence Relationships: Kernel Concept in Combinatorical Topology

Yi Luo
Gábor Lukács

ABSTRACT In this paper we introduce the concept of incidence relationships in combinatorial topology and show that it is the counterpart concept of nearness in point set topology. Starting from the concept of incidence relationships, we investigate several problems in B-rep solid modelling, namely the mathematical models of solids, the representations of non-manifold solids, and the mathematical expressions for regularized Boolean operations on B-rep solid models.

9.1 Introduction

Constructive Solid Geometry (CSG) and Boundary Representation are the two best known approaches in solid modelling. CSG basically starts modelling from volumetric primitives (cubes, pyramids, spheres, cones etc.). It uses Boolean operations (union, intersection, difference, complement) for getting complex objects. The mathemetical basis here is more or less the naïve set theory. If evaluation is necessary it prefers the usage of the well-known De Morgan formulae in order to reduce problems to those concerning the primitives. However if rigorously applied one can get easily irregular results as dangling edges, faces etc. For overcoming this the concept of regularized Boolean operations has been introduced (Requicha & Tilove [11]) where only the closure of the interior of the result is considered. This requires the extension of the naïve set theory with some, basically point set topological notions (interior, boundary, nearness etc).

On the other hand, B-rep starts modelling from objects defined directly by their boundaries. The mathematical theory behind has been Euclidean and descriptive geometry. In order to solve Boolean operations intersection computation of surfaces and curves (planes, quadrics, free-form surfaces etc.) has been used. However irregularity may occur here as well. Thus whether it has been admitted or not regularized operations have been performed as well — as in the CSG case. Because the interest has been focused on the boundary of the result (boundary with inhomogeneous dimensionality) the natural domain for dealing with these problems is combinatorial (or algebraic) topology, where the boundary is partitioned onto subsets of fairly simple geometrical and topological structure (vertices, edges, faces). Moreover combinatorial topology is fortunate because it deals with these finite, countable objects which can be represented unambigously in computers. Instead of the concept "nearness" being used in point-set topology, the application of certain "finite" relations among these primitives are preferred. These relationships are called adjacency relationships.

In the B-rep approach, boundaries of modelled objects are subdivided into vertices, edges and faces which are unified with a single word *simplices* [5]. B-rep systems represent not only the simplices, but also the relationships between them. Baer et al. [1] listed all nine possible classes of adjacency relationships between pairs of the three types of simplices, as shown in Figure 9.1, and summarized several systems storing different subclasses of the relationships. Weiler [10] analysed the adjacency relationships in detail, both in the domain of two-manifolds and in the domain of non-manifolds. Gursoz et al. [4] and Yamaguchi & Kimura [11] also discussed the adjacency relationships in the domain

of non-manifolds. In this paper we shall examine the special subclasses of the adjacency relationships, which are called *incidence relationships.* We shall state that the incidence relationships, just like the concept *nearness* in point set topology, are a kernel concept in combinatorial topology. It is known that subdividing boundaries into simplices so as to have something to count and the reliance on counting alone to found a theory is one of the advantages of combinatorial topology [5]. In order to have something to count for the incidence relationships as well, we shall define basic units for these, too. Applying the concept of incidence relationships, we shall discuss the following problems in B-rep solid modelling.
(1) Definitions of mathematical models for boundaries of solid objects.
(2) B-rep data structures for representing solids, including those with non-manifold boundaries, as shown in Figure 9.2.
(3) Combinatorial boundary expressions of regularized Boolean operators.

9.2 Concept of incidence Relationships

9.2.1 Basic concepts in point set topology

The following definitions are quoted from [5] which are tightly related to the contents of this paper.

DEFINITION 2.1. *A* topological space *is a set* W *together with the choice of a class of subsets* T *of* W *(each of which is called* neighbourhood *of its points) such that*

1. *Every point is in some neighbourhood;*
2. *The intersection of any two neighbourhoods of a point contains a neighbourhood of that point.*

DEFINITION 2.2. *Let* W *be a topological space. Let* X *be a subset of* W *and* p *be a point of* W. p *is near* X *if every neighbourhood of* p *contains a point of* X.

DEFINITION 2.3. *Let* W *be any topological space. A subset* X *of* W *is called* closed *if* X *contains all its near points,* open *if every point in* X *is not near the complement of* X *(denoted* cX*),* compact *if every sequence in* X *has a near point in* X*, and connected if whenever* X *is divided into two nonempty disjoint parts then one of these parts contains a point near the other. Associated with each set* X *are the* interior *of* X *(denoted* iX*) consisting of all points of* X *not near* cX*, the closure of* X *(denoted* kX*) consisting of* X *plus all points near* X*, and the* boundary *of* X *(denoted* bX*) consisting of the points near both* X *and* cX.

9.2.2 Definition of Incidence relationships

Semantically, *adjacency* is used if two things are near to (but not necessarily touching) each other. While *incidence* is used if two things are naturally connected to each other. Among the nine classes of adjacency relationships shown in Figure 9.1, FF, EE, VV are the relationships between two simplices of same type. Obviously, two adjacent faces are separated by either a vertex or an edge, two adjacent edges are separated by a vertex, and two adjacent vertices are separated by an edge. Therefore, FF, EE, VV do not belong to incidence relationships. Incidence relationships are the other six classes of adjacency relationships.

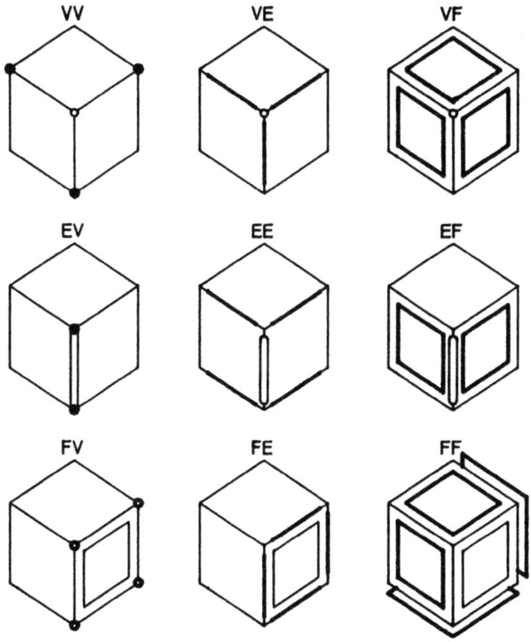

FIGURE 9.1. Nine classes of adjacency relationships.

DEFINITION 2.4. *The incidence relationship for a simplex is the relationship between the simplex and all other simplices which are naturally connected with the simplex.*

9.2.3 Combinatorial interpretations of other elementary concepts using incidence

We have seen from Definition 2.1 – 2.3 that in point set topology *nearness* is a kernel concept from which other elementary concepts can be defined. In combinatorial topology, compactness and connectedness are interpreted in terms of triangulations. However, the other concepts are not convenient to interpret in terms of triangulations. Here we show that the concept of *incidence relationships* is a counterpart concept of *nearness* and all concepts defined using nearness can be interpreted using incidence. By the definition in combinatorial topology, a *complex* is a two-dimensional topological space subdivided into simplices, where 0-simplices are synonymous with vertices, 1-simplices with edges and 2-simplices with faces. (The 0, 1, or 2 is the rank of the simplices). So the concept of simplex sets is a counterpart concept of point sets. With the concepts of complexes, simplex sets and incidence, we are able to interpret those defined in Definition 2.3 combinatorially.

STATEMENT. *Let W' be a complex. A simplex subset Γ of W' is* closed *if and only if Γ contains all simplices with lower rank than, and* incident *to, the highest rank simplices in Γ,* open *if and only if every simplex in Γ is not* incident *to the highest rank simplices in the complement of Γ ($c\Gamma$),* compact *if and only if every 2-simplex in Γ is* incident *to loops of lower rank simplices, and* connected *if and only if whenever Γ is divided into two non-empty disjoint parts then both of these parts contain a simplex* incident *to the other. Associated with each set Γ are the* interior *of Γ consisting of all simplices not* incident *to the highest rank simplices in $c\Gamma$, the* closure *of Γ consisting of Γ plus all* *simplices* incident *to the highest rank simplices in* Γ, *and the* boundary *of* Γ *consisting of all simplices* incident *to the 2-simplices in both* Γ *and* $c\Gamma$.

For our applications, it is necessary to introduce a new concept, the *surrounding* of Γ, which is a set consisting of all simplices *incident* to the closure of Γ. We call the interior, closure, boundary and surrounding of Γ *associate sets* of Γ.

9.2.4 Basic units of incidence relationships

The types of simplices which can be incident to a face are edges and vertices. These edges and vertices may be in several closed chains, which have been named *loops* by Braid [2]. Each loop composes a basic unit of incidence relationship to the face.

Similarly, the types of simplices which can be incident to a vertex are edges and faces. Suppose that the vertex is on the boundary of a pure 3-D object. Taking the vertex as the center and drawing a ball with a very small radius, we can see that the interior of the object around the vertex is in one or several components in the ball; and each component can be surrounded by several disks of edges and faces. Each disk composes a basic unit of the incidence relationship to the vertex. We name the unit a *bundle.* In Figure 9.2, v_1 has three bundles; v_4 has two; and each of v_2, v_3 and v_5 has one.

The types of simplices which can be incident to an edge are faces and vertices. Suppose that the edge is on the boundary of a pure 3-D object. Taking the edge as the axis and drawing a cylinder with a very small radius, we can see that the interior of the object around the edge is in one or several components in the cylinder. Each component is surrounded by at most two faces and the length of the component is ended by at most two vertices, The faces and vertices relating to each component compose a basic unit of the incidence relationship to the edge. We name the unit a *wedge.* In Figure 9.2, each of e_1 and e_2 has two wedges.

Note that the concepts of bundles and wedges are related to the interior of the solid. If we take the object in Figure 9.2 as a negative solid, then the exterior of the positive solid now is the interior of the negative solid. Correspondingly, v_1 has three bundles; v_4 has one; and each of v_2, v_3 and v_5 has two. The faces composing a wedge are also rearranged. As a limit case, a sheet has no interior and each bundle or wedge on its boundary is a *degenerate* one.

Loop, wedge and bundle give complete units of incidence relationships. There should be a uniform terminology for the three terms, just like the word *simplex* is used for face, edge and vertex. We propose a new word, *inciplex* (*inciplices* for plural), where *0-inciplices* stand for bundles, *1-inciplices* for wedges and *2-inciplices* for loops.

9.3 Applications of Incidence Relationships in Solid Modelling

9.3.1 Definitions of mathematical models of solid boundaries using inciplices

By the definition in point set topology, surfaces, or two-manifolds are topological spaces in which every point has a neighbourhood that is topologically equivalent to an open disk. The combinatorial definition of two-manifolds is in terms of triangulations and identifications of edges and vertices of the triangles [5]. In the literature of B-rep solid modelling, the solid boundaries have been abstracted as two-manifolds satisfying certain conditions [8] and their definition has followed the way presented in [5]. Using inciplices, a two-manifold can be combinatorially defined as below.

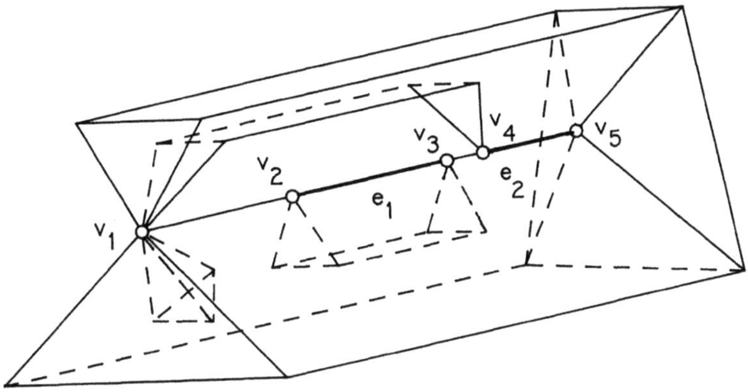

FIGURE 9.2. A non-manifold solid object.

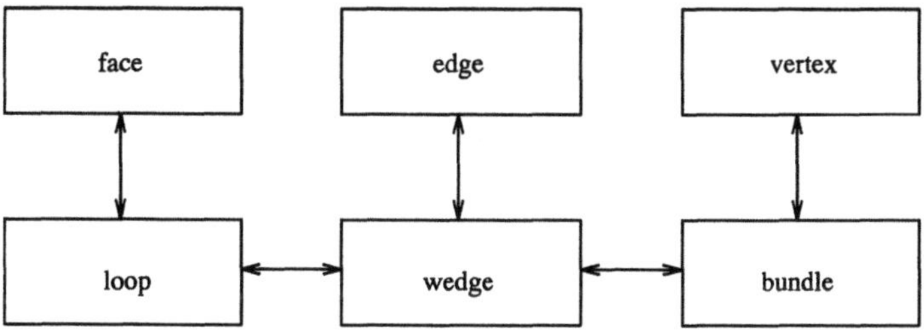

FIGURE 9.3. Relations of all types of records in the new B-rep data structure.

DEFINITION 3.1. *An orientable two-manifold is a topological space constructed from simplices and satisfying the following conditions:*

1. *Each vertex is incident to only one bundle.*
2. *Each edge is incident to only one wedge.*
3. *Each face is incident to a finite number of loops.*

The concept of *orientability* can also be found in [5].

Two-manifolds do not cover all cases of solid boundaries. For example, the object in Figure 9.2 is a solid, but some vertices on its boundary are incident to more than one bundle and some edges are incident to more than one wedge. Definition 3.1 can be modified to include the non-manifold solid boundaries as below.

DEFINITION 3.2. *A solid boundary is an orientable topological space constructed from simplices and satisfying the condition that every simplex is incident to a finite number of inciplices.*

9.3.2 B-rep data structure

The B-rep data structure for representing solids should be designed to reflect the mathematical models. According to Definition 3.2, we design a data structure with the three types of simplex records and the three types of inciplex records. The relations of all types of records are illustrated in Figure 9.3. With wedges and bundles, non-manifold vertices and edges, such as v_1, v_4 and e_1, e_2 in Figure 9.2, can be represented in the same way as the multiply connected faces are represented with loops [2]. See [6] and [7] for details. When applications require only a two-manifold domain, wedge records and bundle records can be cancelled and their contents can be merged into edge records and vertex records respectively. The data structure for two-manifolds then happens to be consistent with that proposed by Falcidieno & Giannini [3].

9.3.3 Combinatorial boundary expressions of regularized Boolean operators

Boolean operations are convenient tools to obtain more complicated solid models from simple ones. In general, a Boolean algorithm should include three steps. The first step finds all coincident parts between two operand solid boundaries. The second step determines which part of a solid boundary is inside, outside, or lying on the other solid boundary. The third step determines which parts of the two boundaries belong to the resultant boundary. The first two steps mainly deal with geometric computation and the last step deals with topological computation. Up to now, although there are boundary expressions for regularized Boolean operations derived by Requicha & Tilove [9], the algorithm for the topological computation is designed relying only on the space imagination and case analysis, not on the expressions. That is because the expressions are based on point set topology, which use the Boolean relations of closure, interior, boundary of subsets of the operand boundaries to express the boundary result of Boolean operations. However, these associate sets of a point set, such as its closure, interior and boundary, have not been able to be determined algorithmically thus the expressions cannot be applied directly for designing the topological computation in Boolean algorithms.

Now we show that the combinatorial concepts of *closure, interior, boundary*, and *surrounding* of a simplex subset Γ of a complex W' can be determined through algorithms if the inciplices to each simplex is available.

The following are the concrete descriptions of the algorithms. In the algorithms, Γ is a buffer used for storing simplices of a subset of complex; and A and B are working buffers.

ALGORITHM 3.1. *Determination of boundary of* Γ.

1. Γ is the input of the algorithm.
2. Copy all 0-simplices and 1-simplices from Γ to A.
3. For each 1-simplex in A do
    If all 2-simplices in its 1-inciplices $\in \Gamma$, then
      remove the 1-simplex to B from A.
4. For each 2-simplex in Γ do
    For each of its 2-inciplices do
      For each simplex in the 2-inciplices do
        If the simplex is in neither A nor B, then
          copy it to A.
5. A is the output of the algorithm.

ALGORITHM 3.2. *Determination of interior of* Γ.

1. Γ and the output of Algorithm 3.1, A, are the input of the algorithm.
2. Copy Γ to B.
3. For each 0- and 1-simplex in Γ do
   If the simplex is in A, then
   delete it from B.
4. B is the output of the algorithm.

ALGORITHM 3.3. *Determination of closure of* Γ.

1. Γ is the input of the algorithm.
2. Copy Γ to A.
3. For each 2-simplex in A do
   For each 2-inciplex of the simplex do
   For each 0- and 1-simplex in the inciplex do
   If the simplex is not in Γ, then
   copy it to A.
4. A is the output of the algorithm.

ALGORITHM 3.4. *Determination of surrounding of* Γ.

1. Γ and the output of Algorithm 3.1, A, are the input of the algorithm.
2. For each simplex in A do
   For each inciplex of the simplex do
   For each simplex in the inciplex do
   If the simplex is neither in Γ nor in A, then
   copy it to B.
4. B is the output of the algorithm.

We can classify the two intersecting solid boundaries into several pieces mainly using the concept of *surrounding.* Then we can apply the *closure, interior* and *boundary* of these pieces to get the combinatorial boundary expressions for regularized Boolean operations. Since the surrounding, closure, interior and boundary of these pieces can be determined algorithmically, the expressions should be directly available for designing the topological computation in Boolean algorithms. See the appendix (Section 9.6) for details.

9.4 Conclusions

We have argued that incidence relationships of simplices are the kernel concept in combinatorial topology using which the other elementary concepts defined in the language of point set topology can be combinatorially interpreted. We have also created a new word *inciplices* to unify the terms *loops, wedges* and *bundles*, which are the names of basic units of incidence relationships to faces, edges and vertices respectively. Applying the concepts of incidence relationships and inciplices, many theoretical and practical problems concerning the topological aspects of B-rep solid modelling, some of which have been mentioned in this paper, can be readily and satisfactorily solved. We use Table 1 to illustrate the main points in this paper. The items marked with $*$ are the new contributions.

Point set topology	Combinatorial topology
2-D topological space	complex
point set	simplex set (face, edge, vertex)
nearness	(old) no counterpart concept * incidence/inciplex (loop, wedge, bundle)
elementary concepts defined using nearness	(old) part of them interpreted using triangulations. * all of them interpreted using incidence.
definition of two-manifolds based on 2-D topological space, point set and nearness	(old) two-manifolds defined using triangles and the identification of their edges and vertices. * two-manifolds defined using complex, simplex, and inciplex.
definition of regular sets based on 3-D topological space, point set and nearness	(old) no definition of boundaries of regular sets * boundaries of regular sets defined using complex, simplex set and inciplex.
boundary expressions for regularized Boolean operators in terms of associate sets not determinable algorithmically	(old) no rigorous boundary expressions of Boolean operators * boundary expressions for regularized Boolean in terms of associate sets determinable algorithmically

Table 1: Corresponding concepts in point set topology and combinatorial topology

Acknowledgements:

This work was supported by the National Research Foundation (OTKA) of the Hungarian Academy of Sciences. We would like to thank Dr. Tamás Várady for his management support to this work.

9.5 References

[1] A. Baer, C. Eastman and M. Henrion. Geometric Modelling: a Survey *Computer Aided Design* 11,5:253–272, 1979.

[2] I.C. Braid. *Notes on a Geometric Modeller.* CAD Group Document No. 101, Computer Laboratory, University of Cambridge, 1979.

[3] B. Falcidieno and F. Giannini. Automatic Recognition and Representation of Shape-Based Features in a Geometric Modeling System *Computer Vision, Graphics, and Image Processing*, 48:93–123, 1989.

[4] E.L. Gursoz, Y. Choi and F.B. Prinz. Vertex-based Representation of Non-Manifold Boundaries. In Wozny, M.J., Turner, J.U. and Preiss, K., editors, *Geometric Modeling for Product Engineering*, North-Holland, New York, 107–130, 1990.

[5] M. Henle. *A Combinatorial Introduction to Topology* W.H.Freeman and Company, San Francisco, 1979.

[6] Y. Luo and G. Lukács. Regularized Set Operators for Non-Manifold Boundary Solid Modelling. In *Proc. of International conference on CAD & CG '89*, Beijing, 281–287, August 1989.

[7] Y. Luo and G. Lukács. A Boundary Representation for Form Features and Non-Manifold Solid Objects. In *Proc. of the first ACM/SIGGRAPH Symposium on Solid Modeling Foundations and CAD/CAM Applications*, Austin, Texas, 45–60, June 1991.

[8] M. Mäntylä. *An Introduction to Solid Modelling.* Computer Science Press, Rockville, Maryland, 1988.

[9] K. Weiler. *Topological Structure for Geometric Modeling.* PhD thesis, Rensselaer Polytechnic Institute, Troy, New York, 1986.

[10] Y. Yamaguchi and F. Kimura. Boundary Neighborhood Representation for Non-Manifold Topology. In *Proc. of the 1990 IFIP/RPI Geometric Modeling Conference*, Rensselaerville, New York 1990.

[11] A.A.G. Requicha and R.B. Tilove. *Mathematical Foundations of Constructive Solid Geometry: General Topology of Closed Regular Sets.* Production Automation Project TM-27a, University of Rochester, New York, 1978.

9.6 Appendix: Derivation of Combinatorial Boundary Expressions for Regularized Boolean Operators

9.6.1 Notations

$\cup$	set union
$\cap$	set intersection
$\setminus$	set difference
W	3-D topological space
W'	complex
X, Y	subsets of W
iX	interior of X
cX	complement of X
bX	boundary of X
b_iX, Δ_iX, $\Delta_{ij}X$	pieces of bX (i and j are figures)
p, q	coincident components of Δ_iX and Δ_iY
$a(p)$	area of p
$n(p)$	orientation of p
$s'(p)$	surrounding of p
$i'(p)$, $i'p$	interior of p
$k'(p)$, $k'p$	closure of p
$b'(p)$, $b'p$	boundary of p
bX *on* bY	pieces of bX satisfying $bX \cap bY$
bY *on* bX	pieces of bY satisfying $bX \cap bY$

9.6.2 Concepts concerned

Regularization of X [9]: $rX = kiX$. A set X is **regular** if $X = rX$. If a regular set X can be put into a box with finite volume, then X is bounded and is named an **r-set**. R-sets in W are consistent with the general concept of **solid**. If X is regular, then rcX is an unbounded regular set which corresponds to a **negative solid**.
Regularized Boolean operations [9]:

$$\begin{aligned} X \cup^* Y &= r(X \cup Y) \\ X \cap^* Y &= r(X \cap Y) \\ X \setminus^* Y &= r(X \setminus Y) \\ c^* &= rcX \end{aligned}$$

Related properties of regular sets and regularized Boolean operations [9]:
PROPERTY 1.
If X and Y are regular then
(1) (De Morgan's laws for regular sets)

$$\begin{aligned} c^*(X \cup^* Y) &= c^*X \cap^* c^*Y; \\ c^*(X \cap^* Y) &= c^*X \cup^* c^*Y; \\ c^*(X \setminus^* Y) &= c^*X \cup^* Y. \end{aligned}$$

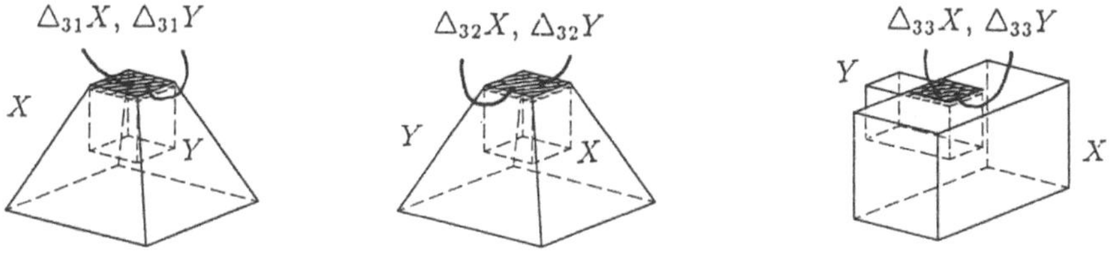

FIGURE 9.4. Subclasses of pair $\Delta_3 X$ and $\Delta_3 Y$.

(2) $ic^*X = cX$.
(3) $c^*c^*X = X$.
(4) $cc^*X = iX$.

Reversing operation [2]: obX. The orientation of a solid boundary is used to determine if the solid is positive or negative. The reversing operation is applied to reverse the orientation of a solid boundary, i.e., $n(obX) = -n(bX)$.

Properties of reversing operation:

PROPERTY 2.

If X is regular, then

(1) $obX = bc^*X$. (2) $oobX = bX$. (3) if two pieces of bX, $\Delta_1 X$ and $\Delta_2 X$, satisfy the condition that

$$\Delta_1 X \cap \Delta_2 X = \emptyset \text{(the empty set)},$$

then

$$o(\Delta_1 X \cup \Delta_2 X) = o\Delta_1 X \cup o\Delta_2 X.$$

Notice that (1) gives the boundary expression of regularized complement operation.

9.6.3 Boundary classification

Let X and Y be two intersecting regular sets (not necessarily bounded). Their boundaries bX and bY can be classified into the following pairs of subsets:

(1) $\Delta_1 X = bX \cap cY$ and $\Delta_1 Y = cX \cap bY$.
(2) $\Delta_2 X = bX \cap iY$ and $\Delta_2 Y = iX \cap bY$.

All the following subsets belong to the case "on".

(3) $\Delta_3 X \subset (bX \text{ } on \text{ } bY)$ and $\Delta_3 Y \subset (bY \text{ } on \text{ } bX)$. Let p and q be coincident connected components of simplices; and let $p \subset \Delta_3 X$ and $q \subset \Delta_3 Y$. Then p and q satisfy the conditions that

$$a(p) = a(q) \text{ and } n(p) = n(q).$$

The pair $\Delta_3 X$ and $\Delta_3 Y$ can be further classified according to their surroundings as below:

$\Delta_{31} X$ and $\Delta_{31} Y$: $s'(p) \subset \Delta_1 X$ and $s'(q) \subset \Delta_2 Y$;
$\Delta_{32} X$ and $\Delta_{32} Y$: $s'(p) \subset \Delta_2 X$ and $s'(q) \subset \Delta_1 Y$;
$\Delta_{33} X$ and $\Delta_{33} Y$: $s'(p) \subset (\Delta_1 X \cup \Delta_2 X)$ and $s'(q) \subset (\Delta_1 Y \cup \Delta_2 Y)$ (See Figure 9.4).

(4) $\Delta_4 X \subset (bX \text{ } on \text{ } bY)$ and $\Delta_4 Y \subset (bY \text{ } on \text{ } bX)$. Let p and q be coincident connected components of simplices; and let $p \subset \Delta_4 X$ and $q \subset \Delta_4 Y$. Then p and q satisfy the

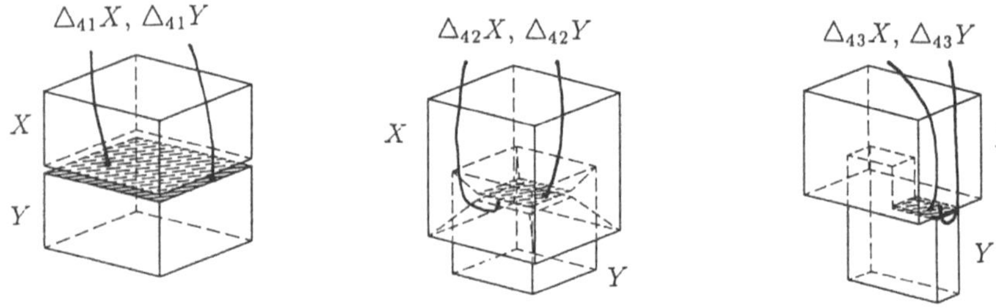

FIGURE 9.5. Subclasses of pair $\Delta_4 X$ and $\Delta_4 Y$.

conditions that

$$a(p) = a(q) \text{ and } n(p) = -n(q).$$

The pair $\Delta_4 X$ and $\Delta_4 Y$ can be further classified according to their surroundings as below:

$\Delta_{41}X$ and $\Delta_{41}Y$: $s'(p) \subset \Delta_1 X$ and $s'(q) \subset \Delta_1 Y$;
$\Delta_{42}X$ and $\Delta_{42}Y$: $s'(p) \subset \Delta_2 X$ and $s'(q) \subset \Delta_2 Y$;
$\Delta_{43}X$ and $\Delta_{43}Y$: $s'(p) \subset (\Delta_1 X \cup \Delta_2 X)$ and $s'(q) \subset (\Delta_1 Y \cup \Delta_2 Y)$ (See Figure 9.5).

(5) $\Delta_5 X \subset (bX \textit{ on } bY)$ and $\Delta_5 Y \subset (bY \textit{ on } bX)$. $\Delta_5 X$ and $\Delta_5 Y$ consist of only faces such that every connected subset in $\Delta_5 X$ or $\Delta_5 Y$ is an individual face. Let p and q be coincident connected components of simplices; and let $p \subset \Delta_5 X$, $q \subset \Delta_5 Y$. Then p and q satisfy the conditions that

$$a(p) \neq a(q) \text{ and } b'(p) \text{ does not touch } b'(q).$$

The pair $\Delta_5 X$ and $\Delta_5 Y$ can be further classified according to their areas and surroundings as below:

$\Delta_{51}X$ and $\Delta_{51}Y$: $a(p) < a(q)$; $s'(k'p) \subset \Delta_1 X$ and $s'(k'q) \subset \Delta_1 Y$;
$\Delta_{52}X$ and $\Delta_{52}Y$: $a(p) < a(q)$; $s'(k'p) \subset \Delta_2 X$ and $s'(k'q) \subset \Delta_2 Y$;
$\Delta_{53}X$ and $\Delta_{53}Y$: $a(p) < a(q)$; $s'(k'p) \subset \Delta_1 X$ and $s'(k'q) \subset \Delta_2 Y$;
$\Delta_{54}X$ and $\Delta_{54}Y$: $a(p) < a(q)$; $s'(k'p) \subset \Delta_2 X$ and $s'(k'q) \subset \Delta_1 Y$;
$\Delta_{55}X$ and $\Delta_{55}Y$: $a(p) > a(q)$; $s'(k'p) \subset \Delta_1 X$ and $s'(k'q) \subset \Delta_1 Y$;
$\Delta_{56}X$ and $\Delta_{56}Y$: $a(p) > a(q)$; $s'(k'p) \subset \Delta_2 X$ and $s'(k'q) \subset \Delta_2 Y$;
$\Delta_{57}X$ and $\Delta_{57}Y$: $a(p) > a(q)$; $s'(k'p) \subset \Delta_1 X$ and $s'(k'q) \subset \Delta_2 Y$;
$\Delta_{58}X$ and $\Delta_{58}Y$: $a(p) > a(q)$; $s'(k'p) \subset \Delta_2 X$ and $s'(k'q) \subset \Delta_1 Y$ (See Figure 9.6).

Class (5) can be put into class (3) or (4) by adding a loop of edges and vertices in the center of a face before classification. To simplify the representation, however, this class has been retained.

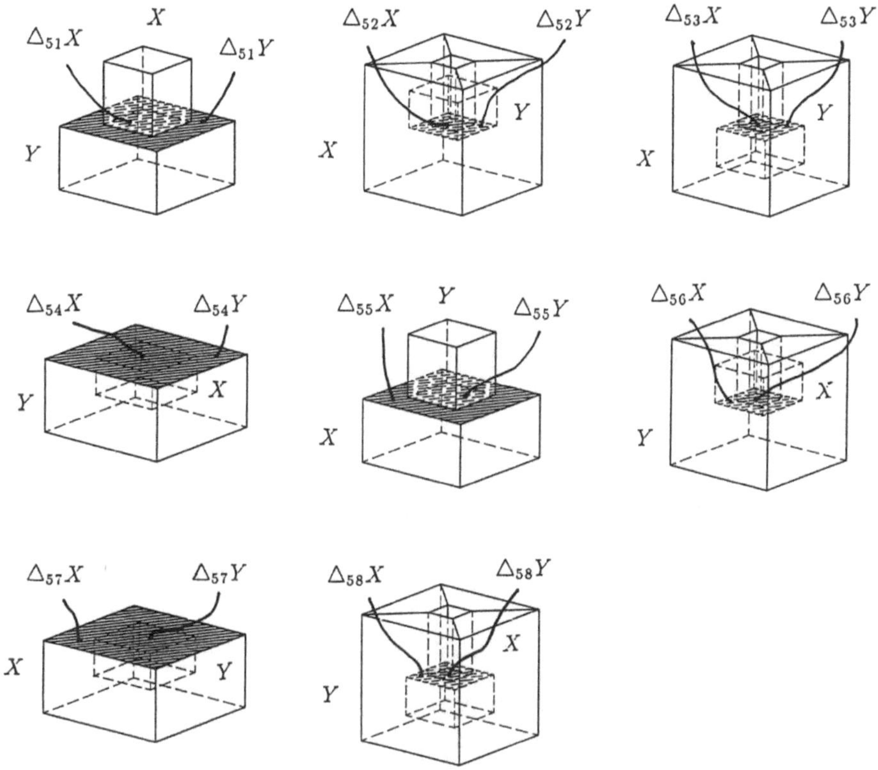

FIGURE 9.6. Subclasses of pair $\Delta_5 X$ and $\Delta_5 Y$.

(6) $\Delta_6 X \subset (bX \textit{ on } bY)$ and $\Delta_6 Y \subset (bY \textit{ on } bX)$. $\Delta_6 X$ and $\Delta_6 Y$ consist of only edges and vertices. Let p and q be coincident connected components of simplices; and let $p \subset \Delta_6 X$ and $q \subset \Delta_6 Y$. The pair $\Delta_6 X$ and $\Delta_6 Y$ can be further classified according to their surroundings as below:

$\Delta_{61}X$ and $\Delta_{61}Y$: $s'(p) \subset \Delta_1 X$ and $s'(q) \subset \Delta_1 Y$;
$\Delta_{62}X$ and $\Delta_{62}Y$: $s'(p) \subset \Delta_2 X$ and $s'(q) \subset \Delta_2 Y$;
$\Delta_{63}X$ and $\Delta_{63}Y$: $s'(p) \subset \Delta_1 X$ and $s'(q) \subset \Delta_2 Y$;
$\Delta_{64}X$ and $\Delta_{64}Y$: $s'(p) \subset \Delta_2 X$ and $s'(q) \subset \Delta_1 Y$;
$\Delta_{65}X$ and $\Delta_{65}Y$: $s'(p) \subset (\Delta_1 X \cup \Delta_2 X)$ and $s'(q) \subset (\Delta_1 Y \cup \Delta_2 Y)$ (See Figure 9.7).

The classification is independent of the specific set operation ($\cup^*$, $\cap^*$, $\setminus^*$). In the following sections, we reclassify the boundaries according to the regularized set operations for obtaining the boundary expressions of the operations.

9.6.4 Boundary expression of regularized union

For $X \cup^* Y$, bX and bY can be reclassified using the operators k', i' and b' on Δ's such that each boundary is divided into two half pieces ($b_1 X$ and $b_2 X$) plus the common boundary between them ($b_3 X$) and the faces each of which will be, after the operation, incident to

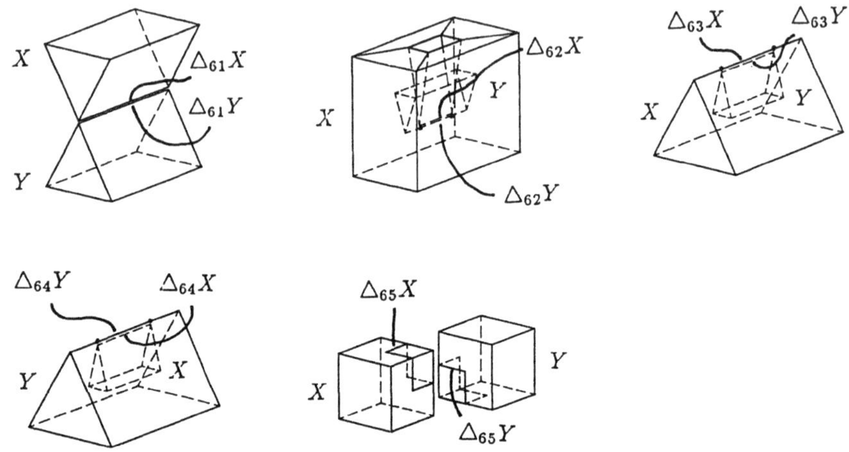

FIGURE 9.7. Subclasses of pair $\Delta_6 X$ and $\Delta_6 Y$.

two loops (b_4X and b_5X), the edges to several wedges and the vertices to several bundles (b_6X). The reclassification is as follows:

$$
\begin{aligned}
b_1X &= i'(\Delta_1X \cup \Delta_{31}X \cup \Delta_{33}X \cup k'\Delta_{53}X \cup k'\Delta_{57}X \cup \Delta_{63}X) \cup \\
&\quad b'(\Delta_{51}X \cup \Delta_{52}X \cup \Delta_{55}X \cup \Delta_{56}X) \\
b_1Y &= i'(\Delta_1Y \cup \Delta_{32}Y \cup k'\Delta_{54}Y \cup k'\Delta_{58}Y \cup \Delta_{64}Y) \cup \\
&\quad b'(\Delta_{51}Y \cup \Delta_{52}Y \cup \Delta_{55}Y \cup \Delta_{56}Y) \\
b_2X &= i'(\Delta_2X \cup \Delta_{32}X \cup \Delta_4X \cup k'\Delta_{54}X \cup k'\Delta_{58}X \cup \\
&\quad \Delta_{62}X \cup \Delta_{64}X) \\
b_2Y &= i'(\Delta_2Y \cup \Delta_{31}Y \cup \Delta_{33}Y \cup \Delta_4Y \cup k'\Delta_{53}Y \cup k'\Delta_{57}Y \cup \\
&\quad \Delta_{62}Y \cup \Delta_{63}Y) \\
b_3X &= k'(b_1X) \cap k'(b_2X) \\
b_3Y &= k'(b_1Y) \cap k'(b_2Y) \\
b_4X &= \Delta_{51}X \cup \Delta_{52}X \\
b_4Y &= \Delta_{51}Y \cup \Delta_{52}Y \\
b_5X &= \Delta_{55}X \cup \Delta_{56}X \\
b_5Y &= \Delta_{55}Y \cup \Delta_{56}Y \\
b_6X &= \Delta_{61}X \\
b_6Y &= \Delta_{61}Y.
\end{aligned}
\tag{9.1}
$$

The boundary of the result of the regularized union on X and Y then can be expressed by:

$$
\begin{aligned}
b(X \cup^* Y) &= b_1X \cup b_1Y \cup b_3X \cup b_4Y \cup b_5X \cup b_6X \\
&= [bX \setminus (b_2X \cup b_4X)] \cup [bY \setminus (b_2Y \cup b_3Y \cup b_5Y \cup b_6Y)]
\end{aligned}
\tag{9.2}
$$

9.6.5 Boundary expression of regularized intersection

By Property 1 and Property 2 in Section 9.6.2 we have

$$b(X \cap^* Y) = bc^*c^*(X \cap^* Y) = bc^*(c^*X \cup c^*Y) = ob(c^*X \cup c^*Y). \tag{9.3}$$

So we can utilize the relations in the regularized union operation to express $b(X \cap^* Y)$. Let X_1 denote c^*X and Y_1 denote c^*Y. By Classification (1) and (2) in Section 9.6.3, Property 1 and Property 2 in Section 9.6.2, we have

$$\Delta_1 X_1 = bX_1 \cap cY_1 = bc^*X \cap cc^*Y = obX \cap iY = o\Delta_2 X;$$

and

$$\Delta_1 Y_1 = cX_1 \cap bY_1 = cc^*X \cap bc^*Y = iX \cap obY = o\Delta_2 Y.$$

Following the above examples, we can obtain the classification of bX and bY in Section 9.6.3 for $X \cap^* Y$ as listed below:

(1) $\Delta_1 X_1 = o\Delta_2 X$, $\quad \Delta_1 Y_1 = o\Delta_2 Y$.
(2) $\Delta_2 X_1 = o\Delta_1 X$, $\quad \Delta_2 Y_1 = o\Delta_1 Y$.
(3) $\Delta_3 X_1 = o\Delta_3 X$, $\quad \Delta_3 Y_1 = o\Delta_3 Y$;
$\Delta_{31} X_1 = o\Delta_{32} X$, $\quad \Delta_{31} Y_1 = o\Delta_{32} Y$;
$\Delta_{32} X_1 = o\Delta_{31} X$, $\quad \Delta_{32} Y_1 = o\Delta_{31} Y$;
$\Delta_{33} X_1 = o\Delta_{33} X$, $\quad \Delta_{33} Y_1 = o\Delta_{33} Y$.
(4) $\Delta_4 X_1 = o\Delta_4 X$, $\quad \Delta_4 Y_1 = o\Delta_4 Y$;
$\Delta_{41} X_1 = o\Delta_{42} X$, $\quad \Delta_{41} Y_1 = o\Delta_{42} Y$;
$\Delta_{42} X_1 = o\Delta_{41} X$, $\quad \Delta_{42} Y_1 = o\Delta_{41} Y$;
$\Delta_{43} X_1 = o\Delta_{43} X$, $\quad \Delta_{43} Y_1 = o\Delta_{43} Y$.
(5) $\Delta_5 X_1 = o\Delta_5 X$, $\quad \Delta_5 Y_1 = o\Delta_5 Y$;
$\Delta_{51} X_1 = o\Delta_{52} X$, $\quad \Delta_{51} Y_1 = o\Delta_{52} Y$;
$\Delta_{52} X_1 = o\Delta_{51} X$, $\quad \Delta_{52} Y_1 = o\Delta_{51} Y$;
$\Delta_{53} X_1 = o\Delta_{54} X$, $\quad \Delta_{53} Y_1 = o\Delta_{54} Y$;
$\Delta_{54} X_1 = o\Delta_{53} X$, $\quad \Delta_{54} Y_1 = o\Delta_{53} Y$;
$\Delta_{55} X_1 = o\Delta_{56} X$, $\quad \Delta_{55} Y_1 = o\Delta_{56} Y$;
$\Delta_{56} X_1 = o\Delta_{55} X$, $\quad \Delta_{56} Y_1 = o\Delta_{55} Y$;
$\Delta_{57} X_1 = o\Delta_{58} X$, $\quad \Delta_{57} Y_1 = o\Delta_{58} Y$;
$\Delta_{58} X_1 = o\Delta_{57} X$, $\quad \Delta_{58} Y_1 = o\Delta_{57} Y$.
(6) $\Delta_6 X_1 = o\Delta_6 X$, $\quad \Delta_6 Y_1 = o\Delta_6 Y$;
$\Delta_{61} X_1 = o\Delta_{62} X$, $\quad \Delta_{61} Y_1 = o\Delta_{62} Y$;
$\Delta_{62} X_1 = o\Delta_{61} X$, $\quad \Delta_{62} Y_1 = o\Delta_{61} Y$;
$\Delta_{63} X_1 = o\Delta_{64} X$, $\quad \Delta_{63} Y_1 = o\Delta_{64} Y$;
$\Delta_{64} X_1 = o\Delta_{63} X$, $\quad \Delta_{64} Y_1 = o\Delta_{63} Y$.

By the relations of b's in (9.1) and Δ's above, we have

$$\begin{aligned} b_1 X &= ob_1 c^* X = ob_1 X_1 = o[i'(\Delta_1 X_1 \cup \Delta_{31} X_1 \cup \Delta_{33} X_1 \cup k'\Delta_{53} X_1 \cup k'\Delta_{57} X_1 \cup \\ &\quad \Delta_{63} X_1) \cup b'(\Delta_{51} X_1 \cup \Delta_{52} X_1 \cup \Delta_{55} X_1 \cup \Delta_{56} X_1)] \\ &= i'(\Delta_2 X \cup \Delta_{32} X \cup \Delta_{33} X \cup k'\Delta_{54} X \cup k'\Delta_{58} X \cup \Delta_{64} X) \cup \\ &\quad b'(\Delta_{52} X \cup \Delta_{51} X \cup \Delta_{56} X \cup \Delta_{55} X) \end{aligned}$$

Following the above step, we get the reclassification of bX and bY for $X \cap^* Y$ as below:

$$\begin{aligned}
b_1X = & \; i'(\Delta_2X \cup \Delta_{32}X \cup \Delta_{33}X \cup k'\Delta_{54}X \cup k'\Delta_{58}X \cup \Delta_{64}X) \cup \\
& \; b'(\Delta_{52}X \cup \Delta_{51}X \cup \Delta_{56}X \cup \Delta_{55}X) \\
b_1Y = & \; i'(\Delta_2Y \cup \Delta_{31}Y \cup k'\Delta_{53}Y \cup k'\Delta_{57}Y \cup \Delta_{63}Y) \cup \\
& \; b'(\Delta_{52}Y \cup \Delta_{51}Y \cup \Delta_{56}Y \cup \Delta_{55}Y) \\
b_2X = & \; i'(\Delta_1X \cup \Delta_{31}X \cup \Delta_4X \cup k'\Delta_{53}X \cup k'\Delta_{57}X \cup \\
& \; \Delta_{61}X \cup \Delta_{63}X) \\
b_2Y = & \; i'(\Delta_1Y \cup \Delta_{32}Y \cup \Delta_{33}Y \cup \Delta_4Y \cup k'\Delta_{54}Y \cup k'\Delta_{58}Y \cup \\
& \; \Delta_{61}Y \cup \Delta_{64}Y) \\
b_3X = & \; k'(b_1X) \cap k'(b_2X) \\
b_3Y = & \; k'(b_1Y) \cap k'(b_2Y) \\
b_4X = & \; \Delta_{52}X \cup \Delta_{51}X \\
b_4Y = & \; \Delta_{52}Y \cup \Delta_{51}Y \\
b_5X = & \; \Delta_{56}X \cup \Delta_{55}X \\
b_5Y = & \; \Delta_{56}Y \cup \Delta_{55}Y \\
b_6X = & \; \Delta_{62}X \\
b_6Y = & \; \Delta_{62}Y.
\end{aligned} \tag{9.4}$$

Equation (9.3) then can be expressed by:

$$\begin{aligned}
b(X \cap^* Y) & = ob(c^*X \cup^* c^*Y) \\
& = o(b_1c^*X \cup b_1c^*Y \cup b_3c^*X \cup b_4c^*Y \cup b_5c^*X \cup b_6c^*X) \\
& = b_1X \cup b_1Y \cup b_3X \cup b_4Y \cup b_5X \cup b_6X \\
& = (bX \cup bY) \setminus [(b_2X \cup b_4X) \cup (b_2Y \cup b_3Y \cup b_5Y \cup b_6Y)]
\end{aligned} \tag{9.5}$$

9.6.6 Boundary expression of regularized difference

By Property 1 and Property 2 in Section 9.6.2 we have

$$b(X \setminus^* Y) = bc^*c^*(X \setminus^* Y) = bc^*(c^*X \cup^* Y) = ob(c^*X \cup^* Y). \tag{9.6}$$

So we can utilize the relations in the regularized union operation to express $b(X \setminus^* Y)$. Let X_2 denote c^*X and Y_2 denote Y. By Classification (1) and (2) in Section 9.6.3, Property 1 and Property 2 in Section 9.6.2, we have

$$\Delta_1X_2 = bX_2 \cap cY_2 = bc^*X \cap cY = obX \cap cY = o\Delta_1X$$

and

$$\Delta_1Y_2 = cX_2 \cap bY_2 = cc^*X \cap bY = iX \cap bY = \Delta_2Y.$$

Following the above examples, we can obtain the classification of bX and bY in Section 9.6.3 for $X \setminus^* Y$ as listed below:

(1) $\Delta_1X_2 = o\Delta_1X,$	$\Delta_1Y_2 = \Delta_2Y.$
(2) $\Delta_2X_2 = o\Delta_2X,$	$\Delta_2Y_2 = \Delta_1Y.$
(3) $\Delta_3X_2 = o\Delta_4X,$	$\Delta_3Y_2 = \Delta_4Y;$
$\Delta_{31}X_2 = o\Delta_{41}X,$	$\Delta_{31}Y_2 = \Delta_{41}Y;$
$\Delta_{32}X_2 = o\Delta_{42}X,$	$\Delta_{32}Y_2 = \Delta_{42}Y;$
$\Delta_{33}X_2 = o\Delta_{43}X,$	$\Delta_{33}Y_2 = \Delta_{43}Y.$
(4) $\Delta_4X_2 = o\Delta_3X,$	$\Delta_4Y_2 = \Delta_3Y;$
$\Delta_{41}X_2 = o\Delta_{31}X,$	$\Delta_{41}Y_2 = \Delta_{31}Y;$

$\Delta_{42}X_2 = o\Delta_{32}X,$ $\Delta_{42}Y_2 = \Delta_{32}Y;$
$\Delta_{43}X_2 = o\Delta_{33}X,$ $\Delta_{43}Y_2 = \Delta_{33}Y.$
(5) $\Delta_5 X_2 = o\Delta_5 X,$ $\Delta_5 Y_2 = \Delta_5 Y;$
$\Delta_{51}X_2 = o\Delta_{53}X,$ $\Delta_{51}Y_2 = \Delta_{53}Y;$
$\Delta_{52}X_2 = o\Delta_{54}X,$ $\Delta_{52}Y_2 = \Delta_{54}Y;$
$\Delta_{53}X_2 = o\Delta_{51}X,$ $\Delta_{53}Y_2 = \Delta_{51}Y;$
$\Delta_{54}X_2 = o\Delta_{52}X,$ $\Delta_{54}Y_2 = \Delta_{52}Y;$
$\Delta_{55}X_2 = o\Delta_{57}X,$ $\Delta_{55}Y_2 = \Delta_{57}Y;$
$\Delta_{56}X_2 = o\Delta_{58}X,$ $\Delta_{56}Y_2 = \Delta_{58}Y;$
$\Delta_{57}X_2 = o\Delta_{55}X,$ $\Delta_{57}Y_2 = \Delta_{55}Y;$
$\Delta_{58}X_2 = o\Delta_{56}X,$ $\Delta_{58}Y_2 = \Delta_{56}Y.$
(6) $\Delta_6 X_2 = o\Delta_6 X,$ $\Delta_6 Y_2 = \Delta_6 Y;$
$\Delta_{61}X_2 = o\Delta_{63}X,$ $\Delta_{61}Y_2 = \Delta_{63}Y;$
$\Delta_{62}X_2 = o\Delta_{64}X,$ $\Delta_{62}Y_2 = \Delta_{64}Y;$
$\Delta_{63}X_2 = o\Delta_{61}X,$ $\Delta_{63}Y_2 = \Delta_{61}Y;$
$\Delta_{64}X_2 = o\Delta_{62}X,$ $\Delta_{64}Y_2 = \Delta_{62}Y.$

By the relations of b's in (9.1) and Δ's above, we have

$$\begin{aligned} b_1X = &\ ob_1c^*X = ob_1X_2 = o[i'(\Delta_1X_2 \cup \Delta_{31}X_2 \cup \Delta_{33}X_2 \cup k'\Delta_{53}X_2 \cup k'\Delta_{57}X_2 \cup \\ &\ \Delta_{63}X_2) \cup b'(\Delta_{51}X_2 \cup \Delta_{52}X_2 \cup \Delta_{55}X_2 \cup \Delta_{56}X_2)] \\ = &\ i'(\Delta_1X \cup \Delta_{41}X \cup \Delta_{43}X \cup k'\Delta_{51}X \cup k'\Delta_{55}X \cup \Delta_{61}X) \cup \\ &\ b'(\Delta_{53}X \cup \Delta_{54}X \cup \Delta_{57}X \cup \Delta_{58}X) \end{aligned}$$

Following the above step, we get the reclassification of bX and bY for $X \setminus^* Y$ as below:

$$\begin{aligned} b_1X = &\ i'(\Delta_1X \cup \Delta_{41}X \cup \Delta_{43}X \cup k'\Delta_{51}X \cup k'\Delta_{55}X \cup \Delta_{61}X) \cup \\ &\ b'(\Delta_{53}X \cup \Delta_{54}X \cup \Delta_{57}X \cup \Delta_{58}X) \\ b_1Y = &\ i'(\Delta_2Y \cup \Delta_{42}Y \cup k'\Delta_{52}Y \cup k'\Delta_{56}Y \cup \Delta_{62}Y) \cup \\ &\ b'(\Delta_{53}Y \cup \Delta_{54}Y \cup \Delta_{57}Y \cup \Delta_{58}Y) \\ b_2X = &\ i'(\Delta_2X \cup \Delta_{42}X \cup \Delta_3X \cup k'\Delta_{52}X \cup k'\Delta_{56}X \cup \\ &\ \Delta_{64}X \cup \Delta_{62}X) \\ b_2Y = &\ i'(\Delta_1Y \cup \Delta_{41}Y \cup \Delta_{43}Y \cup \Delta_3Y \cup k'\Delta_{51}Y \cup k'\Delta_{55}Y \cup \\ &\ \Delta_{64}Y \cup \Delta_{61}Y) \\ b_3X = &\ k'(b_1X) \cap k'(b_2X) \\ b_3Y = &\ k'(b_1Y) \cap k'(b_2Y) \\ b_4X = &\ \Delta_{53}X \cup \Delta_{54}X \\ b_4Y = &\ \Delta_{53}Y \cup \Delta_{54}Y \\ b_5X = &\ \Delta_{57}X \cup \Delta_{58}X \\ b_5Y = &\ \Delta_{57}Y \cup \Delta_{58}Y \\ b_6X = &\ \Delta_{63}X \\ b_6Y = &\ \Delta_{63}Y. \end{aligned} \tag{9.7}$$

Equation (9.6) then can be expressed by:

$$\begin{aligned} b(X \setminus^* Y) &= ob(c^*X \cup^* Y) \\ &= o(b_1c^*X \cup b_1Y \cup b_3c^*X \cup b_4Y \cup b_5c^*X \cup b_6c^*X) \\ &= b_1X \cup ob_1Y \cup b_3X \cup ob_4Y \cup b_5X \cup b_6X \\ &= (bX \cup obY) \setminus [(b_2X \cup b_4X) \cup (ob_2Y \cup ob_3Y \cup ob_5Y \cup ob_6Y)] \end{aligned} \tag{9.8}$$

In summary, Equations (9.1) and (9.2) give combinatorial boundary expression for regularized union; (9.4) and (9.5) for regularized intersection; (9.7) and (9.8) for regularized difference; and (1) in Property 2 for regularized complement operation.

10 A Taxonomy on Geometric and Topological Models

Tapio Takala

ABSTRACT Mathematical fundamentals of point set topology and combinatorial topology are reviewed. All geometric models of practical interest are found to be finite collections of manifold elements of different dimensionalities, with boundary incidence as the most important relation among the elements. Based on this observation a classification scheme is proposed, characterizing a model topologically according to the compositional structure, boundary relations, dimensionality and connectivity of its elements. Geometric shapes are characterized as embeddings of manifolds into the Euclidian space, giving emphasis to homotopy groups. Accidental inconsistencies between embeddings and topological data are discussed, along with surgery operations on them. In focus of criticism is the inexact use of common terms like "boundary representation", "regularization" and "non-manifold model", which are here reconsidered mathematically.

"There is nothing new under the sun ...
... everything has been written before."
Sinuhe, the Egyptian[29]

10.1 Introduction

In its history of over thirty years, the art of computerized geometric modeling has developed from simple wire frames and polyhedra used in graphics, to various forms of curved parametric surfaces and rigid solid volumes utilized in simulation and manufacturing [18, 13]. The three major representation schemes for solids are Boolean expressions on algebraic half-spaces (CSG), closing boundary surfaces (B-rep), and regular space subdivisions into rectangular cells (octree). Additional schemes, used for deformable objects, include irregular subdivisions into finite elements (FEM), non-linear spatial mappings, and particle systems.

A major effort to precisely define the mathematical domain of solid modeling was done by the Rochester group [21], leading to the concept of regularized r-sets. Later, generalizations of solid modeling schemes that would include wire-frames and open surfaces have gained growing interest[30, 31, 32], with a recent boom on so-called non-manifold models (several authors in [28, 5, 24]. Some generalizations have also been made to higher dimensions[9]. However, most of these studies concentrate on the topological data structures, having their theoretical background in combinatorial topology only [14, 8]. Little attempt has been made to carry out comparable theoretical generalizations based on point set topology and Euclidian geometry.

Solid objects with manifold boundary, so-called "manifold solids", have become popular since they allow simple topological data structures that can be manipulated with a fairly compact set of Euler Operators [3, 20]. A more general class of solids can be represented with the same scheme as degenerated "pseudo-manifolds". However, straightforward extrapolation into the realm of mixed-dimensional models is rather artificial.

The fact that a basic scheme has been generalized in different, sometimes incompatible directions, has lead to a conceptual and terminological confusion. With authors having slightly different meanings even for "manifold solid", and developing different extensions to it, the generic complementary term "non-manifold" hardly carries any information content at all.

The purpose of this paper is to clarify the basic mathematical concepts underlying all major geometric modeling schemes, especially the various forms of boundary models. A novel conceptual taxonomy will be presented, where various modeling schemes can be classified based on the types of constraints they have on their geometric domain. Within this framework it is possible to more precisely characterize and assess each different "non-solid" modeling approach and the associated terminology.

Detailed analysis of implementation methods and geometric/topological data structures is out of the scope of this presentation. Models are characterized according to their mathematical form only, which may take various representations, each possibly optimized for a different user profile[1].

An often neglected part of geometric modeling is the embedding or immersion of abstract topological models into Euclidian space or its submanifolds. Homotopy groups, i.e. equivalence classes of geometric objects that can be continuously deformed in space to each other, are introduced as an unexploited tool to characterize different homeomorphic mappings. A note is made on how inconsistent mappings (immersions) can be handled topologically.

The paper is organized in two main parts. First (section 2) is a review of necessary mathematical background, introducing key concepts and terminology. In the second part a taxonomy on the topological form of models is given (section 3), followed by treatment of topological relations between abstract and embedded models (section 4), and discussion of some open problems of contemporary interest.

10.2 Mathematical Background

In this section the basic mathematical concepts are given, that are later needed for the taxonomy. Though most of these have been discovered by mathematicians long ago (see, e.g., [12, 4] and should be well-known in the geometric modeling community [22, 18, 13], there is an evident need to review and illuminate them once more in a concurrent context. The main theme of this paper will be topology, which is considered a special viewpoint to geometry, not as a separate abstract branch of mathematics.

The subsection on point set topology reviews the familiar concepts of open and closed sets, neighborhoods, interior, exterior, boundary and closure of a set, and the concept of n-dimensional manifold. Special attention is given to relative topology of submanifolds, which appears to be important in mixed-dimensional models.

The regularization (of r-sets) is reconsidered in a generalized setting and found to be more useful if defined in a set's relative topology, and if handling the interior and boundary as separate sets. A new concept of inherently non-manifold parts of a model will be introduced.

Combinatorial topology, a branch restricted from general topology to that of triangulable polytopes, is of particular interest in geometric modeling. The basic concepts of simplex, complex, combinatorial summing and other topological surgery operations are reviewed. Connectivity of complexes, homology groups and variations of the famous Euler formula are also studied.

Combinatorial and point set topology are two distinguished approaches in mathematics, which are often pursued separately. In geometric modeling this is reflected as the separation of topological and geometric data structures. However, since both representations have to be consistent, the combinatorial structure of abstract elements should everywhere

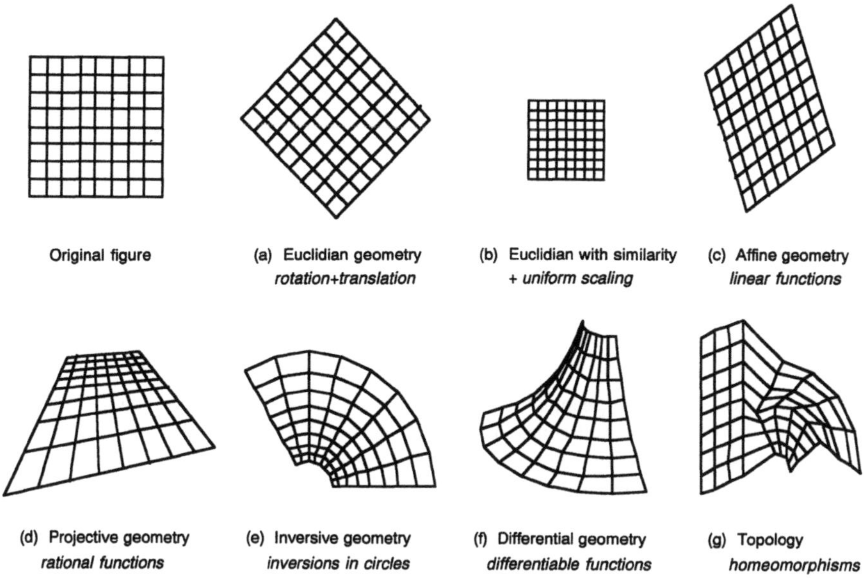

FIGURE 10.1. A rectangular grid with its (a) congruent, (b) similar, (c) affine, (d) projective, (e) inversive, (f) diffeomorphic, and (g) homeomorphic images.

match with the set-theoretical topology of their geometric counterparts.

Throughout the presentation it is assumed that geometric objects of interest are well-behaved. This means that all kinds of wildly embedded, fractal-like structures are excluded. Though such constructions are common examples in general topology, their inclusion would unnecessarily complicate the more practical purposes of geometric modeling.

10.2.1 Aspects of geometry

Since the days of Euclid, geometry has undergone many extensions and generalizations leading to various branches. A clear taxonomy of different geometries, known as the Erlangen program by Felix Klein [15, 6], characterizes each kind with the group of motions that keeps geometric figures equivalent.

In Euclidian geometry, figures are congruent and considered the same if they can be transformed to each other by rigid motions (translation and rotation) only. Another, more general version of it equates similar figures also, adding uniform scaling to the group of motions. The next generalization, called affine geometry, allows all linear (affine) transformations, extending the previous one with shearing or non-uniform scaling. Allowing perspective transformations brings us to projective geometry, for which the geometric space has to be extended with so-called ideal points at infinity. Inversive geometry considers equal all figures generated from each other by inversions in circles and reflections in lines. Finally the most general group of motions are diffeomorphisms (continuously differentiable one-to-one mappings) for differential geometry, and homeomorphisms (continuous and one-to-one, not necessarily differentiable mappings) for topology. Figure 10.1 illustrates the different geometries with an example.

In this view different geometries can be seen as generalizations of each other. Topology is the most abstract form of geometry, and there is a continuum of alternatives between it

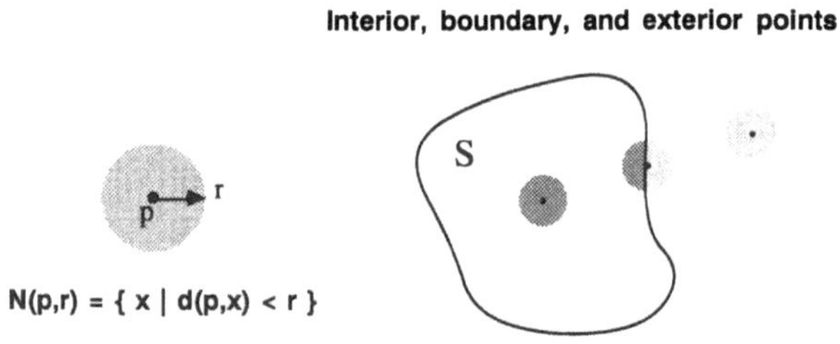

FIGURE 10.2. Neighborhoods of points

and the classical Euclidian geometry. It is important to note that though one may study topological objects as abstract constructions of their own interest, in geometric modeling we are interested in the topology of Euclidian objects, i.e. homeomorphic realizations (embeddings) of topological objects within Euclidian space.

10.2.2 Point set topology

10.2.2.1 Neighborhoods

Out of various cryptomorphic definitions of topology [11] §39, the one most suitable for Euclidian geometry is based on open spherical neighborhoods in n-dimensional space E_n, defined as

$$N(p,r) = \{x|d(p,x) < r\},$$

where $d(p,x)$ is the usual Euclidian distance between points p and x, and r is the radius of the neighborhood (Figure 10.2).

These neighborhoods are the open subsets of E_n serving as a *basis* of topology, satisfying the following axioms:

1. the whole space E_n and the empty set ϕ are open sets,
2. the union of any number of open sets is open, and
3. any finite intersection of open sets is open.

A point x is said to be *near* (or an accumulation point of) set S, if any neighborhood $N(x,r)$ contains at least some point of S. A set is *closed* if it contains all its near points — or equivalently, if its complement is open. A set's *boundary* consists of points near both the set itself and its complement. *Interior* are those points near the set but not its complement, and exterior points are near the complement only. The *closure* of a set S is the union of S with all its near points. Note that a set can be both open and closed, called *closing* (or clopen). In our Euclidian topology, open sets consist of interior points only, and only the trivial open sets ϕ and E_n are closing.

10.2.2.2 Relative topology

A subset X of a topological space S can be topologized with the relative topology, using as a basis not the open sets of S but their intersections with X. All the concepts defined above for the whole space (near points, open/closed sets, interior/exterior/boundary and closure) can be used in this subtopology as well (Figure 10.3).

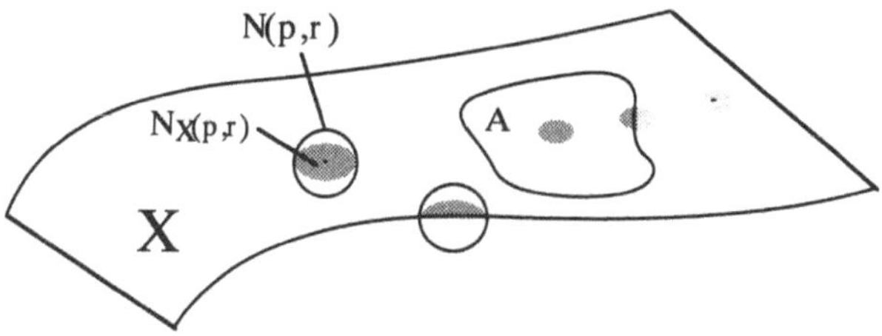

FIGURE 10.3. Relative topology on a subset X of space.

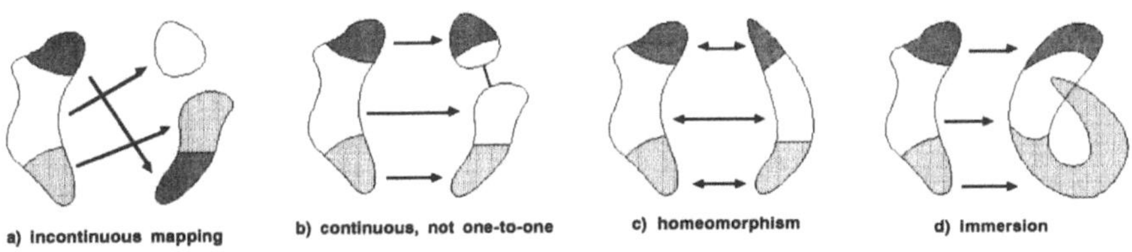

FIGURE 10.4. Kinds of mappings.

Note that a set X is always open (has only interior points) in its own relative topology. In order to speak about exterior and boundary, a larger subtopology is needed. Below, a manifold's relative interior and relative boundary will be distinguished using the topology of its closure.

10.2.2.3 Mappings

A mapping h from set A to set B, written $h : A \rightarrow B$, is a relation associating points of A to points of B. It is called injective if every point of A has a counterpart in B (A forms the domain of h), and surjective if each point of B is associated with some A (the image of h is B). If each domain point in A has a unique image point in B, and vice versa, the mapping is both injection and surjection, or *one-to-one.* A *continuous* mapping preserves nearness, i.e. the images of near points are near each other. The mapping is a *homeomorphism* if it is continuous in both ways (continuous and one-to-one). If the mapping is locally homeomorphic (the neighborhood of each point in A has a homeomorphic image in B) but not necessarily one-to-one as a whole, it is called an *immersion* (Figure 10.4).

Two point sets are called homeomorphic and considered topologically equivalent if there exists a homeomorphism between them. Properties that are retained in homeomorphic mappings are called *topological invariants,* or just topological properties. Most basic of them are nearness of points and the structure of neighborhoods (including interior/exterior/boundary relations). Other invariants, to be introduced later, are connectedness of sets, orientability of manifolds, and the Euler characteristic number.

10.2.2.4 Manifolds

A set is called n-manifold if for each point there is a (relative) neighborhood equivalent to an open neighborhood in E_n. Thus n is the "inherent" dimensionality of the set (Figure 10.5). Clearly, all open sets of E_n are n-manifolds. The simplest of them are open neighborhoods (thus named n-simplexes or simple n-manifolds), which are all homeomorphic to each other and to the whole space E_n.

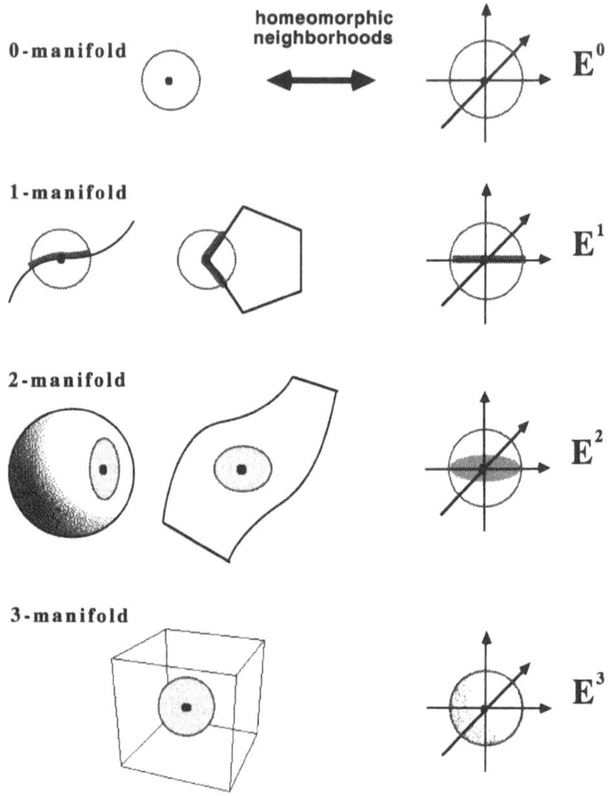

FIGURE 10.5. Manifolds.

Since all homeomorphic sets are considered topologically the same, we can speak about an abstract or topological manifold as of the whole equivalence class. An abstract manifold may be *embedded* onto another one (particularly into an Euclidian space) by a homeomorphic mapping.

Manifold subset of a manifold (often of the Euclidian space), but perhaps of lower dimensionality, is called a *submanifold*. Useful on a submanifold is the relative topology of its closure. In this particular topology it is possible to distinguish between the relative interior (the manifold itself) and its *relative boundary* (points of the closure not belonging to the manifold).

In geometric modeling we are interested in sets, which are decomposable into a finite number of manifold components (of different dimensionalities), but are not necessarily manifolds themselves. An important kind is *n-manifold-with-boundary*, the closure of a manifold, consisting of an n-dimensional relative interior and its lower-dimensional boundary components.[1] Examples are an open surface with its closing boundary curve (Figure 10.6a) and a closing surface (Klein bottle, which cannot be embedded in E_3) im-

[1]In literature "k-manifold-with-boundary" is defined as a set where every point's neighborhood is homeomorphic either to a k-dimensional disk or to ahalf-disk. The latter kind of points form a k-1-manifold, which is the relative boundary of the former former [4]. In this paper, a more liberal definition is adopted.

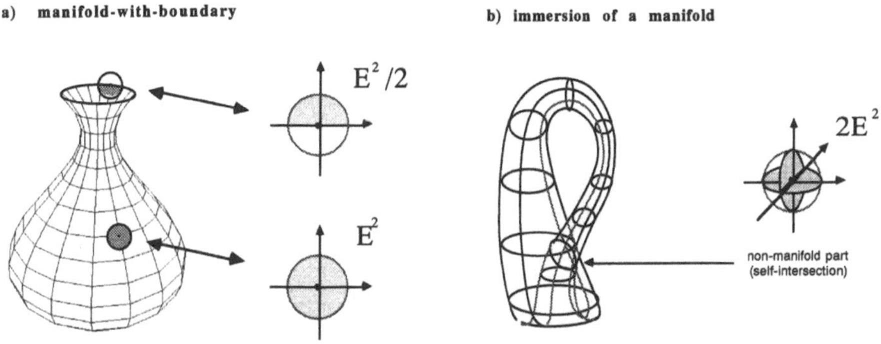

FIGURE 10.6. Some non-manifolds.

mersed in space with self-intersection (Figure 10.6b). In the latter case, the intersection curve's neighborhood is not manifold but homeomorphic to two crossing planes.

10.2.2.5 Regularization

The practical purpose of regularization of solid models is to make them dimensionally homogeneous. Traditionally the operation is defined as

$$regularized(S) = closure(interior(S)),$$

resulting in closed r-sets [21]. However, only the interior of an r-set is full-dimensional, whereas the boundary consists of lower-dimensional parts. In this respect, the dualistic definition

$$regularized(S) = interior(closure(S)),$$

is more natural [2]. Relations of the two alternatives are illustrated in Figure 10.7. With the latter definition we get a nice division of space into three disjoint components: a regularized set, its regularized complement, and the common boundary separating the two.

Traditionally also, the interior and closure are understood to be full-dimensional, i.e. defined in the topology of the whole space. In this paper the same concept will be applied to lower-dimensional components as well. For this purpose, *relative regularization* on a submanifold is introduced here, defined like the open regularization above, but in the relative topology of a submanifold's closure. An example is shown in Figure 10.8.

10.2.2.6 Inherently non-manifold boundaries

The Jordan curve theorem states that in E_2, every closing curve of finite extent separates the plane into exactly two components, one of which is the bounded interior. Its generalization to higher dimensions is the topological basis of boundary representation. Every closing n–1-manifold of limited extent in E_n is the boundary of some n-manifold interiors, separating them from the infinite exterior.

However, the converse it not necessarily true; not every n-manifold's boundary is an n–1-manifold. For example, the interior of two touching cubes (Figure 10.9a–b) is a 3-manifold, but the boundary consists of two manifold surfaces separated by a vertex or by an edge. A similar one is the 8-like figure (Figure 10.9c) consisting of two 1-manifolds joining at a single point.

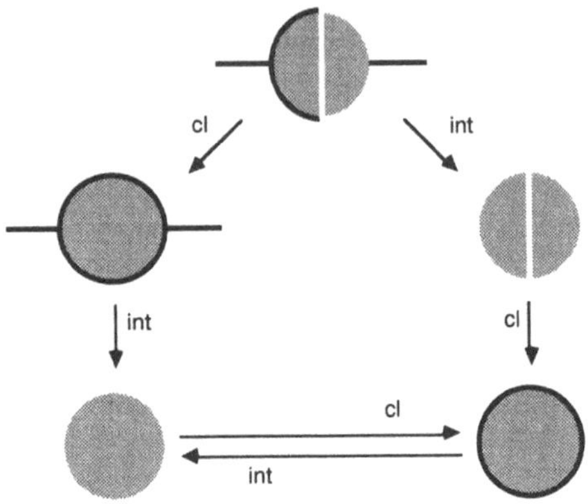

FIGURE 10.7. Open and closed regular sets.

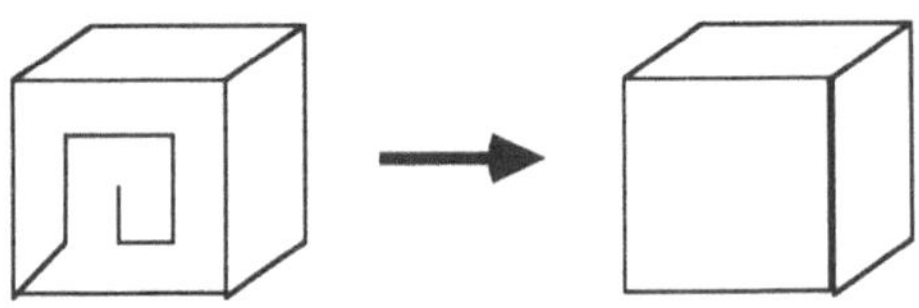

FIGURE 10.8. Relative (2D) regularization of a polyhedral face.

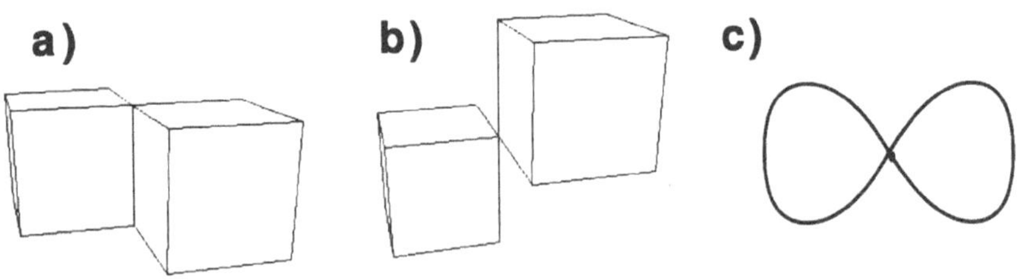

FIGURE 10.9. Inherently non-manifold objects.

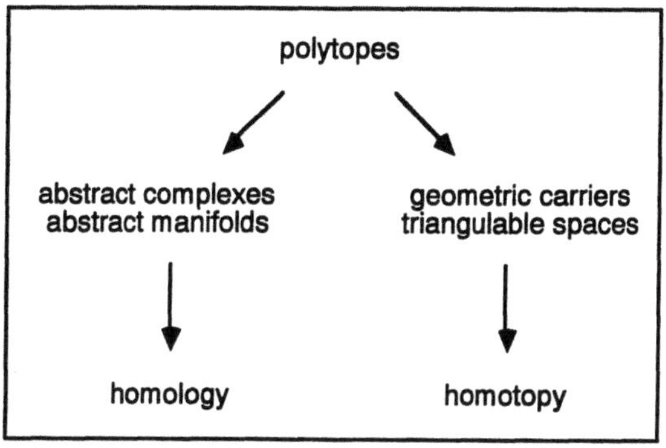

FIGURE 10.10. Abstract and geometric aspects of polytopes.

The essential parts of any n-manifold's boundary are a set of n–1-manifolds, locally separating its interior and exterior. If, however, the relative neighborhood of a boundary point is not a halfspace, it is not a proper boundary point but a non-manifold point. These points form lower-dimensional manifolds, called the *inherently non-manifold* components of the boundary.

10.2.3 Combinatorial topology

Combinatorial topology is an algebraic approach to study the topology of *polytopes* or triangulable point sets, i.e. those which can be formed by combining simple manifolds (chapter 5 of [12, 11]). This is a restricted domain, but sufficient for the purposes of geometric modeling.[2]

There are two major branches in algebraic topology to study polytopes (Figure 10.10). The purely combinatorial approach studies the homology of abstract manifolds, independently of their possible embeddings into Euclidian space. The other more restrictive branch is concerned with various kinds of geometric carriers of polytopes, characterizing them with homotopies.

The two simple but powerful techniques used in combinatorial topology are (1) *surgery* operations, cutting and pasting of manifolds along their boundaries, and (2) *counting* the numbers of elements formed with surgery.

10.2.3.1 Simplexes

The basic combinatorial elements are generalized triangles (this is why to speak about triangulable sets) called simplexes. An *n-simplex* is the region of n-dimensional affine space spanned by n linearly independent vectors emanating from the origin (or any set homeomorphic to this region). Thus 0-simplex is a single point, 1-simplex is a line segment, 2-simplex a triangle, 3-simplex a tetrahedron, etc., as shown in Figure 10.11. Clearly the interior of an n-simplex is a simple n-manifold, homeomorphic to an open n-ball. By construction, the boundary of a simplex is closed and homeomorphic to the n-sphere, and is triangulated into a set of lower simplexes. The closure of an n-simplex consists of k-simplexes with $k = 0 \dots n$, each of which (including the n-simplex itself) is called a k-face of the n-simplex.

[2]As noted in [12], p.206, we cannot even exactly characterize which spaces are triangulable, but at least all 3-manifolds and all differentiable manifolds are included.

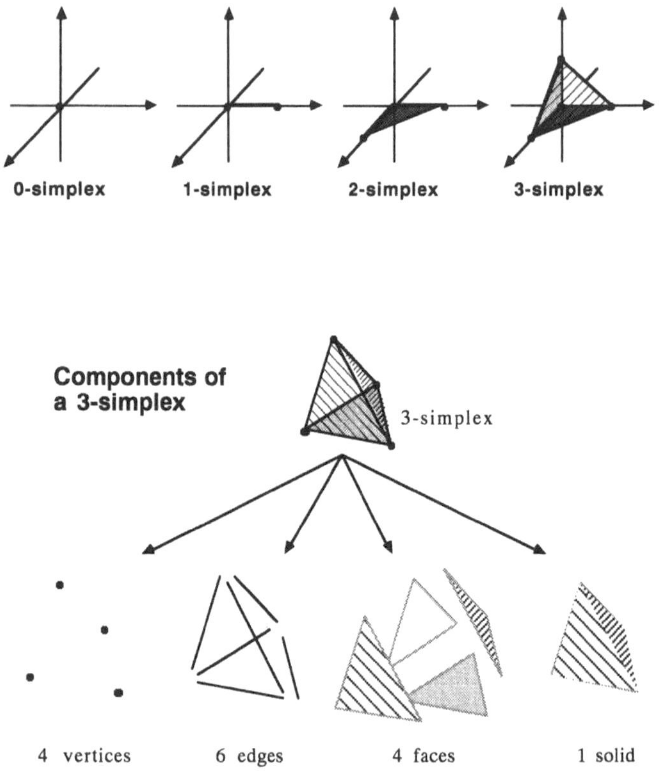

FIGURE 10.11. Simplexes.

10.2.3.2 Complexes

Simplicial complex is a collection of simplexes, such that any two simplexes can meet only along their faces. Examples are shown in Figure 10.12 (first is a simplicial complex, second one has faces that are not simplexes, in the third some simplexes overlap other ways than along their simplicial faces). The geometric carrier of a complex, or simply a *geometric complex*, is a polyhedral structure satisfying these requirements, or any homeomorphic image of it. The equivalence class of homeomorphic geometric complexes is the corresponding abstract complex.

If a simplex s is part of (the boundary of) another simplex S, then S is said to be *incident* to s. Two simplexes are *adjacent* if they are incident to a common simplex [3] (Figure 10.13).

The dimension of a complex is the maximum of its simplexes' dimensions. A dimensionally homogeneous complex containing only k-simplexes is called a *k-chain*. Such a complex obviously is a k-manifold. If the chain is homeomorphic with a k-simplex it will be called a *k-arc*. A k-chain's *boundary* consists of those (k–1-)faces that are incident to

[3] The meanings of terms "adjacency" and "incidency" are somewhat vague in literature. Sometimes adjacency means any kind of combinatorial neighborhood, including incidency. Here the more strict definitions according to [13], p.53, are adopted. However, even then the concept of adjacency is subject to some ambiguity.

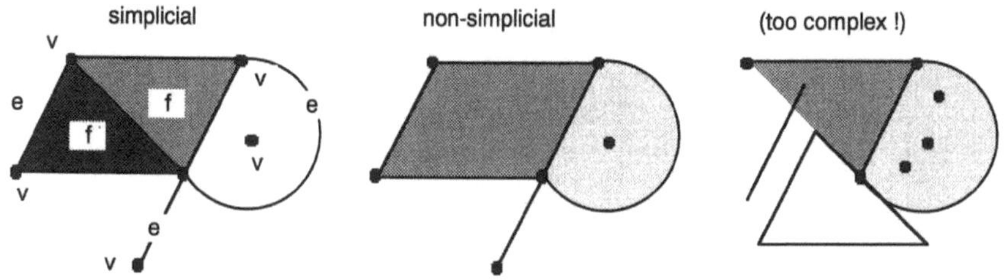

FIGURE 10.12. Complexes.

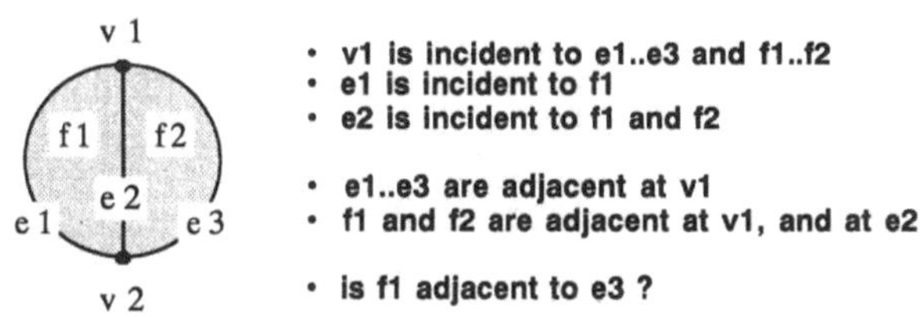

FIGURE 10.13. Incidences and adjacencies.

an odd number of k-simplexes. If every face is shared by an even number of simplexes, the chain has no boundary and is called a *k-cycle*. Examples are shown in Figure 10.14.

The usual concept of connectedness is extended here as follows: two simplexes *a* and *b* of a complex are kD-connected (k-dimensionally connected) if there is a k-arc subcomplex that has incident k–1-faces to both *a* and *b*. A 1D-connected complex is called arcwise connected or just connected (within triangulable spaces these two concepts are identical).

10.2.3.3 Surgery Operations

Surgery operations are used to construct and modify combinatorial complexes. With them, any complex can be subdivided into simplicial elements or reconstructed from them. Two basic types of operations are available: those changing the global topological shape of a complex but not modifying its parts, and those only affecting its internal structure and the topology of individual elements. Figure 10.15 shows the general idea with examples.

One of the former type of operations is *joining* (also known as pasting, sewing or identification). Two simplicial elements are brought geometrically coincident, and topologically identified as a single element. In order to do this consistently, one must join the boundaries of the elements as well, recursively in all dimensions.

The inverse operation of joining is *splitting* or duplication. Geometrically the resulting parts are assumed to be pulled away from each other. If the element bounds other elements, they are to be divided among the duplicates.

Both joining and splitting may change the global topological form. The point set topology of a complex does not change if we *subdivide* an element with an internal boundary. Generally we can make local subdivisions of an n-element by inserting any lower-dimensional elements within it. Before inserting an element, its boundaries must already exist in the environment (in the example, two vertices have to be inserted in a circle's perimeter before halving its area).

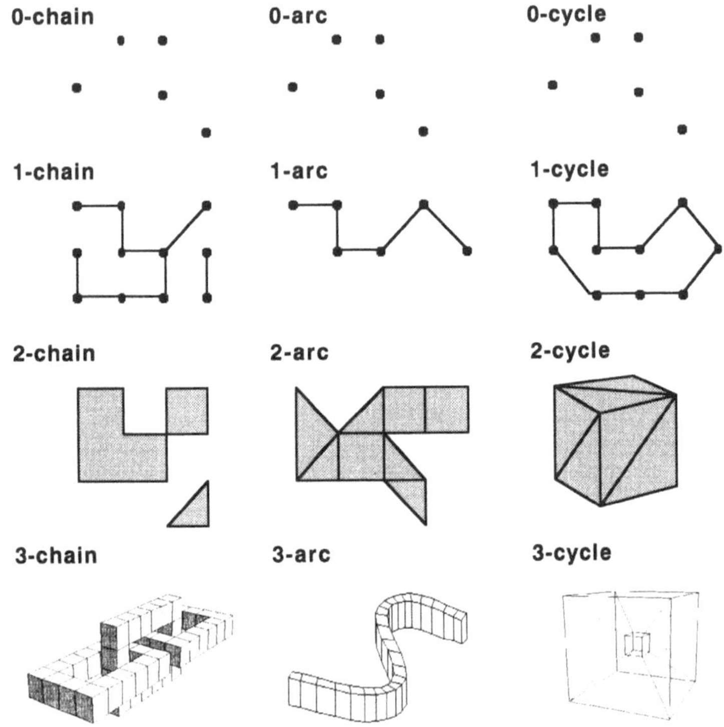

FIGURE 10.14. Chains, arcs and cycles.

Annihilation of an internal boundary is the inverse of subdivision, and also keeps the complex homeomorphically the same. Internal boundary is an n–1-simplex bounding exactly two n-dimensional elements When it is removed, the two adjacent elements together form a non-simplicial element . Annihilation is the combinatorial counterpart of n-dimensional regularization. It can be generalized to remove internal boundaries within a single element also. Then the dimension of removed elements may be lower than n–1.

10.2.3.4 Homology groups and connectivity numbers

Simplicial homology theory is an algebraic approach to characterize topological properties of triangulable objects. It is based on forming various abstract groups on a complex.

Generally a group is a set g of elements and a group operator $\circ$ defined on them, satisfying the following axioms

1. closure: $a, b \in g \Rightarrow a \circ b \in g$,
2. associativity: $(a \circ b) \circ c = a \circ (b \circ c)$,
3. neutral element $0 : \forall a, a \circ 0 = 0 \circ a = a$,
4. inverse element $-a : -a \circ a = a \circ -a = 0$.

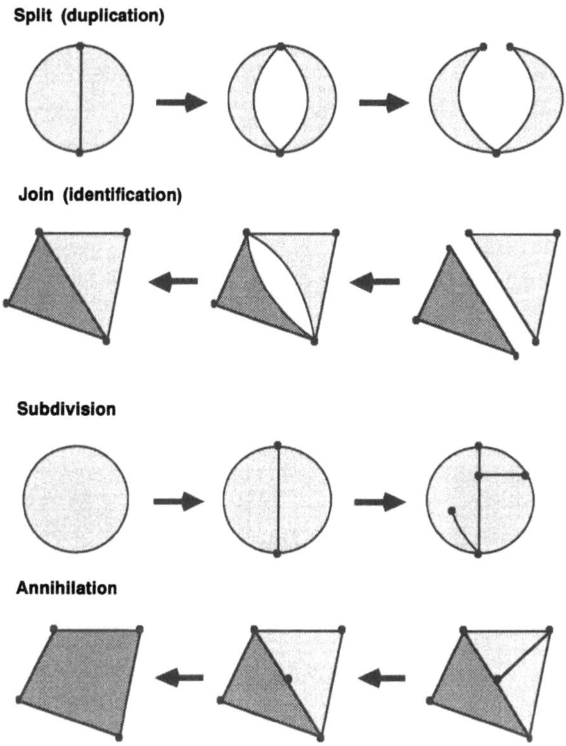

FIGURE 10.15. Surgery operations.

Combinatorial (modulo-2) addition of simplicial chains is the basic operator used in homology theory. The sum of two chains consists of those simplexes appearing in either chain but not in both (Figure 10.16). With this group operator, the neutral element is an empty chain and every chain is the inverse of itself.

Within a complex K, the set of all m-chains (C_m) forms a group. All m-cycles (Zm), and m-cycles bounding m+1-chains (B_m), are also groups under combinatorial addition, and are subgroups of the chain group ($C_m \supseteq Z_m \supseteq B_m$). Examples with m=1 are illustrated in Figure 10.17.

Two cycles are *homologous* ($\sim$), if their sum is a boundary. Homologous m-cycles in a complex K form equivalence classes that are elements in a group called the mth *homology group* (H_m) The generators of this group are the non-bounding m-cycles of K (those cycles that are not boundaries of any m+1-chain). Therefore it can be written $H_m = Z_m - B_m$.

A set of m-cycles is *linearly dependent*, if their combinatorial sum is zero, or equivalently, if any one of them can be formed as the sum of the others. Similarly, a set of elements in H_m are linearly dependent classes, if there are any linearly dependent cycles belonging to the respective classes. For example, the three loops (1-cycles) within a double annulus in Figure 10.18 are representatives of the three classes in its homology group H_1. All loops around the same holes (left, right or both) are homologous. Since any of the three representative loops can be formed as the sum of two others, they are linearly dependent. The maximum number of linearly independent elements in H_m is called its rank, or the mth *Betti number* (p_m) of complex K.

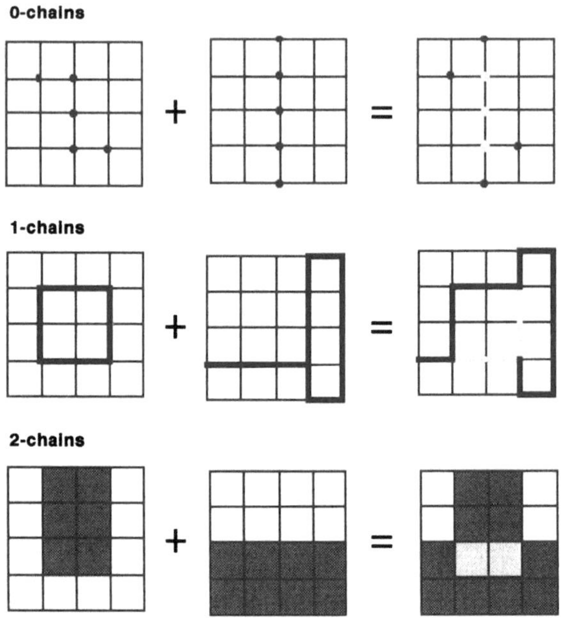

FIGURE 10.16. Summing of chains.

The Betti numbers are also called connectivity numbers of a complex. The lowest one of them, p_0, equals the number of the complex'es connected components[4]. The next is p_1, often simply referred as *the* connectivity number, which indicates how many cuts are needed to make the complex simply connected. For example, a torus must be cut both latitudinal and longitudinal loops in order to open it and make a simple surface patch (Figure 10.19). The second number, p_2, is the number of hollows enclosed by 2-dimensional surfaces. Intuitively the mth number corresponds to the number of m-dimensional holes in the object (i.e. holes surrounded by m-dimensional loops).

10.2.3.5 Euler formula and its variants

The famous formula, formulated by Euler for simple polyhedra, relates the numbers of vertices (V), edges (E) and faces (F) as

$$V - E + F = 2$$

A more general version, derived by Poincaré, relates the numbers of simplicial elements to the connectivity numbers of the complex as

$$\chi = \sum_{m=0}^{n} (-1)^m \alpha_m = \sum_{m=0}^{n} (-1)^m p_m$$

where α_m is the number of m-simplexes and p_m is the mth Betti number. The Eu-

[4]Assuming the so-called non-augmented homology.

Groups on a complex K:

Group: $A, B \in g \rightarrow A + B \in g$

Cm(K) = group of m-chains in K

basis of Cm = all m-simplexes

Zm(K) = group of all m-cycles of K

subgroup of Cm

Bm(K) = group of m-boundaries in K

(m-cycles bounding any m+1-chain);
subgroup of Zm;
basis of Bm = boundary cycles of all m+1-simplexes

Homology: $z^a \sim z^b$ iff $z^a + z^b \in B_m$

Hm(K) = Zm(K) - Bm(K)
= m^{th} homology group of K

equivalence classes of homologous (~)
non-bounding m-cycles
subgroup of Zm;

Pm(K) = m^{th} Betti number of K

number of independent,
non-trivial components of Hm

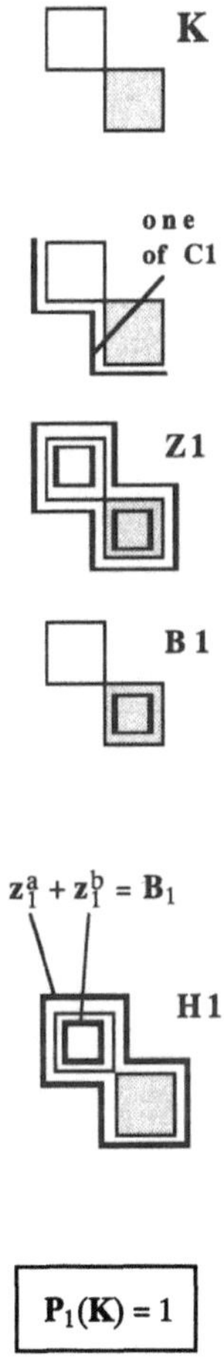

FIGURE 10.17. Homology groups.

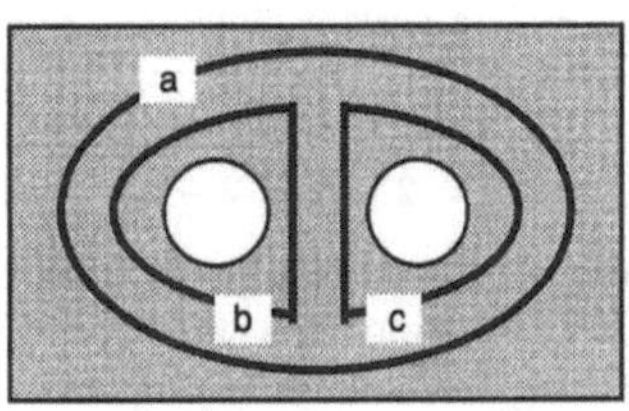

FIGURE 10.18. Linear dependency of cycles.

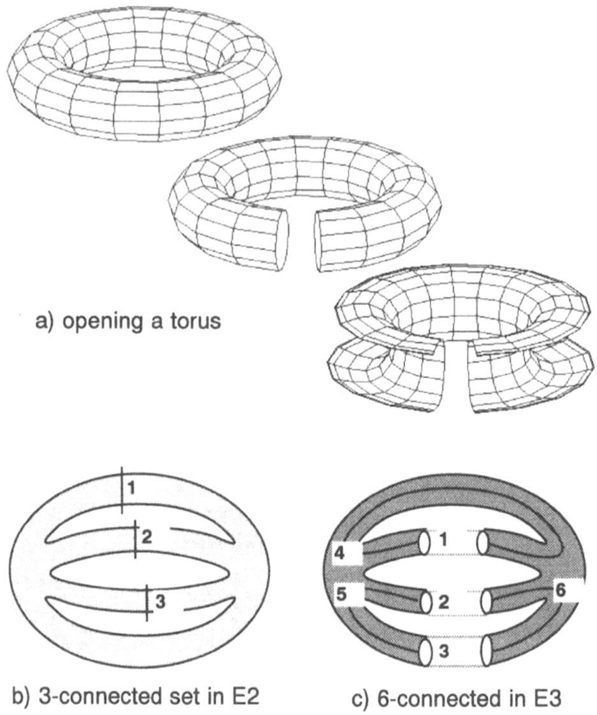

FIGURE 10.19. One- dimensional connectivity number.

ler characteristic χ is a topological invariant of a geometric complex, the same for all its triangulations. Originally the formula was derived for simplicial complexes, but can be extended with surgery operations to hold for all complexes with simply connected faces (homeomorphic to simplexes). Multiply connected faces may change the connectivity numbers on the right side and thus the Euler characteristic, too.

A topological classification statement tells that every closing orientable 2-manifold is homeomorphic to a sphere with a number of through holes or "handles" in it. The number g of holes is also known as the *genus* of the surface. This special case is of particular interest in solid modeling, since often the boundaries of solids are 2-manifolds, and all closing 2-manifolds in E_3 are indeed orientable. For a single closing surface we have $p_0 = 1$ (one connected component) and $p_2 = 1$ (a surface alone is a 2-cycle not bounding a volume). Then the formula yields to

$$\chi = 2 - p_1 = 2 - 2g$$

giving a simple way to compute the surface genus from a purely combinatorial structure.If multiply connected faces are allowed in a surface, they increment p_1 by one for each "ring" or hole in a face. Denoting the number of rings by r, and the number of separate surface components by s, we get a formula with six variables, familiar from solid modeling

$$v - e + f = 2s - 2g - r$$

FIGURE 10.20. Original Euler Operators.

Euler originally proved his formula using two surgery operations (Figure 10.20), which are sufficient to construct any simplicial polyhedral surface. Later, other sets of Euler Operators have been deviced, based on the formula above or simple extensions of it. Note however, that this formula was derived for the special case of closing manifold surfaces and may not be extended in a natural way to other combinatorial complexes. Instead, one should rely on the general Euler-Poincaré form in all non-manifold cases. Moreover, the operations to manipulate such models should not be based on the restricted formula, but on the general surgery techniques.

10.3 Topological Taxonomy

The mathematical concepts introduced above will now be used to construct a taxonomy framework, where various geometric models can be characterized (Figure 10.21).

Considering only the geometric part of a general model, the first categorization in the taxonomy is due to triangulability. "Wild" geometries with infinitely complex structure, like fractals, are ruled out from further analysis. What is then left are collections of manifolds, or geometric complexes. This is well justified, since all practically interesting models are finite collections of smooth manifolds — even fractals are usually evaluated as finite regular approximations.

Out of all possible manifold compositions, the most important subgroup are tessellations of a topological space. Tessellation is a finite collection of mutually disjoint open manifolds of different dimensions (a geometric complex), the union of which is the whole space. To this group belong most non-procedural geometric models without mutually intersecting parts, including open r-sets with explicitly distinguished boundary. Though usually only a single point set is of interest, the only other possible set being its implicitly defined complement, arbitrary selections can be made within a tessellation. Such frameworks are Selective Geometric Complexes, or SGC's [25].

Tessellative models are sometimes called "non-regularized" models, though they are a proper superset of regularized models. Thus the expression is not complementary, but means rather "not-necessarily-regularized" models. Many recent works on "non-manifold" boundary representations fall into this category category [31, 32, 34, 33, 7, 10, 23, 17].

The alternative branch of taxonomy consists of models with possibly overlapping components, with attached procedures defining how interactions are handled. Such are for example the CSG with Boolean set operations, or free-form surface modelers with sweeping, skinning or blending operations. An interesting special case are inconsistent immersions, briefly discussed in the next section. When fully evaluated, all these result in tessellative models.

A restriction among tessellations is regularization. In non-regularized models, there may exist components such as dangling faces, which are neither full-dimensional cells nor separating boundaries between different cells. The relative regularization is a local

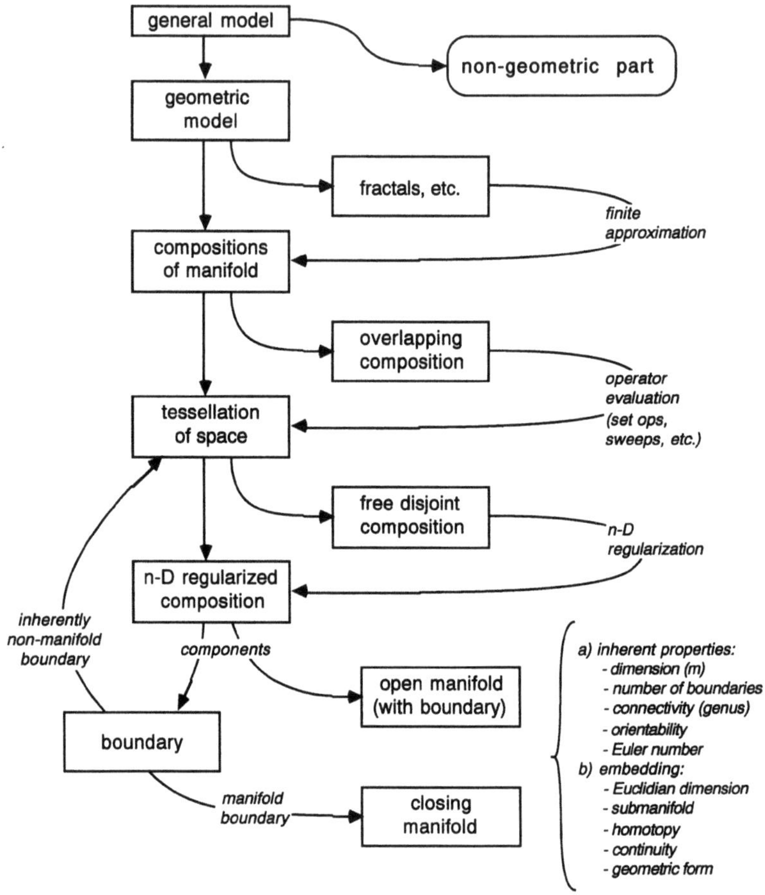

FIGURE 10.21. Taxonomy as a diagram.

operation, which can be applied in different dimensions and separately in the relative topology of a model's each component, though this seems not have been realized yet in any modeler. For example, a 3-dimensionally regularized solid model may have a clean 2-manifold boundary surface, the decomposition of which still contains strut edges dangling on its faces, and thus is not regularized 2-dimensionally (see Figure 10.8). In this respect the taxonomy is recursive.

In so-called "manifold solid" modelers the boundary of a regularized model is usually assumed to be a closing manifold - actually a more accurate name would thus be "solid with manifold boundary". However, a manifold's boundary may also have some inherently non-manifold elements. These may be understood as degenerated regions where two or more manifold boundaries make contact [19], and are thus sometimes called pseudomanifolds [27]. This interpretation is ambiguous, however, and therefore it is more natural to consider coincidental elements as just one element, as is indeed done in a couple of works [33, 16].

Each manifold component of a tessellation can be characterized separately according to its dimension, number of its boundary components, its connectivity number (genus), orientability, Euler characteristic, etc. Extensive studies of this have been done in mathematics (especially for surfaces), which will not be repeated here. Note however, that generally there are more degrees of freedom with open manifolds, than with closed ones. For them there are no classification theorems as simple as that for closing 2-manifolds (see subsection above on the Euler formula). Thus the Euler number is useful for closed 2-manifolds only — it is trivial for lower and ambiguous for higher dimensions, as it is for 2-manifolds with boundary.

10.4 Mappings to Geometry

A topology may be mapped into Euclidian space not only in various geometric shapes, but also in a number of essentially different ways topologically. In this section some problems associated with mappings will be pointed out. Many of these issues are actually still open research subjects.

A geometric object should be topologically similar to its abstract representation, in order to be consistent. The two mappings of interest here are embeddings and immersions. Continuity of mapping is essential to preserve the topological properties. It is very misleading to speak about "discontinuous embeddings" [14]. One topological entity cannot be in two or more geometrically apart places. In such cases the entity should be explicitly split with topological surgery, perhaps resulting in piecewise embeddings.

Following the previous classification scheme, a first distinction would rule out all "wild" mappings which, though continuous and perhaps even differentiable, fold the geometric shape into infinitely small details. Theoretically a homeomorphic mapping is known to be tame if it can be extended to the whole space. However, practical characterization of these mappings is still an open issue.

Whether all embeddings of an abstract model are topologically alike or even possible, depends on the dimensionality, structure and completeness of the model, relative to the embedding space. It is obvious that a tessellation homeomorphic to an open simplex can be embedded onto any manifold of the same dimension (which may be either an Euclidian space or a submanifold in a higher-dimensional one), and that the dimension of any embedding space should be at least that of the topological complex (Figure 10.22). Also it is known that a closing n-manifold cannot be embedded in Euclidian space of the same dimension, and if embedded in Euclidian n+1-space it must be orientable, then becoming the separating boundary of disjoint regions (open n+1-manifolds). For example, the Klein bottle is a non-orientable 2-manifold and thus cannot be embedded in E_3.

10.4.0.6 Homotopy groups

A topology may be embedded in various ways which are not *homotopic*, i.e. deformable to each other through a continuous family of homeomorphisms in the same space. For example, a closed curve or a torus can take the form of various knots, which are indeed homeomorphic but not deformable to a simple loop. Deformability divides a set of homeomorphic objects into smaller subsets, called *homotopy groups.*

If the topological model is not a complete tessellation of the Euclidian space, its embedding is always homotopically ambiguous. For example, the neighborhood of a vertex inside a face, with two strut edges emanating from it, can be embedded in E_3 with the two edges either on the same side or on different sides of the face (Figure 10.23). Both are homeomorphic (and so are their complements in 3-D space) but not homotopic. Another

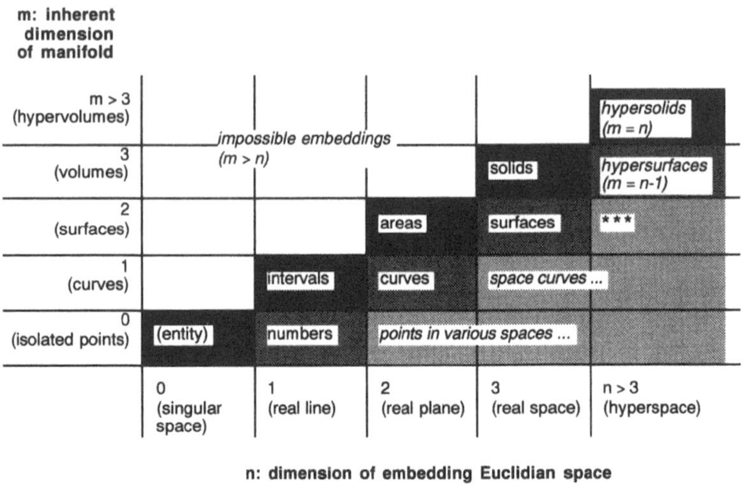

FIGURE 10.22. Dimensions of objects and spaces

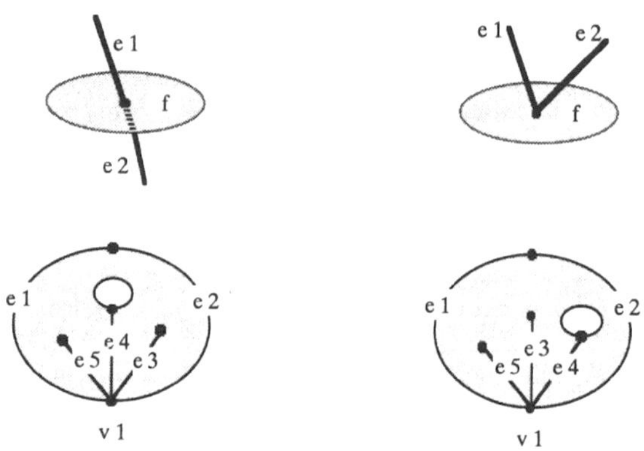

FIGURE 10.23. Examples of homotopy groups.

example is the ordering of strut edges emanating from a common vertex on a face. Two homotopy groups are shown, which cannot be distinguished by general adjacency relations alone. Usually this kind of situations are solved by adjacency ordering of edges along a face loop.

What information is generally needed to distinguish between homotopy groups is yet unclear. As shown by examples, incidences and adjacencies between topological elements are not enough, neither is the topological form (genus etc.) of the model's complement always sufficient. Local topological ordering of elements, like the tree structure of surfaces at a vertex by Weiler [31], seems to be necessary, but a comprehensive study of the issue is yet to be done.

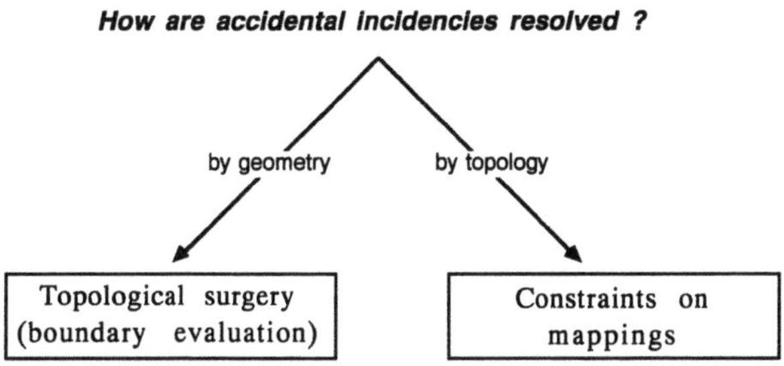

FIGURE 10.24. Complementary uses of topological models.

10.4.0.7 Surgery imposed by inconsistent mappings

An immersion may map a model into a non-homeomorphic image, deformed such that it touches or penetrates itself. The original topology is then violated by additional incidences. In the taxonomy this is a special case of overlapping combinations, and could be handled with topological surgery, splitting the interacting elements into disjoint pieces and possibly unifying coincident elements.

The problem arising here is, that often a topological model is assumed to be correct regardless of its associated geometry. In that case the overlapping would be merely accidental, and should be avoided by constraining the allowable embeddings of the model.

Figure 10.24 illustrates the two complementary strategies, whether geometric relations or topological data structures are considered the primary information from which the other is derived [26]. Unfortunately this distinction is often not made clear in modeling systems.

10.5 Conclusions

The main aim of this paper has been the clarification of basic concepts and terminology, giving a mathematically sound framework for a taxonomy where various types of geometric objects or modeling domains can be characterized.

Useful novelties in the mathematical review are the definitions of relative boundary and relative regularization. With these concepts the various mixed-dimensional models, including inherently non-manifold ones, can be more accurately characterized.

The leading theme throughout the paper has been two basic issues. First, manifolds are the basic building blocks, out of which all models of practical interest are composed. Thus tessellations of space into manifold compositions (SGC) are considered a basic framework, on which all others can be assessed. Second, the embeddings are important and have to be consistent with topological models. Depending on whether combinatorial structure or geometric shape is considered primary information, the consistency is forced by constraints or by implied surgery at accidental incidencies.

The meaningfulness of Euler operators on non-regularized models is questioned. Since the Euler characteristic of such a model cannot be used to classify it topologically, is there any reason to maintain the Euler equation? Instead, the basic surgery operations are suggested as a more general and intuitively easily understandable approach.

Similar question is posed concerning the term "non-manifold", the rejection of which is strongly proposed. The term is a negative complementary expression, lumping together too many possibilities — even pure manifold models. It is like using "non-integers" to mean rational, real and imaginary numbers all together. The proposed taxonomy helps to distinguish between various alternatives called by the same name.

A subject pointed out for future research is the study of homotopy groups and their representations with topological data structures.

Acknowledgements:

The idea of a topological taxonomy first arose during informal group meetings at the IFIP WG 5.2 workshop in Rensselaerville, NY[28], where the ambiguity of term "non-manifold" was recognized. Later, the paper has strongly been affected by discussions with Mike O'Connor, Jarek Rossignac and Neil Stewart. Expanding these ideas to a full paper was made possible in part by my sabbatical leave at the George Washington University, Washington, DC.

10.6 References

[1] S.R. Ala. Design Methodology of Boundary Data Structures. In J. Rossignac and J. Turner, editors, *Proc. of Symposium on Solid Modeling Foundations*, ACM, pp.13–23, 1991.

[2] F. Arbab. Set Models and Boolean Operations for Solids and Assemblies. *IEEE Computer Graphics and Applications*, 10(6), November 1990.

[3] B. Baumgart. *Geometric Modeling for Computer Vision.* Stanford Artificial Intelligence Report No. CS-463, October 1974.

[4] Boothby. *An Introduction to Differentiable Manifolds and Riemannian Geometry.* Academic Press, 1975.

[5] J. Rossignac, editor. CAD Special Issue, Beyond Solid Modeling. *CAD*, 23(1), Janunary/February 1991.

[6] H.M.S. Coxeter. *Introduction to Geometry* John Wiley, 1961.

[7] G. Crocker and W. Reinke. An Editable Nonmanifold Boundary Representation. *IEEE Computer Graphics and Applications*, 11(2):39–51, March 1991.

[8] J-F. Dufourd. An OBJ3 Functional Specification for boundary Representation. In J. Rossignac and J. Turner, editors, *Proc. of Symposium on Solid Modeling Foundations*, ACM, pp.61–72, 1991.

[9] V. Ferrucci and A. Paoluzzi. Extrusion and Boundary Evaluation for Multidimensional Polyhedra, CAD, (Special Issue: Beyond Solid Modeling), 23(1):40–50, January/February 1991.

[10] E.L. Gursoz, Y. Choi and F. Prinz. Boolean Set Operations on Non-Manifold Boundary Representation Objects. CAD, (Special Issue: Beyond Solid Modeling). 23(1):33–39, January/February 1991.

[11] M Henle. *A Combinatorial Introduction to Topology.* W.H.Freeman and Company, 1979.

[12] J. Hocking and G. Young. *Topology.* Addison-Wesley, 1961.

[13] Chr. Hoffman. *Geometric and Solid Modeling.* Morgan Kaufmann, 1989.

[14] P. Lienhardt. Topological Models for Boundary Representation: A Comparison with N-Dimensional Generalized Maps. CAD, (Special Issue: Beyond Solid Modeling), 23(1):59–82, January/February 1991.

[15] E.A. Lord and C.B. Wilson. The Mathematical Description of Shape and Form. Ellis Horwood, 1986.

[16] Y. Luo and G. Lukács. A Boundary Representation for Form Features and Non-Manifold Solid Objects. In J. Rossignac and J. Turner, editors, *Proc. of Symposium on Solid Modeling Foundations*, ACM Press, pp.45–60, 1991.

[17] A. Maulik. An Efficient Intersection Algorithm for Polyhedral Cellular Decompositions. In J. Rossignac and J. Turner, editors, *Proc. of Symposium on Solid Modeling Foundations*, ACM, pp.109-118, 1991.

[18] M. Mortenson, Geometric Modeling. John Wiley, 1985.

[19] S. Murabata and M. Higashi. Non-Manifold Geometric Modeling for Set Operations and Surface Operations. In J. Turner, Pegna and M. Wozny, editors, *Product Modeling for Computer-Aided Design*, North Holland (in press).

[20] M. Mäntylä. *An Introduction to Solid Modeling.* Computer Science Press, 1988.

[21] A. Requicha. Representation of Rigid Solid Objects. Tech.Memo.No.29, Production Automation Project, Univ. of Rochester, 1977/78. Also in J. Encarnaçao, editor, *Computer-Aided Design,* Lecture Notes in Computer Science No. 89, Springer-Verlag, 2–78, 1980.

[22] A. Requicha, *Mathematical models of Rigid Solid Objects.* Tech.Memo.No.28, Production Automation Project, Univ. of Rochester, 1977.

[23] J. Rossignac and A. Requicha. Constructive Non-Regularized Geometry. CAD, (Special Issue: Beyond Solid Modeling), 23(1):21–32, January/February 1991.

[24] J. Rossignac and J. Turner, editors. *Proc. Symposium on Solid Modeling Foundations (Austin, Texas, June 5–7, 1991).* ACM, 1991.

[25] J. Rossignac and M. O'Connor. SGC: A Dimension-Independent model for Pointsets with Internal Structures and Incomplete Boundaries. In M. Wozny, J. Turner and K. Preiss, editors, *Geometric Modeling for Product Engineering*, North-Holland, 1990.

[26] T. Takala, The Role of Multidimensional Models in CAD Systems. In M. Wozny, H. McLaughlin and J. Encarnaçao, editors, *Geometric modeling for CAD Application*, North-Holland, 115–121, 1988.

[27] K. Tang and T. Woo, Algorithmic Aspects of Alternating Sum of Volumes. Part 1: Data Structure and Difference Operation. *CAD*, 23(5):357–366, June 1991.

[28] J. Turner, J. Pegna and M. Wozny, editors. *Product Modeling for Computer-Aided Design (papers from the IFIP WG 5.2 Working Conference in Rensselaerville, USA, 1990)*, North-Holland, (in press).

[29] M. Waltari *The Egyptian.* G.P.Putnam's Sons, 1949.

[30] K. Weiler, *Topological Structures for Geometric Modeling.* PhD Dissertation, Rensselaer Polytechnic Institute, USA, August 1986.

[31] K. Weiler, Boundary Graph Operators for Non-Manifold Geometric Boundary Modeling Topology Representations. In H. McLaughlin and J. Encarnaçao, editors, *Geometric modeling for CAD Application*, North-Holland, 37–66, 1988.

[32] K. Weiler, The Radial Edge Structure: A Topological Representation for Non-Manifold Geometric Boundary Modeling. In H. McLaughlin and J. Encarnaçao, editors, *Geometric modeling for CAD Application*, North-Holland, 3–36, 1988.

[33] S-T. Wu. Considerations about a Minimal Set of Non-Manifold Operators. In J. Turner, Pegna and M. Wozny, editors, *Product Modeling for Computer-Aided Design*, North Holland (in press).

[34] Y. Yamaguchi and F. Kimura, Boundary Neighborhood Representation for Non-Manifold Topology. In J. Turner, Pegna and M. Wozny, editors, *Product Modeling for Computer-Aided Design*, North Holland (in press).

11 Multilevel Generation of Fractal Images Using the Associated Markov Process

Marc Coevoet

ABSTRACT Using a set of contractive affine transformations, one can describe an image with intricate detail. However, the process of generating fractals with these transformations is an expensive operation. In this paper, it is shown that it can become significantly less expensive, when looking at the iteration of a set of affine transformations as a Markov process. Using the special structure of the state transition matrix for the solution of the large set of equations, some multilevel scheme-algorithms emerge.

11.1 Introduction

Complex natural objects are very hard to describe in Euclidean geometric. How do you model clouds, mountains, coastlines by using points, lines and circles? Mandelbrot studied the structure of natural objects and came to what he called "fractal geometry". Objects constructed with this geometry can represent the shape of natural objects.

Fractals occur when studying (complex) dynamical systems. A dynamical system is any process evolving in time. Such a dynamical system is e.g. population growth, the weather, the economy, a chemical reaction .... Hutchinson [7] and Barnsley [1] showed that even simple mathematical expressions such as sets of transformations, iteration of polynomials, etc. yield fractals. Fractals owe their success mainly to the fact that they are easily generated and they simulate natural phenomena. Fractal images can be generated which are almost indistinguishable from natural images; natural scenes, containing mountain landscapes, planets, clouds or ferns have already been generated as well as pictures of a Brazilian girl.

There is however a major disadvantage in generating fractals with a set of affine transformations as Barnsley proposes: the algorithms to render an image are rather slow. If the process of generating is viewed as a stochastic process, more exactly a Markov chain, the image can also be seen as the stationary distribution of that stochastic process. The state transition matrix is used to calculate the stationary distribution. This leads to a (large) set of linear equations. Because of the special structure of the state transition matrix, a nested iteration scheme is applied. It will also be shown in this paper why the classical multi-grid method is unnecessary.

11.2 Iterated function systems

The term 'Iterated function system' was introduced by Michael Barnsley. Earlier, in [7], Hutchinson proved a theorem about self-similar sets.

11.2.1 Hutchinson's Theorem

DEFINITION 11.2.1 (INVARIANT SET). *$A \subset \mathcal{R}^d$ is invariant with respect to a set Δ of contraction maps $\{S_i \mid i = 1, \ldots, N\}$ if*

$$A = \bigcup_{i=1}^{N} S_i(A).$$

THEOREM 11.2.2 (HUTCHINSON'S THEOREM).
Let a set of contraction mappings $\Delta = \{S_i \mid i = 1, \ldots, N\}$, $S_i : \mathcal{R}^d \rightarrow \mathcal{R}^d$, be given.

(i) There is a unique closed bounded and compact *subset K which is invariant with respect to Δ:*

$$K = \bigcup_{i=1}^{N} S_i(K).$$

(ii) K is the closure of the set of fixed points of the mappings $S_{i_1 i_2 \ldots i_k}$ where the $S_{i_1 i_2 \ldots i_k}$ denote the products $S_{i_1} S_{i_2} \ldots S_{i_k}$ for any random sequence $i_1 i_2 \ldots i_k$ of length k with $i_j \in \{1..N\}$.

(iii) Let $\Omega = \prod_{k=1}^{\infty} \{1, \ldots, N\}$. Ω is the set of all mappings of $\mathcal{N}$ into the finite set $\{1, \ldots, N\}$ equipped with the product topology. For an element $\omega \in \Omega$, $\omega = (i_1, i_2, \ldots, i_k, \ldots)$ and any $x \in \mathcal{R}^d$

$$\lim_{k \rightarrow \infty} S_{i_1} S_{i_2} \ldots S_{i_k} (x)$$

exists and is independent of x. If we denote $\pi(\omega)$ this limiting value, we have a map

$$\pi : \Omega \rightarrow \mathcal{R}^d.$$

(iv) The map $\pi : \Omega \rightarrow \mathcal{R}^d$ is continuous and has image ($F[\Delta]$ is the fractal object associated to the set of contraction maps Δ.)

$$F[\Delta] := Im(\pi) = \bigcup_{k=1}^{\infty} \bigcup_{i_1=1}^{N} \cdots \bigcup_{i_k=1}^{N} \{F_{i_1 i_2 \ldots i_k} \mid F_{i_1 i_2 \ldots i_k}\, fixed\, point\, of\, S_{i_1} S_{i_2} \ldots S_{i_k}\}.$$

(v) Let $A \subset \mathcal{R}^d$ be any *non-empty bounded set, then*

$$\lim_{k \rightarrow \infty} \bigcup_{i_1=1}^{N} \cdots \bigcup_{i_k=1}^{N} S_{i_1} S_{i_2} \ldots S_{i_k} (A) = K$$

in the Hausdorff metric.
For a proof of this theorem, see [7].

11.2.2 Generating fractals

Hutchinson's Theorem suggests several ways to approximate a fractal.

- *Fixed point method.* It suffices to calculate the set of fixed points of the products $S_{i_1 i_2 \ldots i_k}$ up to a certain level k where enough points are plotted.
- *Recursive rendering.* Evaluate the expression mentioned in (v) up to a certain k, where the diameter of the transformed set becomes smaller than the pixel size (the diameter will decrease because of the contractivity of the affine transformations). Whatever the set A (in (v)) looks like, the same image will emerge, as shown in Figure 11.1.

FIGURE 11.1. The sierpinski, with different starting sets A.

Both the fixed point and recursive method are expensive as far as memory and timeare concerned: in the calculation of the fractal, a recursion stack is involved. In each step of the recursion, the number of objects involved in the computation grows exponentially with the depth of the recursion; see for example the fern (levels 0, 1, 2, ... ,8) in Figure 11.2.

- *Random iteration algorithm (RIA).* Start from any point in K, select at random one of the transformations, apply it to the point, plot the result. Repeat this procedure with this new point as input.

Up to now, we supposed that all the transformations had the same probability. To fill in different parts of the image with a certain rate, you can assign also a probability to a transformation. When generating black and white images, you will always get the same image, whatever probabilities you assign, as long as you do not exclude one of the transformations by assigning it *zero* probability. There is no difference whether the random dancing point hits a pixel one or several times. With gray-value or coloured images, the visits the random dancing point pays to a pixel will be cumulated, in this way defining a gray-scale or colour to a pixel, after *many* iterations. This pixel value is proportional to the chance the random dancing point will visit the pixel. To display the image on the computer screen, the value is rescaled according to the gray-spectrum or colour-spectrum of the computer.

Let X be the random process defined by the affine transformations, the chance of visiting pixel p is

$$P[X = p] \approx \frac{pixel\ value\ of\ p}{number\ of\ iterations}.$$

If we assign to every point of $\mathcal{R}^d$ a visiting probability, this leads to a probability measure on $\mathcal{R}^d$. Of course we have

$$\frac{pixel\ value\ of\ p}{number\ of\ iterations} \geq 0,$$

$$\sum_{p \in \mathcal{R}^d} \frac{pixel\ value\ of\ p}{number\ of\ iterations} = 1.$$

For a black and white generation with the RIA, see Figure 11.3.

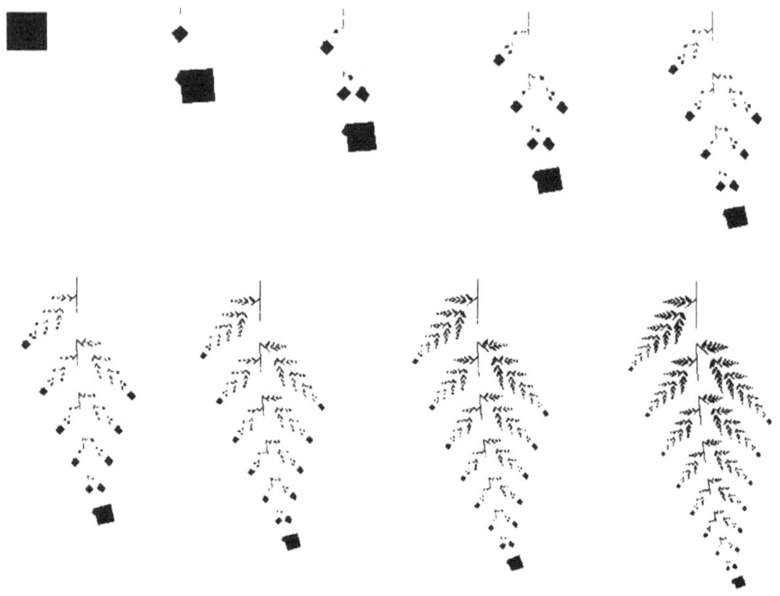

FIGURE 11.2. Levels 0, 1, 2, ... ,8 of the recursive Fern ($N = 4$).

FIGURE 11.3. RIA for the Fern for increasing numbers of iterations

Before citing the theorem that proves convergence of the RIA, let us have the definition of *Iterated Function System.*

DEFINITION 11.2.3 (IFS).

$$\begin{aligned} IFS &= \{\Delta \times p \mid \Delta \text{ set of affine transformations, } p \text{ probability measure on } \Delta\} \\ &= \{(S_i, p_i) \mid S_i \text{ affine transformation, } p_i \text{ associated probability}(i = 1, \ldots, N)\} \end{aligned}$$

THEOREM 11.2.4 (THE COLLAGE THEOREM). *Let $\mathcal{X}$ be a complete metric space. Let $\mathcal{H}(\mathcal{X})$ denote the space of compact subsets of $\mathcal{X}$, other than $\emptyset$. Let $K' \in \mathcal{H}(\mathcal{X})$ be given, and let $\epsilon \geq 0$ be given. Choose a set of transformations $\Delta = S_1, S_2, \ldots, S_N$ with contractivity $0 \leq s \leq 1$, so that*

$$d_{\mathcal{H}(\mathcal{X})}(K', \bigcup_{i=1}^{N} S_i(K')) \leq \epsilon \ (in\ the\ Hausdorff\ metric)$$

Then

$$d_{\mathcal{H}(\mathcal{X})}(K', K) \leq \frac{\epsilon}{1-s}$$

where K is the set determined by Δ. Equivalently,

$$d_{\mathcal{H}(\mathcal{X})}(K', K) \leq \frac{d_{\mathcal{H}(\mathcal{X})}(K', \bigcup_{i=1}^{N} S_i(K'))}{1-s} \quad \forall K' \in \mathcal{H}(\mathcal{X}).$$

This is an important property used in image compression: if one can identify parts of an image as transformed copies of the whole image, the affine transformations which generate the image can be calculated yielding a high compression factor. Some pictures such as the Brazilian girl on the cover of [1], were hand-coded. Automatic techniques already existing are based on moment matching, padé approximants, iterative improvement via gradient or simulated annealing ...etc. These methods are based on the application of the Collage Theorem. Another way of fractal-like compression is the one elaborated in Arnaud Jacquin's PhD ([8]). Jacquin does not map the whole image to transformed parts, but defines a partition. Each element of the partition is then mapped to another one, in this way defining a Markov operator.

THEOREM 11.2.5 (ELTON'S ERGODIC THEOREM). *Let $\mathcal{X}$ be a compact metric space. Let*

$$\Delta = \{(S_i, p_i) \mid i = 1, \ldots, N\}$$

be an IFS. Let $\{x_n\}_{n=0}^{\infty}$ denote an orbit of the RIA. That is

$$x_n = S_{\omega_n} S_{\omega_{n-1}} \ldots S_{\omega_0}(x_0) \quad (\omega \in \Omega),$$

where the map S_i is chosen at random from $\{S_1, S_2, \ldots, S_N\}$ with probability p_i. Let μ be the measure for the IFS. Then, with probability one

$$\lim_{n\to\infty} \frac{1}{n+1} \sum_{k=0}^{n} f(x_k) = \int_{\mathcal{X}} f(x) d\mu(x)$$

for all continuous functions

$$f : \mathcal{X} \to \mathcal{R}\ and\, for\, all\, x_0.$$

FIGURE 11.4. Some 2D fractals

COROLLARY 11.2.6. *Let B be a Borel subset of $\mathcal{X}$ and let μ(boundary of B) $= 0$. Let $\mathcal{N}(B,n) = \#\{x_i\}_{i=0}^{n} \cap B$. Then , with probability one,*

$$\mu(B) = \lim_{n\to\infty} \frac{\mathcal{N}(B,n)}{n+1},$$

for all starting points x_0. That is, the 'mass' of B is the proportion of iteration steps, when running the RIA, which produces points in B.

Elton's Ergodic Theorem proves the existence of a unique measure. An extension to this theorem can be found in [2]: the existence of a unique, invariant measure is demonstrated, without the affine transformations being contractive and without the space $\mathcal{X}$ being compact.

11.3 RIA as Markov Process

Recall the Random Iteration Algorithm: having arrived at a point x in $\mathcal{R}^d$, there are N possible next points

$$S_i(x) = y_i,$$

where the process can arrive. Each next point y_i has probability p_i, independent of the arrival time at point x, the process only depends on the most recent known value, not on preceding ones. Looking in this way to the RIA leads to the definition of the RIA as a homogeneous Markov process.

DEFINITION 11.3.1 (RIA AS MARKOV PROCESS).
$x_t = S_{i_{t-1}} S_{i_{t-2}} \dots S_{i_1} S_{i_0}(x_0)$
$P[X(t+1) = S_{i_t}(x_t) = x_{t+1} \mid X(0) = x_0; X(1) = x_1; \dots, X(t) = x_t]$

$$= P[X(t+1) = S_{i_t}(x_t) = x_{t+1} \mid X(t) = x_t]$$
$$= p_i$$

Since the aim is to produce images on a discrete computer screen, take a finite number of states. A state will then correspond to a pixel. We can now define the homogeneous Markov chain.

DEFINITION 11.3.2 (RIA AS MARKOV CHAIN). *Let T be the mapping function of the continuous space into the discrete space, T corresponds to truncate.*

$$P[X(t+1) = T(S_{i_t}(x_t)) = T(x_{t+1}) \mid X(0) = T(x_0); X(1) = T(x_1); \ldots, X(t) = T(x_t)]$$
$$= P[X(t+1) = T(S_{i_t}(x_t)) = T(x_{t+1}) \mid X(t) = T(x_t)]$$
$$= p_i$$

Let $X_K = \{x_1, x_2, \ldots, x_{K^d}\} \subset \mathcal{X}$ be a discrete set, consisting of the truncated points from $\mathcal{X}$, $\#X_K = K^d$. We know the transition functions (the affine transformations) and the transition probabilities, so we can define the state transition matrix. Let us take an IFS on a one-dimensional grid of points $\{0, \ldots, K\}$. So

$$IFS = \{(a_i x + b_i, p_i) \mid i = 1, \ldots, N\}.$$

The transition for a point $x_j \in X_K$ will be $x_l := T(a_i x_j + b_i)$ with probability p_i. Hence the element p_{jl} of the state transition matrix P will get a contribution p_i from the ith affine transformation. Clearly, P is a sparse matrix, in which each affine transformation is represented by a line of non-zero elements.

Elton's Ergodic Theorem and its extensions yield that the fractal (and thus its Markov chain) has a stationary distribution μ. In [4] the equivalence between the irreducibility of P and the existence of its unique invariant measure μ is proved. It can be solved by the set of linear equations $\mu P = \mu$ with the additional constraint $\sum \mu_i = 1$. This is a set of equations which is as large as the number of pixels in the image. In a d-dimensional space, taking a d-dimensional cube with side K, the number of equations is K^d.

The state transition matrix on a $(2n)^d$-grid being a refinement of the matrix on an n^d-grid, we can take advantage of this structure by applying a multi-grid iteration scheme. It will be shown later on that the fast contraction of the affine transformations can be used in the prolongation efficiently, such that the iterations need not backtrack to coarser grids and a plain nested iteration suffices.

11.4 Multi-grid method

Before explaining multi-grid, let us start with the two-grid method; this will make things easier.

11.4.1 Two-grid method

The idea behind the two-grid method is to solve a $(n * n)$ set of equations using the solution of the $(m * m)$ $(m < n)$ set. Let the index l denote the level we are working at. The steps involved in two-grid iteration for solving $L_l \mu_l = f_l$ (see also Figure 11.5) are as follows.

μ_l^j given iterate.
$\mu_l := \mathcal{S}(\mu_l, f_l)$ (smoothing)
$d_l := L_l \mu_l - f_l$ (calculation of defect)
$d_{l-1} := r\, d_l$ (restriction of defect)
$v_{l-1} := L_{l-1}^{-1}\, d_{l-1}$ (solution of coarse-grid equation)
$\mu_l^{j+1} := \mu_l - p v_{l-1}$ (correction of μ_l)

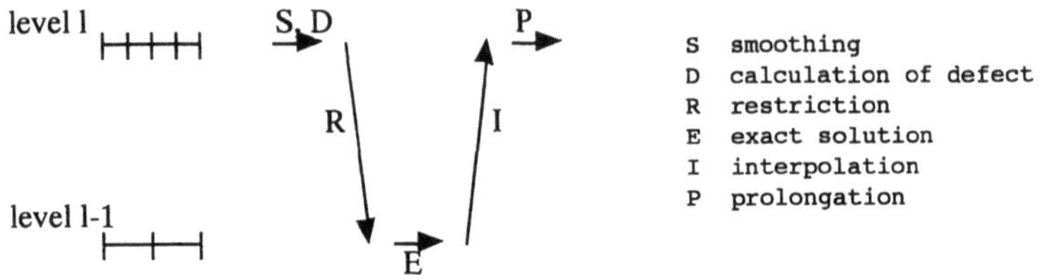

FIGURE 11.5. The two-grid method

The coarse grid correction can be condensed to

$$\mu_l^{j+1} := \mu_l - pL_{l-1}^{-1}r(L_l\mu_l - f_l)$$

We will not focus on the individual steps of the algorithm and would rather refer to [5].

11.4.2 Multi-grid method

In the two-grid method, an approximation of the solution of level l is found using the exact solution of level $l-1$. When the exact solution is recursively deferred to a lower level, the multi-grid method is obtained. Here is a PASCAL version of the algorithm (see also Figure 11.6).

```
procedure MGM(l, μ, f);
var
  l, j: integer;
  μ, f, v, d: array;
begin
  if (l = 0) then
    μ := L_0^{-1} * f
  else
    begin
      μ := S(μ, f);
      d := r(L_l μ - f);
      v := 0;
      for j := 1 to γ do MGM(l - 1, v, d);
      μ := μ - pv;
    end
end;
```

The least efficient step of the classic multi-grid method is the interpolation step. A frequently used interpolation step is the piecewise linear one: (h is the mesh size of the fine grid)

$$\mu_l(x) = \begin{cases} \mu_{l-1}(x) & if\ x\ \in\ coarse\ grid \\ \frac{\mu_{l-1}(x-h)+\mu_{l-1}(x+h)}{2} & otherwise \end{cases}$$

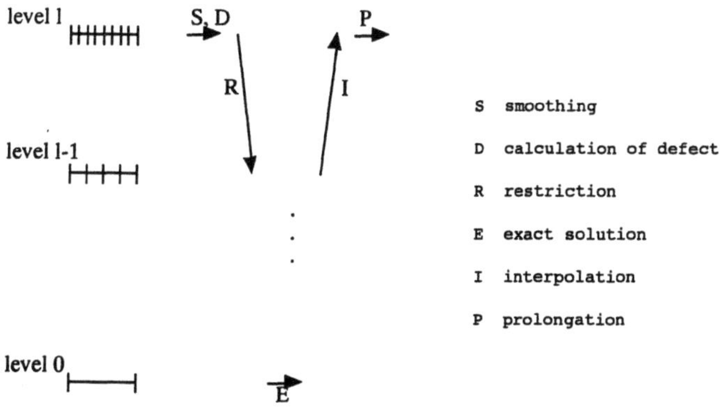

FIGURE 11.6. The multi-grid method

This is a very unsuccessful operation when, e.g., a chequerboard pattern is to be generated!

Some of the classical smoothing steps are the Jacobi-iteration, the Gauss-Seidel iteration and their SOR-variants. In the multi-grid implementation the Jacobi-iteration has been used.

In the next section, some weaknesses of the multi-grid method will be avoided by using specific properties of the problem.

11.5 Nested Iteration

The nested iteration consists of interpolation and iteration steps. The interpolation step consists of a transition from level l to level $l+1$, where the number of unknowns is respectively 2^l and 2^{l+1}. Most likely, the interpolation step does not destroy the structure already calculated, however, most of the times it is an averaging operation. The iteration step calculates a better approximation on the same level l.

- *Interpolation of coarse to fine grid.* Recall the fact that the state transition matrix expresses some affine transformations. These affine transformations could be used to transform the coarse grid into the fine grid. This corresponds to defining a state transition matrix from the coarse grid to the fine grid, not only embodying affine transformations, but also an interpolation between succeeding levels. Let us define $J_{l-1\to l}$ to be an $(m \times n)$ $(m < n)$ transition matrix. The interpolation step can be defined as:

$$\mu_l' = \mu_{l-1}\, J_{l-1\to l}.$$

- *Iteration at a certain level.* The iteration step corresponds to the application of the affine transformations.

$$\mu_l' = \mu_l\, P_l.$$

- *Summary.* The equation at level 0 is

$$x = 1.$$

A cascade of interpolation and iteration steps gives:

$$\mu_l = \mu_0 \; J_{0\to 1} \; P_1^k \; \ldots \; P_{l-1}^k \; J_{l-1\to l} \; P_l^k. \tag{11.1}$$

where k is the number of iteration steps, $\mu_0 = [1]$

- *Markov chains.* When P is the state transition matrix of a homogeneous irreducible Markov chain, the stationary distribution μ^∞ can be calculated as

$$\mu^\infty = \lim_{n\to\infty} \mu^0 P^n, \tag{11.2}$$

independently of the initial distribution μ^0.

Formula (11.1) is, in fact, an approximation of (11.2) calculated on several levels.

11.5.1 Why not Multi-grid

We had first implemented the classical multi-grid method for a 1D-fractal. However, looking for a better interpolation step (instead of the *naive* averaging step), we adapted the MGM using the affine transformations to transform from the coarse grid into the fine grid. The smoothing step – originally a Jacobi iteration – was also replaced by the application of the affine transformations. In the context of Markov chains, this corresponds to the calculation of (11.2) on a cascade of grids. Experiments have shown that the convergence of the Jacobi iteration is *not* faster than the application of the state transition matrix, they both tend to the solution with the same speed. The convergence is related to the contractivity of the affine transformations.

With the adapted interpolation and smoothing steps, the restriction does not improve the solution. Why? Looking at the condensed 2-grid iteration (see [5]):

$$\mu_l^{j+1} := \mu_l - pL_{l-1}^{-1} r(L_l \mu_l - f_l)$$

Defined in terms of the adapted steps:

Defect:	d_l	$:= \mu_l P_l - \mu_l$
Restriction:	r_{l-1}	$:= d_l J_{l\to l-1}$
Exact solution:	μ_∞	$:= \lim_{n\to\infty} (\mu_l P_l - \mu_l) J_{l\to l-1} P_{l-1}^n$
		$= \lim_{n\to\infty} v_{l-1} P_{l-1}^n$

Suppose we always work with positive vectors. Since the exact solution is calculated as the invariant distribution of an irreducible Markov chain using the state transition matrix (which expresses affine transformations) it will always be the same. When the affine transformations are applied a number of times through the product $v_{l-1} P_{l-1}^n$, the mass of v_{l-1} will at any level be spred in a similar structure. This means that the calculated correction always is proportional to the exact solution. Backtracking makes no sense; instead of arriving at level 0 with a *mass* δ (the sum of the elements of the vector, $0 \le \delta \le 1$) and spreading this small mass in a structure proportional to the exact solution, we can as well start with mass 1 and apply nested iteration without backtracking.

11.6 Implementation

11.6.1 Sparse matrix implementation

For each affine transformation, there is a vector of states with at index i the image of x_i. The images of the pixels are calculated using a DDA-algorithm (Digital Differential Analyzer, see [6]), no multiplications are involved, only additions. The set of vectors implements the *state transition matrix*. It is zero everywhere, except at the $(j,\ l)$th element $(T(S_i(x_j)) = l)$, where the value is p_i.

For the 1D-IFS $= \{(\frac{1}{2}x, p_1), (\frac{1}{2}x + \frac{1}{2}, p_2) \mid p_2 = 1 - p_1\}$ where the states are equally spaced in the interval [0, 1], the state transition matrix looks like

$$\begin{pmatrix} p_1 & 0 & & 0 & p_2 & 0 & & 0 & 0 \\ 0 & p_1 & & 0 & 0 & p_2 & & 0 & 0 \\ 0 & p_1 & & 0 & 0 & p_2 & & 0 & 0 \\ 0 & 0 & & 0 & 0 & 0 & & 0 & 0 \\ \vdots & \vdots & \ddots & \vdots & \vdots & \vdots & \ddots & \vdots & \vdots \\ 0 & 0 & & p_1 & 0 & 0 & & p_2 & 0 \\ 0 & 0 & & p_1 & 0 & 0 & & p_2 & 0 \\ 0 & 0 & & 0 & p_1 & 0 & & 0 & p_2 \end{pmatrix}.$$

When $p_1 = p_2 = \frac{1}{2}$, this yields the uniform distribution.

This method was also implemented for fractals in a 2D space. It generates a gray-value picture of size 256×256, which involves a set of $256 * 256 = 65536$ linear equations. The pixels of the image are ordered lexicographically, i.e. let $(n \times n)$ be the size of the 2-dimensional grid, then
$\mu[0 \ldots n-1]$ contains the first line of the image,
$\mu[n \ldots 2n-1]$ contains the second line of the image, etc

On a MacIIx, the code uses about 10 seconds to generate a 256×256 pixel image of the fern with 256 gray-values using integer arithmetic only. See some of the generated 2-dimensional gray-value pictures at the end of this chapter. You can find their coefficients in the Appendix (section 11.8).

For a discussion on the convergence of the gray-values of the pixels, see the section on convergence.

11.6.2 Coarse & Fine grid

The fine grid is generated from the coarse grid using the uniform mesh coarsening by factor 2. In 1D, this corresponds to doubling the number of grid points, in 2D, the factor is 4, in $\mathcal{R}^d$, the factor is 2^d.
Example

$$1D - IFS = \{(\frac{1}{2}\,x,\ \frac{3}{5}),\ (\frac{1}{2}\,x + \frac{1}{2},\ \frac{2}{5})\}$$

See Figure 11.7 for the generation of the stationary distribution starting from level 0, where $\mu_0 = [1]$.

11.6.3 Efficiency

Let $\mathcal{R}^d$ be the space in which fractals are generated. Take a d-dimensional cube of side $n = 2^m$. The IFS has t transformations. When applying the Nested Iteration method, s is the number of iteration steps. The mesh sizes of the coarse and the fine grid are related

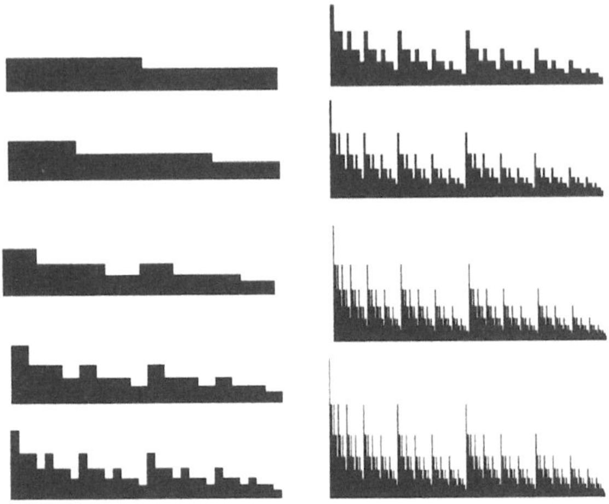

FIGURE 11.7. Several levels of 1D rendering.

by a factor 2. One flat iteration at level l takes $t * 2^{d*l}$ flops. This is also the number of flops needed to interpolate from level $l-1$ to level l. To generate the image from level 0 to level m, the number of flops is

$$\sum_{i=1}^{m}(1+s)*t*2^{d*i} = \frac{2^d}{2^d-1}(1+s)*t*(2^{m*d}-1) = O(n).$$

The computational work is therefore proportional to the number of unknowns in the linear system.

The number of flops required to calculate an approximation of the image by using only interpolation steps ($s=0$) between level 0 and level m is as follows:

$$\frac{2^d}{2^d-1}t*(2^{m*d}-1) = 2^{m*d}\frac{2^d}{2^d-1}t - \frac{2^d}{2^d-1}t.$$

This number is close to the number of flops needed for $\frac{2^d}{2^d-1}$ flat iterations at level m. Starting from any initial image, this number of flat iterations yields a very poor approximation of the image, while m interpolation steps are nearly as good as m flat iterations.

The nested iteration is faster than flat iteration by a factor of

$$\frac{1-\frac{1}{2}^{m*d}}{m(1-\frac{1}{2}^{d})}.$$

Finally, for the 1D-IFS-example above, this factor is

$$\frac{1-\frac{1}{2^9}}{9(1-\frac{1}{2})} = \frac{511}{2304} \approx 0.22$$

11.6.4 Convergence

How fast does a pixel gray-value converge to its limit value? Let C be the overall contractivity of the set of affine transformations. The probabilities of transformations are related to their contractivity C(see [1]), i.e. the maximum of the probabilities is smaller or equal to C. Let us consider the case when we iterate on the same level. A new pixel value is calculated as the sum of products of a probability with some pixel-values; it converges to its limit value with factor at least C.

11.7 Conclusion

Looking at the IFS-generation of fractal images as a Markov process leads to a large set of linear equations. This problem can be solved by the classical Multigrid Method leading to a recursive scheme containing a cascade of mainly smoothing and interpolation steps. However, when we make use of the properties of the IFS, this scheme can be significantly simplified. Instead of the traditional smoothing, we use the affine transformations of the IFS to iterate at a certain level. Likewise, we use the affine transformations to transform the image from a coarse grid to a fine grid, instead of a *naive* interpolation. The convergence of this nested iteration technique depends only on the contractivity of the affine transformations. This way of generating fractals is much faster than point-to-point or recursive generation.

11.8 Appendix

The affine transformations, and their probabilities, which generate the pictures (the top-left corner is the origin!) in Figure 11.8 are:

- *The black square*

$$\begin{pmatrix} 0.5 & 0 \\ 0 & 0.5 \end{pmatrix} + \begin{pmatrix} 0 \\ 0 \end{pmatrix}, \quad p_1 = 0.2$$

$$\begin{pmatrix} 0.5 & 0 \\ 0 & 0.5 \end{pmatrix} + \begin{pmatrix} 0.5 \\ 0 \end{pmatrix}, \quad p_2 = 0.25$$

$$\begin{pmatrix} 0.5 & 0 \\ 0 & 0.5 \end{pmatrix} + \begin{pmatrix} 0 \\ 0.5 \end{pmatrix}, \quad p_3 = 0.25$$

$$\begin{pmatrix} 0.5 & 0 \\ 0 & 0.5 \end{pmatrix} + \begin{pmatrix} 0.5 \\ 0.5 \end{pmatrix}, \quad p_4 = 0.3$$

The efficiency of the multilevel method is very visible in this case: a naive method would calculate 256×256 pixels for each image, the multilevel method only needs 4^i pixels for the ith image. The image i can be considered as the ith nested iteration, magnified with a factor of 2^{8-i}.

- *Fuzzy square 1*

$$\begin{pmatrix} 0.5 & 0 \\ 0 & 1 \end{pmatrix} + \begin{pmatrix} 0 \\ 0 \end{pmatrix}, \quad p_1 = 0.36$$

$$\begin{pmatrix} 0 & 0.5 \\ -1 & 0 \end{pmatrix} + \begin{pmatrix} 0.5 \\ 1 \end{pmatrix}, \quad p_2 = 0.36$$

$$\begin{pmatrix} 0.45 & -0.2 \\ 0.22 & 0.34 \end{pmatrix} + \begin{pmatrix} 0.22 \\ 0.31 \end{pmatrix}, \quad p_3 = 0.14$$

$$\begin{pmatrix} 0.42 & -0.2 \\ 0.23 & 0.32 \end{pmatrix} + \begin{pmatrix} 0.4 \\ 0.1 \end{pmatrix}, \quad p_4 = 0.14$$

- *Fuzzy square 2*

$$\begin{pmatrix} 0.5 & 0 \\ 0 & 1 \end{pmatrix} + \begin{pmatrix} 0 \\ 0 \end{pmatrix}, \quad p_1 = 0.34$$

$$\begin{pmatrix} 0 & 0.5 \\ -1 & 0 \end{pmatrix} + \begin{pmatrix} 0.5 \\ 1 \end{pmatrix}, \quad p_2 = 0.34$$

$$\begin{pmatrix} 0.56 & -0.2 \\ 0.25 & 0.34 \end{pmatrix} + \begin{pmatrix} 0.21 \\ 0.39 \end{pmatrix}, \quad p_3 = 0.16$$

$$\begin{pmatrix} -0.58 & -0.17 \\ -0.27 & 0.32 \end{pmatrix} + \begin{pmatrix} 1 \\ 0.28 \end{pmatrix}, \quad p_4 = 0.16$$

Acknowledgements:

I would like to thank Prof. T. D'Hondt, for his continuing enthusiasm and support, Prof. P. de Groen for his advice on numerical algorithms and Prof. F. Delbaen for his help with probabilistic theory aspects of the subject. Also many thanks to the people at PROG and SCT, and especially to Lieven Verreycken and Michel Tilman for reading early drafts of this text. Also thanks to my sponsors IWONL and SCT.

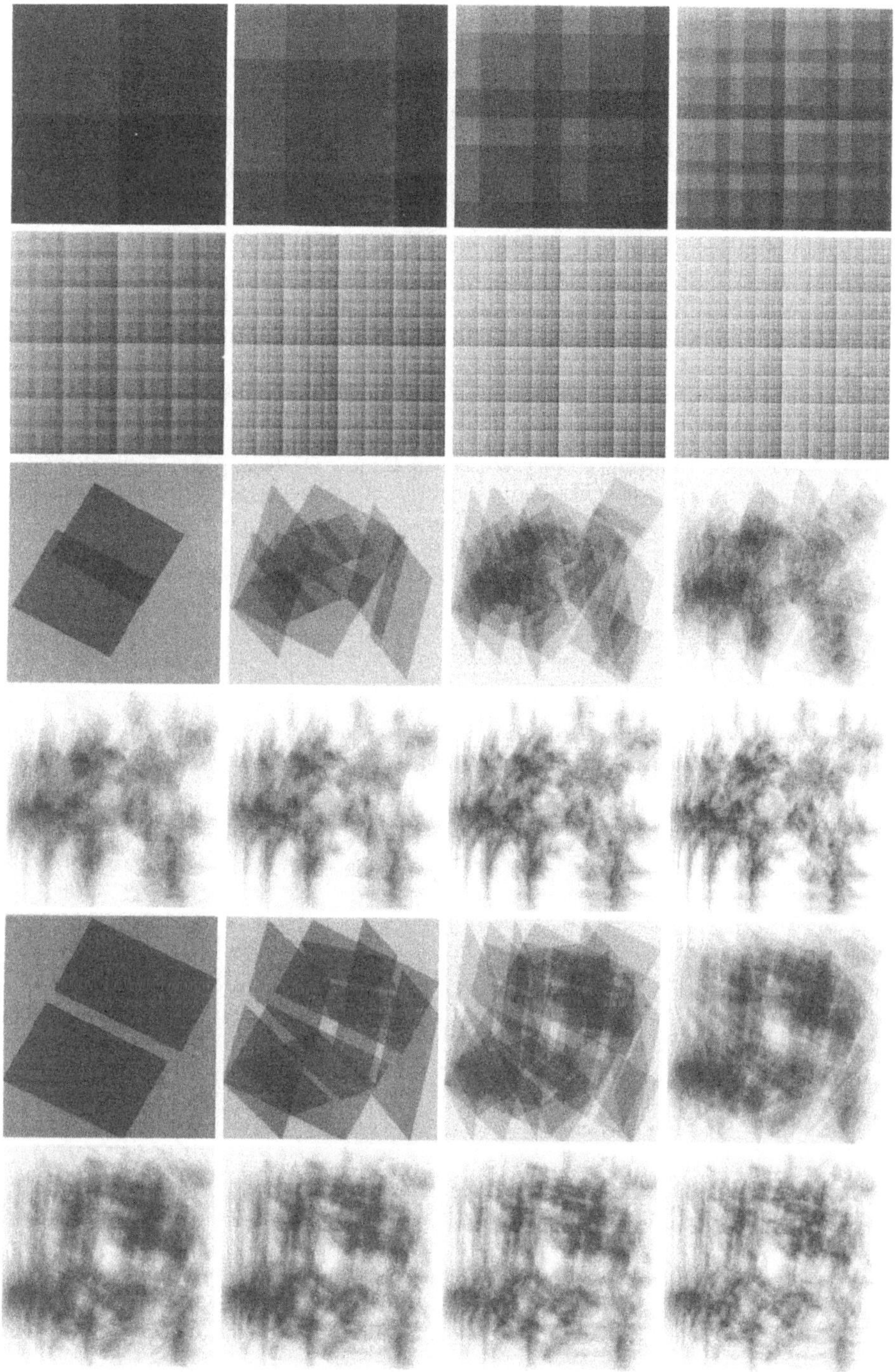

FIGURE 11.8. The gray-value pictures.

11.9 References

[1] M.F. Barnsley. *Fractals Everywhere.* Academic Press, 1988.

[2] M.F. Barnsley and J.H. Elton. A New Class of Markov Processes for Image Encoding. *Adv. Prob.* 20:14–32, 1988.

[3] U. Narayan Bhat. *Elements of Applied Stochastic Processes*, John Wiley, 1984.

[4] P. Billingsley. *Ergodic Theory and Information.* Krieger, 1978.

[5] W. Hackbusch. *Multi-grid Methods and Applications.* Springer-Verlag, 1985.

[6] D. Hearn, *Computer Graphics.* Prentice-Hall International, 1985.

[7] J. E. Hutchinson. Fractals and Self Similarity. *Indiana University Journal*, 30(5), 1981.

[8] A. Jacquin. *A Fractal Theory of Iterated Markov Operators, with Applications to Digital Image Coding.* PhD, Georgia Institute of Technology, School of Mathematics, 1989.

[9] B.B. Mandelbrot. *The Fractal Geometry of Nature.* W. H. Freeman and Co, 1982.

12 Statistical Colour Quantization for Minimum Distortion

Xiaolin Wu

ABSTRACT Colour quantization for frame buffer displays, a basic computer graphics technique for efficient use of colours, is statistically a large scale clustering process and it poses a difficult discrete optimization problem. Many heuristic color quantization algorithms were proposed [7, 12, 22, 24] but they all suffer from various degrees of non-adaptability to the colour statistics of input images. In this article we will introduce an adaptive colour image quantization algorithm that outperforms existing algorithms in subjective image quality and in least square approximation sense. The new algorithm eliminates many current suboptimal treatments of colour quantization such as a prequantization of discarding few lower bits of RGB values, restricted cuts orthogonal to RGB axes, partition criteria based on population or marginal distributions rather than variance minimization. A constrained global optimization scheme is incorporated into a divide-and-conquer clustering process to minimize quantization distortion. The time and space complexities of the new algorithm are $O(N \log K)$ and $O(N)$, where N is the number of pixels in the image and K is the number of colours in quantized image.

12.1 Introduction

To render colour images of photographic quality on raster displays, 24 bits (a byte for each of primary colours red, green, and blue) are required to encode each pixel. This is because average human eyes can distinguish up to ten million different colors. However, a given image with finite resolution will contain only a small subset of the sixteen million colours represented by the above encoding scheme. Very often 256 or even less statistically chosen colour representatives of an image suffice to reproduce the image without serious loss of fidelity. The process of selecting a small number of representative colours for an image of higher colour resolution is called colour image quantization in computer graphics. Colour image quantization is a must to display continuous-tone colour images on displays without 24 bits/pixel full-colour frame buffer. Even for full-colour displays colour image quantization is still important since it can free valuable frame buffer space for animation. High-speed frame buffer will continue to be a premium resource for real-time graphics. Also colour image quantization lessens burden of massive image data on storage and transmission bandwidth which will be bottlenecks if computer graphics is integrated into multimedia systems and when HDTV technology becomes commonplace.

Formally, colour quantization is a large scale clustering problem. A colour image in RGB mode corresponds to a 3-dimensional discrete density $P(\mathbf{c})$, where $\mathbf{c} = (c_r, c_g, c_b)$ in RGB space. The intensity of each primary colour is discrete and encoded by m bits, hence $P(\mathbf{c})$ is defined on a cubic lattice S of $2^m \cdot 2^m \cdot 2^m$ points. In colour image quantization, the point set S is partitioned into K subsets S_k, $1 \leq k \leq K$, $\sum_{\mathbf{c} \in S_k} P(\mathbf{c}) \neq 0$, $\cup_{1 \leq k \leq K} S_k = S$, $S_k \cap_{k \neq j} S_j = \phi$, and all colours $\mathbf{c} \in S_k$ are mapped to a representative colour $\mathbf{q}_k = (q_{kr}, q_{kg}, q_{kb})$. Given such a partition, the expected quantization distortion is defined to be

$$E(S_1, S_2, \cdots, S_K) = \sum_{1 \leq k \leq K} \sum_{\mathbf{c} \in S_k} P(\mathbf{c}) \|\mathbf{c}, \mathbf{q}_k\| \tag{12.1}$$

where $\|\mathbf{c}, \mathbf{q}_k\|$ is a perceptual distance between $\mathbf{c}$ and $\mathbf{q}_k$. To design a subjectively meaningful colour quantizer, we need a perceptually uniform distortion measure $\|\cdot, \cdot\|$. Namely,

$\|\mathbf{c}_0, \mathbf{c}_1\| = \|\mathbf{c}_1, \mathbf{c}_2\|$ if and only if $\mathbf{c}_0$ and $\mathbf{c}_1$ differ as much as $\mathbf{c}_1$ and $\mathbf{c}_2$ in visual sensation, and this quantified difference is independent of $\mathbf{c}$. Much research has been devoted to perceptually uniform colour measurements [13, 18, 15] but no universally accepted solutions have been found yet. This paper will focus on the algorithmic aspect of statistical colour quantization. For simplicity the ubiquitous mean-squared error is used in our development as in many popular colour quantization algorithms [12, 22, 7]. The new color quantization algorithm to be derived is, however, independent of the colour space in which the clustering of trivariant colour data takes place. It can be directly applied in a perceptually uniform colour space for an improved performance than in the RGB space.

Ideally, we would like to minimize the quantization distortion Eq(12.1) over all possible K-partitions of the input data set S given the number of output colours. The problem is one of discrete optimization. The fact that there are approaching K^N different K-partitions of N points [2] makes optimal colour quantization algorithmically intractable. For instance, there exist $2,436,684,974,110,751$ different ways of dividing 25 objects into 5 groups! Proofs for the NP-completeness of optimal K-clustering, for which optimal colour quantization is an instance, for variable K were given for various cost functions [4, 17] [1].

It is interesting to note, however, that for monochrome images, i.e., each pixel is a scalar value, optimal quantization is solvable in polynomial time by dynamic programming. $O(K|S|^2)$ optimal scalar quantization algorithms were proposed by Bruce [3]. This time complexity was reduced to $O(K|S|\log|S|)$ by Wu and Rokne [25, 26], and most recently to $O(K|S|)$ [23]. Unfortunately, our success in optimal scalar quantization cannot be extended to 3-dimensional colour space. Due to the NP-completeness of optimal colour quantization, any algorithms for the problem are necessarily heuristic and approximate solutions.

In the next section we will survey the existing colour quantization algorithms, and expose their weakness. Then to improve the current algorithms, a new technique based on principal component analysis will be developed in section 3, to tailor the partition of the colour space to particular colour distributions. In section 4, it will be shown how the above technique, after being incorporated into a divide-and-conquer framework, leads to a K-clustering algorithm for colour quantization. The details of the new algorithm will be given in section 5 together with its complexity analysis. Finally some empirical evidence will be provided to judge the quantization performance of the new algorithm in practice.

12.2 Historic Review

The research on colour quantization for frame buffers started a decade ago when Heckbert published apparently the first paper on the topic [12]. His algorithm and all other later algorithms are supported by the standard frame buffer architecture equipped with a colour look-up table. For the hardware support of colour quantization algorithms please refer to computer graphics textbooks, if necessary.

12.2.1 Colour histogram equalization

Heckbert's algorithm, commonly referred to as the median cut algorithm, is outlined below.

[1]In [22] an incorrect source was quoted for the NP-completeness proof.

Step 1. Partition the RGB space into K rectangular boxes under two criteria a) each box contains roughly the same number of pixels and b) pixels in each box have roughly equal spreads in R, G, and B directions.

Step 2. The means of K boxes formed in step 1 are chosen as representative colors.

Step 3. Pixels are quantized to representative colors by nearest neighbor rule using Euclidean distance.

Step 1 is basically a *k-d* tree like recursive cutting of boxes at the median point with an orthogonal plane. The algorithm reproduces better colour images than the naive uniform quantization scheme which simply discards the least significant bits of R, G, and B values. Clearly the latter could produce many empty boxes if the colour distribution is biased, wasting the valuable information capacity of the available colours. By giving equal colour quota to equal input colours the median cut algorithm is effectively a three-dimensional histogram equalization process. This idea was also explored by Stevens *et al.* in a less known paper [21]. They first converted the three-dimensional colour histogram to a one-dimensional histogram using peano curve scan, and then carried out a conventional histogram equalization on the colours. A problem with the histogram equalization is that it tends to distort distinctive colours. For instance, if the above approach was used to quantize colour portraits, then it would represent the colours of eyes by some very different chosen colours. This is because the eye colours, due to the small physical size of eyes, do not carry enough weight to be allocated a representative by the histogram equalization process, although being vital information. On a second reflection, we should consider the moments of colour points from their representatives, i.e., the variance, rather than only the colour populations in colour quantization.

12.2.2 Marginal variance minimization

In an internal report we [27] proposed a so-called adaptive bisplit algorithm. The algorithm differed from Heckbert's only in the cutting criterion at step 1. A color box Ω is cut into two subboxes Ω_1 and Ω_2 with an orthogonal plane at the position ξ_{opt} that minimizes

$$E(\xi) = \sum_{\mathbf{c}\in\Omega_1} P(\mathbf{c})[c_j - \mu_{1j}(\xi)]^2 + \sum_{\mathbf{c}\in\Omega_2} P(\mathbf{c})[c_j - \mu_{2j}(\xi)]^2, \tag{12.2}$$

where μ_{1j} and μ_{2j} are the marginal means of Ω_1 and Ω_2 on axis j in which Ω has the widest extent. The reason for minimizing the sum of two marginal variances expressed by Eq(12.2) rather than true variances was to reduce the computational cost. To compete with Heckbert's algorithm in speed, an estimator [2] of ξ_{opt}

$$\xi_{opt} \approx \mu_j + 0.2(\mu_j - \tau_j) \tag{12.3}$$

was used, where μ_j and τ_j are respectively the j coordinates of the mean and the geometric center of the box Ω. To make the algorithm more adaptive, the concept of color quota partition was introduced. At beginning K color quotas are assigned to the set S. After S is partitioned by ξ_{opt} into S_1 and S_2 that are tightly bounded by Ω_1 and Ω_2 (in the sequel the symbols S_i and Ω_i are used interchangeably since no distinction is necessary), the integer K is also partitioned into K_1 and K_2, K_1 color quotas to S_1 and K_2 to S_2, by the ratio of the two local variances for S_1 and S_2, namely,

$$K_1 = \left\lfloor \frac{K\sum_{\mathbf{c}\in\Omega_1} P(\mathbf{c})[\mathbf{c} - \mu(\xi_{opt})]^2}{\sum_{\mathbf{c}\in\Omega_1} P(\mathbf{c})[\mathbf{c} - \mu_1(\xi_{opt})]^2 + \sum_{\mathbf{c}\in\Omega_2} P(\mathbf{c})[\mathbf{c} - \mu_2(\xi_{opt})]^2} \right\rfloor, \quad K_2 = K - K_1. \tag{12.4}$$

[2] See [25] for derivation. The formula was misrepresented as the mean in [22].

To avoid computing the variances, Eq(12.4) was approximated by

$$K_1 = \left\lfloor \frac{K|S_1|L(\Omega_1)}{|S_1|L(\Omega_1) + |S_2|L(\Omega_2)} \right\rfloor, \quad K_2 = K - K_1, \tag{12.5}$$

where $L(\Omega_t)$, $t = 1, 2$, is the average of three extents of the box Ω_t.

The adaptive bisplit algorithm is described by the following recursive procedure.

procedure Bisplit(S, K)
begin
  if K=1 **then return**;
  partition S by Eq(12.3) into S_1 and S_2;
  partition K by Eq(12.5) into K_1 and K_2;
  Bisplit(S_1, K_1);
  Bisplit(S_2, K_2);
end.

Smaller quantization distortion and better image fidelity were obtained by replacing step 1 of Heckbert's algorithm with the above procedure, while the time complexity remained the same.

Wan *et al.* extended the above work by reducing the quantization error at the expense of higher time complexity. The differences between their algorithm [27] and ours are 1) the optimal cutting position ξ_{opt} was computed by exhaustive search rather than by Eq(12.3), for all three axes rather than only the axis of the widest color spread, then cutting was done in the direction where the smallest sum of two marginal variances is obtained; 2) boxes were cut in the order of their variances rather than recursive color quota partition.

12.2.3 Other methods

The marginal variance minimization technique was obviously a computational expedience rather than an optimization process by choice. It was not until recently that colour quantization under true variance minimization criterion was made computationally practical for interactive applications [24]. This was achieved by a statistical preprocessing based on the inclusion-and-exclusion rule of combinatorics. Aided by this preprocessing, the variance, mean, and population of any box in the colour space can be reported in $O(1)$ time so that sweeping the colour space in a search for optimal bipartition becomes a linear-time operation. This preprocessing technique also reduces the time complexity of the colour histogram equalization algorithms.

Another interesting colour quantization technique due to Gervautz and Purgathofer [7] is octree-based clustering of pixels in color space.

Optimal colour quantization was formulated in Eq(12.1) as an optimal K-clustering problem. Therefore, the well-known iterative K-means clustering algorithm [14] can be directly used for colour quantization. In terms of optimization, the K-means algorithm is a typical descent method and it can find a local minimum on the distortion function. This algorithm can be used as a postprocessing to improve the optimality of all the other algorithms mentioned above by using the latters' outputs as the initial cluster centroids to start the iterative K-means clustering. However, this improvement should be considered only as an off-line postprocessing because of the inherently high computational cost of the K-means algorithm. Even with its high cost the K-means algorithm is still a heuristic approach since it can only guarantee a local optimum. Simulated annealing technique can be used to further improve the performance of the K-means algorithm in colour quantization [9], but its cost is even higher.

All the above algorithms work in the RGB colour space. But it is possible to cluster colours in image space as well under the assumption that certain locality of colours in image space exists. An attempt in this direction was made by Campbell *et al.* [5]. They used the BTC encoding technique in signal processing to compress colours by blocks of an input image. In this paper we will only study colour clustering in a three-dimensional colour space, and will not elaborate on the other technique any further.

The colour clustering algorithms in [12, 7, 24, 22] compromise the optimality by orthogonal partition of colour space, namely, the clusters are formed by cutting planes perpendicular to one of the R, G, B axes even though the colour distribution $P(\mathbf{c})$ of a real image is seldom orthogonal to the R, G, B axes. This restriction makes these algorithm less adaptable to the colour statistics of individual images. In fact, there is no reason other than algorithm efficiency to employ only orthogonal cuts in clustering. Splitting a space with arbitrary halfplanes in clustering multivariate data can be dated back to 1963 [19]. A natural choice of cutting direction is the principal axis along which the colour points have the maximum variance. The eigenvector transform [16] on the covariance matrix of the data set can be performed to find this principal axis. Due to the small dimensionality of three for colour quantization, the eigenvector transform on a small (3×3) symmetric positive definite covariance matrix is very fast hence practical. In what follows the above idea will develop into a new colour quantization algorithm.

12.3 Globally Optimal Cut on Principal Axes

The new colour quantization algorithm is based on the following simple idea. Instead of optimizing the K-partition of the set S, an extremely difficult task, we try to compute (approximate) the optimal bipartition of S into S_1 and S_2, i.e., to minimize the sum $D(S_1) + D(S_2)$, where

$$D(S_j) = \sum_{\mathbf{c} \in S_j} P(\mathbf{c}) \|\mathbf{c} - \mathbf{u}(S_j)\|_2, \tag{12.6}$$

$\mathbf{u}(S_j)$ is the centroid of the data set S_j, $\| \ \|_2$ stands for the L_2-norm (the squared Euclidean distance), and $D(S_j)$ is the squared error in quantizing all $\mathbf{c} \in S_j$ to $\mathbf{u}(S_j)$. Since the optimal bipartition of S is done by a halfplane under the Euclidean metric, we would like to sweep a plane in search for the optimal cutting position. Still the freedom to place such a plane is too enormous, we need to fix the orientation of this sweeping plane to make the search feasible. Our goal of minimizing $D(S_1) + D(S_2)$ can be best met by cutting S in the direction in which the maximum variance of S occurs, hence the cutting plane should be set perpendicular to the principal axis of the 3-dimensional point set S. As will be demonstrated in section 12.5 it is now computationally feasible to examine all possible bipartitions of S into $\hat{S}_1$ and $\hat{S}_2$ by the family of parallel cutting planes orthogonal to the principal axis, and determine the optimal one that minimizes $D(\hat{S}_1) + D(\hat{S}_2)$. Note optimizing the bipartition of S by cuts orthogonal to the principal axis becomes a scalar optimal quantization problem, and its globally optimal solution can be obtained through a linear search.

Next we lift the restriction on the cutting plane orientation, and start an iterative 2-means clustering process on the resulting subsets $\hat{S}_1$ and $\hat{S}_2$. The 2-means clustering process is a special case of the well-known K-means clustering algorithm [14]. For the $(i+1)^{th}$ iteration the algorithm has as input a bipartition of S into S_1^i and S_2^i, and it cuts S into S_1^{i+1} and S_2^{i+1} by the half plane bisecting the two centroids $\mathbf{u}(S_1^i)$ and $\mathbf{u}(S_2^i)$, and so forth until the sequence $\{S_1^i | i = 0, 1, \cdots\}$ (consequently $\{S_2^i | i = 0, 1, \cdots\}$) converges. It has been proven [20] that the K-means algorithm converges to a local optimum. The

2-means clustering algorithm initialized to $S_1^0 = \hat{S}_1$ and $S_2^0 = \hat{S}_2$ will adjust the optimal cutting plane perpendicular to the principal axis, if necessary, to a locally optimal position in colour space. That is, the result of globally optimal scalar quantization is used to obtain a locally optimal colour (multivariate) quantization through iterative clustering. We have high confidence that this local optimum is very close to, if is not, the globally optimal bipartition since the progressive improvement of the bipartition by the iterative clustering process starts from a bipartition which is globally optimal under the reasonable constraint of cutting perpendicularly to the principal axis.

12.4 From Bipartition to K-Partition

After S is cut into S_1 and S_2 by the above optimized bipartition algorithm, the same algorithm can be applied to S_1 and S_2, and then to the subsequent subsets until K subsets (clusters) are generated. If we treat an intermediate subset Ω of S to be further split as an internal node of a binary tree and Ω_1 and Ω_2, the two subsets of Ω, as the two sons of the node pertaining to Ω, then the entire colour clustering process through a sequence of bipartitions can be represented by the binary tree whose root is S, each of whose $K-1$ internal nodes corresponds to a bipartition, and whose K leaves are the final K subsets S_j, $1 \leq j \leq K$. The K centroids $\mathbf{u}(S_j)$ are the K representative colours to be loaded into the colour look-up table. The above tree structure is in essence a so-called binary spatial partitioning tree (BSP tree), a powerful data structure for many computer graphics problems. The BSP tree was first used by Fuchs *et al.* [6] but for efficient visibility determinations. Of course, our partition criterion for colour quantization is different from theirs. For convenience we will adopt the BSP tree terminology in the sequel.

In general, the order in which the internal nodes of the BSP tree get split matters to the optimality of colour quantization. The simplest order of repeated bipartitions is recursion. Before growing to the next level, each internal node of the binary tree at the current level is split regardless of its statistic characteristic. Smaller quantization errors may be achieved by more elaborated orders of splitting the internal nodes. One obvious alternative is to always split the node with the largest variance; the other choice is to split the node whose bipartition brings the largest reduction in the total mean squared error. Precisely let Ω_j, $1 \leq j \leq k < K$, be the k subsets in the middle of colour clustering process, and $\Omega_{j,1}$ and $\Omega_{j,2}$ be the two subsets of Ω_j if it is split by the optimal cutting plane of the previous section. Then the next subset to be split is Ω_t such that

$$t = \arg \max_{1 \leq j \leq k} \{D(\Omega_j) - D(\Omega_{j,1}) - D(\Omega_{j,2})\}. \tag{12.7}$$

The above two splitting criteria were tested and compared with the blind recursion scheme. The criterion Eq(12.7) performed the best, and the criterion of splitting intermediate subsets in the order of their variances performed the second. The difference between the two criteria in terms of the final quantization distortion is marginal, but they outperformed the blind recursion scheme noticeably with their total mean squared errors being twenty percent smaller. The increase in computational costs is negligible for the maximum variance criterion and is two folds for the criterion of Eq(12.7).

12.5 Algorithm Details and Complexity Analysis

In this section we discuss mathematical and algorithmic issues arose in the proposed colour clustering process. Here we assume that the BSP tree pertaining to the colour clustering process is balanced. This is definitely the case if the recursive splitting criterion is used; the $O(\log K)$ height was also observed for the other two splitting criteria in section 12.4 in our experiments. At the ith level of the binary tree, $0 \leq i < \log K$ (assuming that K is a power of 2), there are 2^i subsets of S: $S_{i,t}$, $1 \leq t \leq 2^i$, to be bipartitioned. To find the principal axes of those subsets, we need to perform an eigenvector transform on each of those 2^i subsets. For each subset $S_{i,t}$, we have to compute its covariance matrix $\mathbf{C}$, the largest eigenvalue λ_{max} and the corresponding eigenvector $\mathbf{v}$ of $\mathbf{C}$, i.e., $\mathbf{Cv} = \lambda_{max}\mathbf{v}$, which determines the principal axis of $S_{i,t}$. The cost of the eigenvector transform is dominated by the construction of covariance matrix $\mathbf{C}$:

$$\mathbf{C} = E\{\mathbf{c}\mathbf{c}^T\} - E\{\mathbf{c}\}[E\{\mathbf{c}\}]^T. \tag{12.8}$$

The matrix $\mathbf{C}$ has 9 covariances each of which requires $O(|S_{i,t}|)$ operations, namely,

$$c_{a,b} = \frac{\sum_{\mathbf{c}\in S_{i,t}} c_a c_b}{|S_{i,t}|} - u_a u_b, \quad 1 \leq a,b \leq 3. \tag{12.9}$$

In the above formula and in the sequel c_r, c_g and c_b are indexed to c_1, c_2 and c_3 respectively for convenience.

Additional $O(|S_{i,t}|)$ operations are needed to compute the mean vector $\mathbf{u}(S_{i,t})$ before its components can be used in Eq(12.9). It follows from the above counting that computing $\mathbf{C}$ for a given $S_{i,t}$ needs $O(|S_{i,t}|)$ operations if we ignore the coefficient 6 for the 3×3 symmetric matrix $\mathbf{C}$. Since

$$\sum_{i=0}^{\log K-1} \sum_{t=1}^{2^i} |S_{i,t}| = |S| \log K \tag{12.10}$$

the total cost of $K-1$ eigenvector transforms for the entire colour clustering process is $O(|S| \log K)$.

Once the principal axis of an $S_{i,t}$ is found, we sweep a cutting plane perpendicular to this axis in a linear search for the optimal cutting position. To this end we sort all colour points $\mathbf{c} \in S_{i,t}$ by the ordinates of their projections onto the principal axis, i.e., establishing the order that $\mathbf{c}_1 \leq \mathbf{c}_2$ if $\mathbf{c_1}^T\mathbf{v} \leq \mathbf{c_2}^T\mathbf{v}$. The projection of $\mathbf{c} \in S_{i,t}$ on the principal axis takes $O(|S_{i,t}|)$ operations (multiplications and additions). Since these projection values are real numbers, the sorting seems to require $O(|S_{i,t}| \log |S_{i,t}|)$ comparisons. But on a second reflection, the real arithmetic is unnecessary here because the intensity resolution of frame buffer devices is discrete and finite anyway, being typically integers from 0 to 255. Therefore, we can put the real projection values into M buckets ($M = 512$ was used in our experiments and was shown to be sufficient not to introduce additional quantization error) so that linear-time bucket sort algorithm can be used to sort the projection values. The order of $\mathbf{c} \in S_{i,t}$ by their projection on the principal axis makes it possible to find the optimal cutting position of the sweeping plane in $O(d|S_{i,t}|)$ operations as explained below.

Let $S_1(\xi)$ and $S_2(\xi)$ be a bipartition of an S formed by the cutting plane perpendicular to and positioned at coordinate ξ of the principal axis of S. Our goal is to find the optimal cutting position ξ_{opt} in the least square sense, i.e.,

$$\begin{aligned} \xi_{opt} &= \arg\min_{\xi}\{D(S_1(\xi)) + D(S_2(\xi))\} \\ &= \arg\min_{\xi}\left\{ \sum_{\mathbf{c}\in S_1(\xi)} [\mathbf{c} - \mathbf{u}(S_1(\xi))]^T[\mathbf{c} - \mathbf{u}(S_1(\xi))] + \right. \\ &\qquad \left. \sum_{\mathbf{c}\in S_2(\xi)} [\mathbf{c} - \mathbf{u}(S_2(\xi))]^T[\mathbf{c} - \mathbf{u}(S_2(\xi))] \right\}. \end{aligned} \tag{12.11}$$

It can be verified after some manipulations that

$$\sum_{\mathbf{c}\in S_1(\xi)} [\mathbf{c}-\mathbf{u}(S_1(\xi))]^T[\mathbf{c}-\mathbf{u}(S_1(\xi))] + \sum_{\mathbf{c}\in S_2(\xi)} [\mathbf{c}-\mathbf{u}(S_2(\xi))]^T[\mathbf{c}-\mathbf{u}(S_2(\xi))] =$$

$$\sum_{k=1}^3 \sum_{\mathbf{c}\in S} c_k^2 - \sum_{k=1}^3 \left[\frac{[\sum_{\mathbf{c}\in S_1(\xi)} c_k]^2}{|S_1(\xi)|} + \frac{[\sum_{\mathbf{c}\in S_2(\xi)} c_k]^2}{|S_2(\xi)|}\right] \quad (12.12)$$

Since the first term is a constant for any given S, the minimization problem of Eq(12.11) is equivalent to the maximization one:

$$\begin{aligned}\xi_{opt} &= \arg\max_{\xi}\left\{\sum_{k=1}^{3}\left[\frac{[\sum_{\mathbf{c}\in S_1(\xi)} c_k]^2}{|S_1(\xi)|} + \frac{[\sum_{\mathbf{c}\in S_2(\xi)} c_k]^2}{|S_2(\xi)|}\right]\right\} \\ &= \arg\max_{\xi}\left\{\sum_{k=1}^{3}\left[\frac{[\sum_{\mathbf{c}\in S_1(\xi)} c_k]^2}{|S_1(\xi)|} + \frac{[\sum_{\mathbf{c}\in S} c_k - \sum_{\mathbf{c}\in S_1(\xi)} c_k]^2}{|S_2(\xi)|}\right]\right\} \qquad (12.13)\end{aligned}$$

where $\sum_{\mathbf{c}\in S} c_k$ is also a constant for any given S. The primitive operation in the above maximization process is $\sum_{\mathbf{c}\in S_1(\xi)} c_k$, and it can be done incrementally. Suppose that ξ_1 and ξ_2 are two consecutive ordinates on the principal axis projected by $\mathbf{c} \in S$, we have

$$\sum_{\mathbf{c}\in S_1(\xi_2)} c_k = \sum_{\mathbf{c}\in S_1(\xi_1)} c_k + \sum_{\mathbf{c}^T\mathbf{v}=\xi_2} c_k \qquad (12.14)$$

Since the data set S is sorted by $\mathbf{c}^T\mathbf{v}$ we can retrieve all those $\mathbf{c}$ satisfying $\mathbf{c}^T\mathbf{v} = \xi_2$ by simply marching in the sorted list from ξ_1 to ξ_2 and evaluate $\sum_{\mathbf{c}^T\mathbf{v}=\xi_2} c_k$ along the way. After this $\sum_{\mathbf{c}\in S_1(\xi_2)} c_k$ can be incrementally obtained in constant time from $\sum_{\mathbf{c}\in S_1(\xi_1)} c_k$. Based on the above analysis we finally conclude that ξ_{opt} defined in Eq(12.13) can be found in $O(|S|)$ time after $\mathbf{c} \in S$ are sorted by $\mathbf{c}^T\mathbf{v}$.

The adjustment of the cutting plane by iterative two-means clustering can be done efficiently. Given two tentative centroids before convergence $\mathbf{c}_1$ and $\mathbf{c}_2$ of the two subsets S_1 and S_2, we have the 3-dimensional halfplane:

$$(\mathbf{c}-\mathbf{c}_h)^T(\mathbf{c}_2-\mathbf{c}_1) = 0 \qquad (12.15)$$

where $\mathbf{c}_h$ is the midpoint of the line segment connecting $\mathbf{c}_1$ and $\mathbf{c}_2$. Then we can determine the cluster membership of a colour point $\mathbf{c} \in S$ by substituting the $\mathbf{c}$ into the plane equation Eq(12.15) and check the sign of the result, if positive, $\mathbf{c}$ is closer to $\mathbf{c}_2$; otherwise, it is closer to $\mathbf{c}_1$. Thus an iteration of the two-means clustering takes $O(|S|)$ operations. Since the starting point of the clustering process is the globally optimized cutting plane perpendicular to the principal axis, it takes very few iterations for the iterative clustering algorithm to converge. Our empirical data showed that on average four iterations are needed.

Adding together the costs of $O(|S|\log K)$ for eigenvector transforms, $O(|S|\log K)$ for sorting the projections of colour points on the principal axes (linear-time bucket sort is used here), $O(|S|\log K)$ for optimizing the cutting planes perpendicular to the principal axes, and $O(|S|\log K)$ (based on empirical data) for two-means clustering processes, we conclude that the proposed colour quantization algorithm runs in $O(|S|\log K)$ expected time. Recently Wu proposed a preprocessing technique [24] to reduce the cost of median-cut algorithm from $O(|S|\log K)$ to $O(|S|)$, and reduce the cost of Wan *et. al.*'s algorithm from $O(K|S|)$ to $O(|S|)$. The new algorithm is more expensive than the improved previous algorithms but no slower than their original versions actually in use. On a 4D/20 personal IRIS workstation our algorithm took on average 31.9, 27.7, and 26.1 seconds for quantizing a 24 bit/pixel 512 × 512 colour image into 256, 64, and 32 colours.

	Median-Cut			Wan *et. al.*			New		
K	Zelda	Toys	Hotel	Zelda	Toys	Hotel	Zelda	Toys	Hotel
16	1328.9	1076.4	832.5	1294.6	1138.3	898.5	311.5	217.7	396.7
32	1228.1	695.7	711.3	581.9	466.6	618.0	161.1	111.2	224.9
64	713.2	343.2	512.7	304.1	253.2	450.6	80.9	60.6	119.8
256	79.5	68.3	186.3	69.6	52.6	140.4	28.2	22.0	37.6

TABLE 12.1. Quantization errors of different algorithms on different images.

The new colour quantization algorithm requires only $O(|S|)$ space. At any given moment, S is partitioned into k subsets, $2 \leq k \leq K$. To dynamically maintain the membership of colour points in subsets, we use an index array of $|S|$ entries each of which is a pointer to a pixel. The pixels' indices of each subset occupy a contiguous sequence of locations in the index array. This sequence can be recorded by an auxiliary pointer to the first position of the sequence and the size of the subset. If this subset is further split into two, we exchange indices, if necessary, to separate the indices of pixels into two subsequences, and then record the new head pointers and sizes of the two newly created subsets. The same technique for partitioning an index array in quicksort algorithm (see p. 96 of [1]) can be used for dynamically maintaining the subsets. Therefore, besides the space to save the input colour image, the new algorithm only needs $O(|S|)$ working space, which is less than the previous algorithms [12, 22, 24] by an order. By quantizing in the principal axis the new algorithm does not use large 3-dimensional arrays to hold input data as required by its predecessors [12, 22, 24], hence it does not need a prequantization of chopping few least significant bits to control the space usage. Consequently quantization precision up to the original colour resolution can be obtained by the new algorithm.

After the final K subsets S_j, $1 \leq j \leq K$, are formed, we may quantize $\mathbf{c} \in S_j$ to $\mathbf{u}_j$. This is done by a simple scan of the index array. Of course, we may explicitly construct a binary search tree as the by-product of the colour clustering process and use a k-d tree-like search technique [10] for nearest neighbor match in the final assignment of pixels to cluster centroids. The improvement of the nearest neighbor search over the simple mapping of $\mathbf{c} \in S_j$ to $\mathbf{u}_j$ was insignificant in our experiments, not justifying the extra computational costs. Note that the nearest neighbor search step after quantization is required by the previous algorithms [12, 22] in order to get good results, incurring additional computational cost.

12.6 Quantization Performance

The proposed colour quantization algorithm was implemented in C language and tested on twenty 24 bits/pixel colour images of various types. It should not be a surprise that the proposed algorithm achieved much smaller quantization errors than previous algorithms on each of these test images without a single exception, because it was based on a better optimization strategy in least-square approximation sense. Quantization errors for different algorithms on three ISO colour test images: Zelda, Hotel and Toys representing the classes of portrait, outdoor and indoor pictures are tabulated in Table 12.1.

For a subjective judgement on image quality, quantized images of Zelda and Hotel of 64 colours by Heckbert's median-cut algorithm, Wan *et al.*'s marginal variance minimization algorithm, and the new algorithm are given in the colour plate. Also given, as references, are the two images of 256 colours quantized by the new algorithm which are virtually indistinguishable from the 24 bit/pixel originals.

It is important to understand the impact of different partition criteria in clustering process on the quality of quantized images. The median-cut criterion does not take the variance into consideration. It gives the same information capacity to groups of equal colour population regardless how colours are distributed in each group. As the result, those groups of diverse colours are underrepresented causing false contours in the output, whereas other groups of much concentrated colours will waste their representation quota. The marginal variance minimization criterion corrects the above problem to certain extent. But it puts heavier weight on intensity than on hue in quantization. This is because the colour component in the direction of the largest marginal variance mainly reflects the intensity, while the ratio of different color components determines the hue. Due to its quantization bias, marginal variance minimization algorithm tends to gain image smoothness at the expense of false hue. To verify this pay attention to the seemingly "washed-out" colours of the image quantized by Wan *et al.*'s algorithm. It appeared that the new algorithm outperformed its predecessors not only in minimizing quantization errors but also in the colour fidelity of the quantized images.

12.7 Conclusions

An optimized colour quantization algorithm was proposed. Higher image fidelity and lower quantization error were obtained by the new algorithm over the current algorithms. The optimization is done at reasonable time and space complexities. No prequantization is necessary for the new algorithm.

Acknowledgements:

This works was supported by grants from the Natural Sciences and Engineering Research Council of Canada.

FIGURE 12.1. Zelda of 256 colours quantized by the new algorithm (a); Zelda of 64 colours quantized by the median-cut (b); the marginal variance minimizations (c) and the new algorithm (d).

FIGURE 12.2. Hotel of 256 colours quantized by the new algorithm (a); Hotel of 64 colours quantized by the median-cut (b), the mariginal variance minimization (c) and the new algorithm (d).

12.8 References

[1] A.V. Aho, J.E. Hopcroft, and J.D. Ullman. *The Design and Analysis of Computer Algorithms*, Addison-Wesley Co., 1974.

[2] M.R. Anderberg. *Cluster Analysis for Applications*, Academic Press, New York, 1973.

[3] J.D. Bruce. Optimum Quantization, Sc. D. thesis, M.I.T., May 14, 1964.

[4] P. Brucker. On the complexity of clustering problems. in *Optimization and Operations Research* (R. Henn, B. Korte, and W. Oettli, eds.), pp. 45-54, Springer-Verlag, 1977.

[5] G. Campbell, T. DeFanti, J. Frederiksen, S. Joyce, L. Leske, J. Lindberg and D. Sandin. Two bit/pixel full colour encoding. in *SIGGRAPH'86 proceedings*, pp. 215-233, 1986.

[6] H. Fuchs, G. Z. M. Kedem and B.F. Naylor. On visible surface generation by a priori tree structure, *Computer Graphics*, vol. 14, no. 3, p. 124-133, July 1980.

[7] M. Gervautz and W. Purgathofer. A simple method for colour quantization: octree quantization, in *Graphics Gems* edited by A. Glassner, Academic Press, 1990.

[8] R.M. Gray. Vector quantization, *IEEE ASSP Mag.*, pp. 4-29, April 1984.

[9] E. Fiume and M. Ouellette. On distributed, probabilistic algorithms for computer graphics. in *Proc. Graphics/Interface'89*, London, Ontario, p. 221-218, June 1989.

[10] J. H. Friedman, J.L. Bentley and R.A. Finkel. An algorithm for finding best matches in logarithmic expected time. *ACM Trans. Math. Software*, Vol. 3, No. 3, pp. 209-226, Sept. 1977.

[11] J. A. Hartigan. *Clustering algorithms*, Wiley, New York, 1975.

[12] P. Heckbert. Colour image quantization for frame buffer display. *SIGGRAPH'82 proceedings*, pp. 297-307, 1982.

[13] J.O. Limb, C.B. Rubinstein, and J.E. Thompson. Digital coding of colour video signals — a review, *IEEE Trans. on Comm.*, vol. COM-25, no. 11, p. 1349-1385, Nov. 1977.

[14] J.B. MacQueen. Some methods for classification and analysis of multivariate observations. *Proc. 5th Berkeley Symposium on Mathematical Statistics and Probability*, Vol. 1, pp. 281-297, 1967.

[15] D.L. MacAdam. Uniform colour scale. *J. Opt. Soc. of Amer.*, vol. 64, pp. 1691-, 1974.

[16] B.F.J. Manly. *Multivariate Statistical Methods*, Chapman and Hall, London, 1986.

[17] N. Megiddo, K.J. Supowit. On the complexity of some common geometric location problems. *SIAM J. Comput.* vol. 13, pp. 182-196, 1984.

[18] G.W. Meyer and D.P. Greenberg, D. P. Perceptual colour space for computer graphics. in *Colour and the Computer*, edited by H. J. Durrett, Academic Press, pp. 83-100, 1987.

[19] J.N. Morgan and J.A. Sonquist. Problems in the analysis of survey data, and a proposal. *J. Amer. Statist. Assoc.*, vol. 58, pp. 415-434, 1963.

[20] S.Z. Selim and M.A. Ismail. 'K-means-type algorithms: a generalized convergence theorem and characterization of local optimality, *IEEE Trans. PAMI.*, Vol. 6, No. 1, pp. 81-87, 1984.

[21] R.J. Stevens, A.F. Lehar and R.H. Preston. 'Manipulation and presentation of multidimensional image data using the Peano scan. *IEEE Trans. PAMI*, Vol. 5, No. 2, pp. 520-, Sept. 1983.

[22] S. Wan, A. Wong and P. Prusinkiewicz. An algorithm for multidimensional data clustering. *ACM Trans. on Math. Software*, Vol. 14, No. 2, pp. 153-162, June 1988.

[23] X. Wu. Optimal quantization by matrix-searching. *Journal of Algorithms*, vol. 12, pp. 663-673, Dec. 1991.

[24] X. Wu. Efficient statistical computations for optimal colour quantization, in *More Graphics Gems* vol. II, edited by J. Arvo, Academic Press, pp. 126-133, 1991.

[25] X. Wu. *Algorithmic approaches to optimal mean-square quantization*, Ph.D dissertation, Dept. of Computer Science, Univ. of Calgary, 1988.

[26] X. Wu and J. Rokne. An $O(KN \lg N)$ Algorithm for Optimal K-level Quantization on Histograms of N Points. *Proc. ACM 1989 Conference of Computer Science*, Louisville, KY., pp. 339-343, Feb. 1989.

[27] X. Wu and I. Witten. A fast k-means type clustering algorithm. *Research Report No. 85/197/10*, Dept. of Computer Science, Univ. of Calgary, 1985.

13 Minimizing the Absolute Value Energy Function: An Application to Geometrical Constraint-Solving

Nami Kin
Yoshiaki Takai
Tosiyasu L. Kunii

ABSTRACT
This paper proposes a new connectionist model to minimize the absolute value energy function based on the Boltzmann machine. In conventional models, n^{th}-order constraints (nonlinear optimization problems) are solved by minimizing the $2n^{th}$-order energy function, which brings forth explosive increase of the number of units and connections among them. By using the absolute value energy function, n^{th}-order constraints can be directly represented by the nth-order energy function. Hence, a very great reduction of the network complexity becomes possible. Our model is particularly suited for applications, such as geometrical constraint-solving, in which nonlinearity plays a definitive role.

13.1 Introduction

Artificial neural networks are an emerging computational technology, which are based explicitly on an abstraction of our current understanding of the information processing properties of actual biological neurons. From a structural point of view, neural networks can be characterized by the massive interconnection of a large number of tiny processing elements called units. Consequently, artificial neural networks are sometimes referred to as connectionist models [2]. There are many different types of neural networks, each of which has different advantages suited to their applications.

One of the most distinctive features of the neural network is that it works as a minimizing machine for a certain energy function. In 1982, Hopfield demonstrated that mutually connected artificial neural networks can be applied to obtain rapidly high quality sub-optimal solutions of NP-complete problems such as the traveling salesman problem [4]. Such a network works toward reaching an optimal solution by searching for a minimum of the energy function associated with the network in a highly parallel and distributed way. In conventional methods, the energy functions of neural networks are defined as a quadratic form with respect to the binary states of units building up the networks. Thus, only linear simultaneous equations are directly converted to least square problems on the neural networks [10, 11].

We have already introduced an extended energy function which includes terms of higher-order than quadratic ones, and applied it to solving the nonlinear constraints denoted as nonlinear simultaneous equations [7]. This extended energy function is a natural extension of the conventional quadratic energy function associated with the Boltzmann machine, which is a kind of mutually connected artificial neural network. Any n^{th}-order constraints (nonlinear optimization problems) can be solved by minimizing the $2n^{th}$-order extended energy function. However, the number of nth-order terms is $O({}_mC_n)$, where m is the number of units. Therefore, as the energy function becomes higher-order, the number of connections between units explodes.

This paper describes another connectionist model for minimizing the energy function, which is defined as the absolute value of the difference between a variable and an objective value. Using the absolute value, n^{th}-order constraints can be directly represented by the nth-order energy function. Thus, a tremendous reduction of the network complexity (the number of units required and connections among them) becomes possible. We first review the extended energy function. Then, a connectionist model of minimizing the absolute value energy function is described. Lastly, we apply the new method to geometrical constraint-solving, and evaluate its validity from the viewpoint of reducing complexity.

13.2 The Extended Energy Function

In this section, we review the extended energy function which is defined on the Boltzmann machine.

A Boltzmann machine is a stochastic artificial neural network consisting of a number of two-state units mutually connected [3, 1]. Formally, the network is defined by a pseudograph (U, C), where U denotes a finite set of units and C is a finite set of unordered pairs of elements of U. These unordered pairs represent connections between the units. A *unit* u is assigned either 1 or 0 as its own state. A *connection* $\{u, v\}$ joins the units u and v. With each connection $\{u, v\}$, a *connection strength* $w_{\{u,v\}} \in \boldsymbol{R}$ is associated. The conventional energy function of the Boltzmann machine is defined by

$$E = - \sum_{\{u,v\} \in C} w_{\{u,v\}} u \cdot v.$$

To extend this energy function, a set of *logical connections* L is defined as $L = 2^U$. Then, we introduce a *connection strength function* $W : L \rightarrow \boldsymbol{R}$, and define an extended energy function as follows:

$$E = - \sum_{c \in L} W_c \prod_{u \in c} u.$$

This sigma-pi multilinear function is a natural extension of the conventional energy function. Liapunov stability of the extended energy function is well described in [7].

To reach a minimum of the extended energy function, a non-deterministic state transition mechanism is introduced. This mechanism allows units to adjust their states by a stochastic function of states of the neighbors and the corresponding connection strengths. Hence, Boltzmann machines can escape from local minima in the same manner as a simulated annealing algorithm. Without a procedure of a cooling schedule, the fundamental algorithm of the state transition is represented by an iteration of three steps as follows:

Step 1: Select one unit u at random.

Step 2: Calculate $g(u)$: the sum of input to u,

$$g(u) = \sum_{c \in L, c \ni u} W_c \prod_{v \in c - \{u\}} v.$$

Step 3: Assign 1 to the state of u with probability P.

$$P = \frac{1}{1 + exp(\frac{-g(u)}{T})}, \qquad \text{T: temperature of the network.}$$

13.3 Minimizing the Absolute Value Energy Function

Let us think of solving an nth-order constraint as follows:

$$\sum_{p \in L} a_p \prod_{k \in p} u_k = b,$$

$$a_p, b\text{: real constants}, \qquad u_k \in \{0,1\}\text{: binary variables},$$
$$L = 2^N, \qquad N = \{1, 2, 3, \cdots, n\}.$$

To obtain an optimal solution of the above formula, we introduce the ***absolute value energy function*** E defined as follows:

$$\begin{aligned} E &= |F|, \\ F &= \sum a_p \prod u_k - b. \end{aligned}$$

In the conventional method described in the former section, the state of a unit u is determined by only the *sum of input* g from the other units to u. If the sign is not considered, g is regarded as the amount of energy change associated with a state transition of the unit. That is,

$$\Delta E = E_{old} - E_{new} = \begin{cases} g & \text{if } u_{old} = 0 \ \& \ u_{new} = 1, \\ -g & \text{if } u_{old} = 1 \ \& \ u_{new} = 0. \end{cases}$$

In the case of the absolute value energy function, a state transition is handled in almost the same way as in the conventional method. The function F is considered as a *pseudo energy function*, and the sum of input g is also calculated. Since the real energy is the absolute value of F, however, there is some cases where g is not equal to the amount of energy change. These cases are:

1. $F < 0$,
2. $u = 0$, $g > E$,
3. $u = 1$, $-g > E$.

In case (1), the actual amount of energy change is $-g$ (Figure 13.1). Thus, in a state transition, the sign of F must be considered. In case (2), g is bigger than the current energy value E. Then the actual amount of energy change is $E - (g - E) = 2E - g$ (Figure 13.2). Similarly, in case (3), the actual amount of energy change is $E - (-g - E) = -(-2E - g)$ (Figure 13.3).

According to the case analysis above, let us introduce the fundamental sequential algorithm of a state transition. Here, *sign* represents the sign of F:

$$sign = \begin{cases} 1 & \text{if } F \geq 0, \\ -1 & \text{if } F < 0, \end{cases}$$

and *flag* represents the alternation of the sign of F:

$$flag = \begin{cases} 1 & \text{if the sign of } F \text{ changes}, \\ 0 & \text{if the sign of } F \text{ does not change}. \end{cases}$$

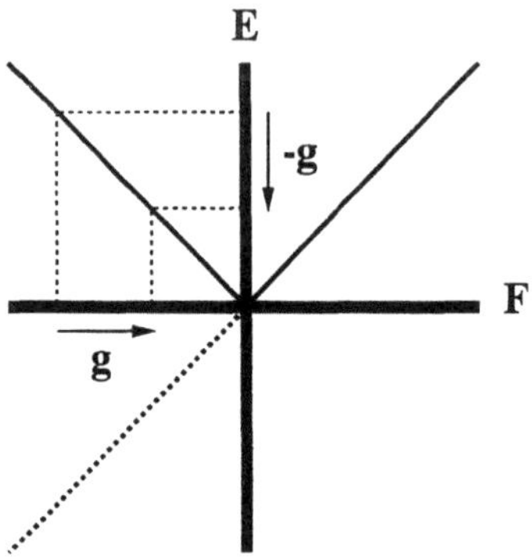

FIGURE 13.1. The case of $F < 0$.

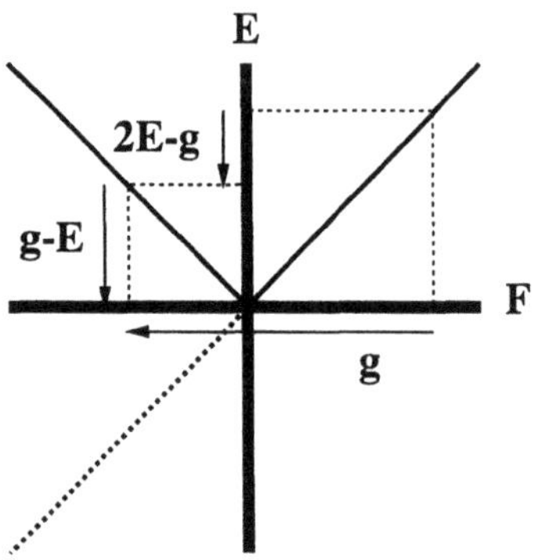

FIGURE 13.2. The case of $u = 0,\ g > E$.

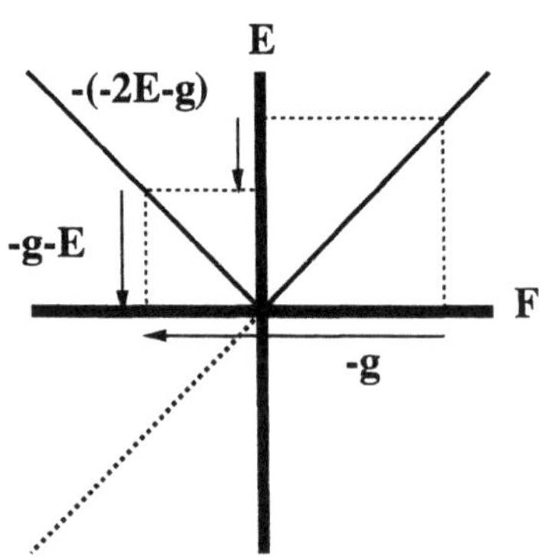

FIGURE 13.3. The case of $u = 1, -g > E$.

The algorithm is as follows:

Initialization: Calculate a pseudo energy function F.
If $F \geq 0$ **then** sign$\leftarrow 1$, $E \leftarrow F$.
else sign$\leftarrow -1$, $E \leftarrow -F$.

Step 1: Select one unit u_i at random.

Step 2: Calculate g_i: the sum of input to u_i,

$$g_i = - \sum_{p \in L, p \ni i} a_p \prod_{k \in p - \{i\}} u_k.$$

Step 3: **If** sign$= -1$ **then** $g_i \leftarrow -g_i$.

Step 4: **If** $u_i = 0$ & $g_i > E$
then $g_i \leftarrow 2E - g_i$, flag$\leftarrow 1$.
else if $u_i = 1$ & $-g_i > E$
then $g_i \leftarrow -2E - g_i$, flag$\leftarrow 1$.
else flag$\leftarrow 0$.

Step 5: Assign 1 to the state of u_i with probability P.

$$P = \frac{1}{1 + exp(\frac{-g_i}{T})}, \qquad \text{T: temperature of the network.}$$

Step 6: **If** the state of u_i has changed **then**
if flag$= 1$ **then** sign$\leftarrow -$sign.
$E \leftarrow E - g_i(2u_i - 1)$.

Step 7: Back to the Step 1 unless a stop criterion is met.

13.4 An Application: Geometrical Constraint-Solving

We have applied our new connectionist model associated with the absolute value energy function to geometrical constraint-solving [8, 6]. As described in [7], a connectionist approach based on neural networks is particularly suitable for solving insufficient and/or inconsistent constraints, which are typical in the fields of data visualization and layout problems [5]. In geometrical constraints, nonlinearity plays an important role. Thus, by using the absolute value energy function, the number of connections between units can be greatly reduced.

As an example, when coordinate values are represented as n-bit binary numbers, the energy function of a constraint "the length of a line segment AB is d on a 2-dimensional Euclidean space" is given by:

$$E = |(\sum_{i=0}^{n-1} 2^i(x_{0i} - x_{2i}))^2 + \sum_{i=0}^{n-1} 2^i(x_{1i} - x_{3i}))^2 - d^2|,$$

$$A : (x_0, x_1), \quad B : (x_2, x_3), \quad x_j = \sum_{i=0}^{n-1} 2^i x_{ji}.$$

Given an arbitrary line segment AB as an initial condition, by iterating state transitions, AB converges to a line segment the length of which is d. In this case, the position of the segment is determined only in a probabilistic way because no stationary constraints are given. This reflects the fact that an initially given constraint is insufficient.

If energy functions are linearly combined as below, a regular pentagram can also be drawn:

$$\begin{aligned} E &= |AB^2 - d^2| + |BC^2 - d^2| + |CD^2 - d^2| + |DE^2 - d^2| + |EA^2 - d^2| \\ &\quad + |AC^2 - e^2| + |CE^2 - e^2| + |EB^2 - e^2| + |BD^2 - e^2| + |DA^2 - e^2|, \end{aligned}$$

$$\text{where } d > e.$$

Figures 13.4 and 13.5 show a result of solving the regular pentagram constraints obtained by the prototype constraint solver.

Figure 13.4 represents an initial state, and Figure 13.5 represents the converged state. The graph below shows the state transition of the energy. This solver was implemented on Sun-4 SPARCstation 330 with XWindows. The coordinate values are represented by 4-bit binary numbers. If the conventional extended energy function is used, it has a 4-th order form, and the number of connections required is about 56000. On the other hand, if the absolute value energy function proposed is used, the number of connections is only 720. Thus, the absolute energy function achieves a tremendous reduction of network complexity, which increases the actual validity of the connectionist approach to geometrical constraint-solving.

13.5 Conclusions

In this paper, we proposed a new method of minimizing the absolute value energy function based on the Boltzmann machine, which achieves a great reduction of the number of connections between units. Our method is particularly valid for solving geometrical constraints, in which nonlinearity plays a definitive role. We expect an appearance of a new world of computer graphics, which can be called *Neurographics*, that is, an integration of neural networks and computer graphics.

As our future work, we consider the parallel state transition algorithm. The method described here utilizes the global information E accessed by all the units equally. Thus, to extend the method to a parallel situation, an asynchronous algorithm of a state transition including the estimation of global information must be considered. Some experimental results are given in [9]. In addition, now geometrical constraints are considered from the viewpoint of an orthogonal coordinate system, which is suitable for the conventional numerical computation. This might not be a good way to fully utilize the advantages of the connectionist models. The evaluation of another geometrical index system must be considered.

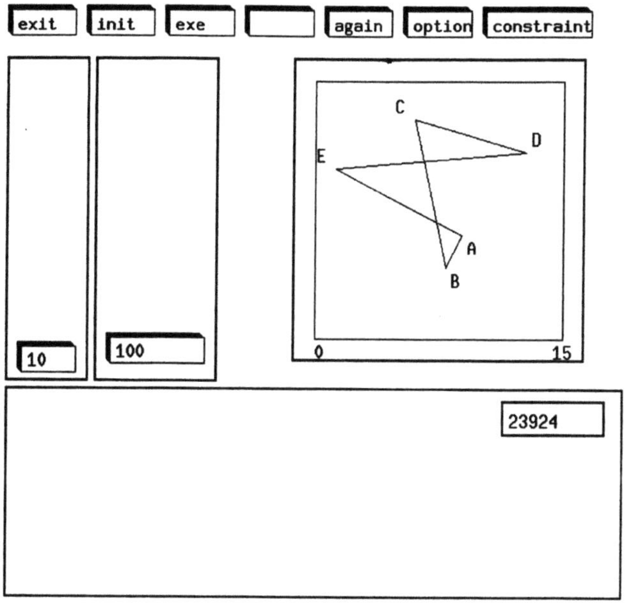

FIGURE 13.4. An example of the prototype system screen: the initial picture.

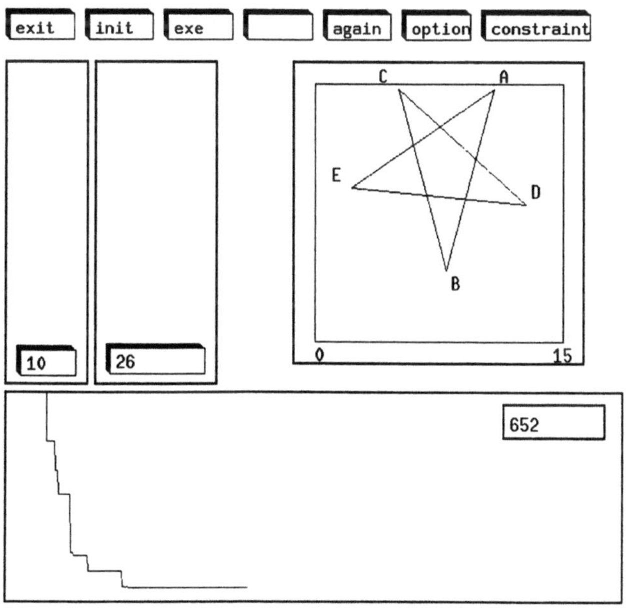

FIGURE 13.5. An example of the prototype system screen: the converged picture.

13.6 References

[1] E. Aarts and J. Korst. *Simulated Annealing and Boltzmann Machines.* John Wiley & Sons, 1989.

[2] J. A. Feldman and D. H. Ballard. Connectionist Models and Their Properties. *Cognitive Science*, 6:205–254, 1982.

[3] G. E. Hinton, T. J. Sejnowski, and D. H. Ackley. Boltzmann Machines: Constraint Satisfaction Networks that Learn. *Technical Report CMU-CS-84-119*, 1984.

[4] J. J. Hopfield. Neural Networks and Physical Systems with Emergent Collective Computational Abilities. *Proceedings of National Academy of Sciences*, 79:2554–2558, 1982.

[5] T. Kamada. *On Visualization of Abstract Objects and Relations.* PhD thesis, Dept. of Information Science, Faculty of Science, The University of Tokyo, 1989.

[6] N. Kin, T. Noma, and T. L. Kunii. PictureEditor: A 2D Picture Editing System Based on Geometric Constructions and Constraints. In R. A. Earnshaw and B. Wyvill, editors, *New Advances in Computer Graphics: Proceedings of CG International '89*, pages 193–207. Springer-Verlag, 1989.

[7] N. Kin, Y. Takai, and T. L. Kunii. A Connectionist Approach to Geometrical Constraint-Solving. In T. L. Kunii, editor, *Modeling in Computer Graphics: Proceedings of IFIP WG 5.10 Working Conference*, pages 367–380. Springer-Verlag, 1991.

[8] W. Leler. *Constraint Programming Languages: Their Specification and Generation.* Addison-Wesley, 1988.

[9] H. Sato. A Parallel Method for the Extended Boltzmann Machine Based on the Absolute Energy Function. Graduation thesis, Dept. of Information Science, Faculty of Science, The University of Tokyo. in preparation.

[10] M. Takeda and J. W. Goodman. Neural Networks for Computation: Number Representations and Programming Complexity. *Applied Optics*, 25(18):3033–3046, 1986.

[11] D. W. Tank and J. J. Hopfield. Simple Neural Optimization Network: An A/D Converter, Signal Decision Circuit, and a Linear Programming Circuit. *IEEE Transactions on Circuits and Systems*, 33(5):533–541, 1986.

14 Geometric Constraint Propagation with Quantum Labels

Remco C. Veltkamp
Farhad Arbab

ABSTRACT
This paper presents an incremental approach to geometric constraint satisfaction that categorizes solutions into so-called quanta. A quantum is a range of solutions with uniform geometric characteristics. This approach is suitable for interactive design because it can handle (perhaps temporarily) incomplete specifications and alternative solutions, and performs satisfaction locally and incrementally. The intermediate solutions are kept in the geometric domain, so that new geometric constraints can be interpreted on the same high level of abstraction, allowing powerful reasoning. Both alternative discrete solutions and continuous ranges of solutions are determined.
Propagation of information is performed by 'propagation of known state', and the information itself results from 'solution set inference', where a solution set is represented in geometrical form by unions and intersections of quantum labels. This representation allows the use of simple logical expressions of constraints, it uniformly handles delayed satisfaction, and it supports 'constraint inference', the derivation of implied constraints by geometric reasoning.

14.1 Introduction

Incremental satisfaction of constraints arises naturally in design applications. A good deal of work has already been done on constraint satisfaction in general [10, 8], and constraint satisfaction in geometry in particular [16, 2, 11, 13, 1, 5] and [14]. Both areas are known to be difficult and no robust methods exist for sufficiently general applications.

Incremental satisfaction of geometric constraints poses additional complications in constraint problem solving. There are two important factors that set incremental satisfaction of geometric constraints (in the context of computer aided design or geometric modeling) apart from other constraint satisfaction methods:

- The constraint satisfaction system must perform useful work and provide *meaningful* feedback to its users, especially, in under-constrained situations.
- The feedback provided by the system must be primarily geometric.

While a design is in progress, it is natural to expect under-constrained situations to be predominant: often not enough information is available to make solutions more specific. Many systems detect over-constrained situations but cannot gracefully handle under-constrained cases. Indeed, many numerical methods cannot solve a set of under-constrained equations. On the other hand, a constrained logic programming system for solving numeric constraints (e.g. [7]) returns a set of deduced equations that forms the solution. However, the solutions and feedback are not in a geometrical form. The operational approach [13, 1] yields only one of all the solutions. This gives rise to two potential problems:

- The proposed solution may be as was intended but the user is not aware of the ambiguity of the specification. This may cause problems for post-processing.

- The solution is not as was intended and the user must interfere because the system cannot propose alternative solutions.

We distinguish two types of alternative solutions:

- A constraint can have several discrete solutions. For example, a circle through two points and a fixed diameter greater than the distance between the points, can have two positions.
- A constraint can have a continuous range of solutions. For example, the locus of a point having a fixed distance to a fixed point is a circle.

Most graphics oriented constraint systems directly translate the (geometric) constraints into a set of numeric constraints or equations. Often, only linear equations can be solved. Like in [1] we want to solve constraints at a high level of geometric abstraction. This allows powerful reasoning which can avoid some of the problems that occur with numeric constraints: recall that even a simple distance constraint results in a quadratic equation.

The rest of this paper is organized as follows. The next section gives an overview of related work in constraint propagation. Section 14.3 introduces our approach using quantum labels, a formal model for which is presented in Section 14.4. The propagation algorithm is given in Section 14.5, and its termination properties are analyzed in Section 14.6. Section 14.7 is about implied constraint inference, and Section 14.8 discusses completeness. A non-trivial example of constraint solving using quantum labels is given in Section 14.9. Some further remarks are made in Section 14.10, and final conclusions are given in Section 14.11.

14.2 Related work

We can classify satisfaction techniques into *structured* and *unstructured* methods. Structured methods impose a structure on the set of constraints by grouping them into sets of dependent constraints. These sets are satisfied independently. Unstructured techniques perform no structuring. An obvious drawback of unstructured techniques in interactive applications is that each single change or addition to the set of constraints results in the re-evaluation of the whole set of constraints, while the real effect might be only local.

An orthogonal classification can be made into *numerical computation* and *deductive* techniques. Numerical methods turn all constraints into equations and solves these numerically by, for example, relaxation or Newton-Raphson iterations, either in a structured or an unstructured way. An example of unstructured numerical constraint satisfaction is Variational Geometry [9].

A set of dependent constraints can be represented as a *network* of constraints. Structured deductive systems use some method to assimilate results of the inference process throughout the network. We mention two such propagation methods.

Propagation of known states, or just local propagation, can be performed when there are parts in the network whose states are completely known (have no degrees of freedom). The satisfaction system looks for one-step deductions that will allow the states of other parts to be known. This is repeated until all constraints are satisfied or no more known states can be propagated. If not all constraints can be satisfied, the remaining constraints must be resolved by, for example, numerical relaxation. Most constraint satisfaction systems use some form of local propagation.

Propagating degrees of freedom amounts to discarding all parts of the network that can be satisfied easily and solving the rest by some other method. This method identifies a part in the constraint network with enough degrees of freedom so that it can be changed to satisfy all its constraints. That part and all the constraints that apply to it are then removed from the network. Deletion of these constraints may give another part enough degrees of freedom so as to satisfy all its constraints. This continues until no more degrees of freedom can be propagated. The part of the network that is left is then satisfied by some other method, if necessary, and the result is propagated towards the discarded parts, which are successively satisfied (propagation of known states). For example, Sketchpad [16], and [12] use propagation of degrees of freedom.

The above techniques are two methods of propagation, independent of *what* is actually propagated. We distinguish the following types of information inference and propagation.

Solution set inference makes deductions on the set of possible solutions, which are restricted by the constraints. Most frequently used are discrete value sets, or intervals of numerical values (see [4, 14]). Only when the full solution set is derived, and the solution does not depend on the order in which the constraints are solved, the declarative semantics of constraints are preserved.

In *single solution inference*, constraint variables get assigned a single value, often numeric. A system with single numeric solution inference performed by algebraic methods and propagation of degrees of freedom is presented by [12]. The *operational approach* [13, 1] performs single *geometric* solution inference. It satisfies constraints sequentially by performing operations (translation, rotation, etc.) on the geometric objects involved. An already satisfied constraint either tolerates an operation on one of its operands, or must propose a transformation to satisfy the constraint again. In this way operations can be propagated through a constraint network until all operations are tolerated. Selection of alternatives based on minimal disturbance [5] is another example of single solution inference.

Local *algebraic expression inference* is a means to deal with loops in propagating numeric values. [15] used powerful algebraic manipulation in the system CONSTRAINTS, but found that these techniques are not powerful enough to solve many interesting problems that people can solve. The way people usually solve these problems is by organizing the solution so that simple canned algebraic solutions suffice.

In *constraint inference*, implied constraints are derived and explicitly added to the network. Implied constraints can be recognized by a unification mechanism or by the use of multiple redundant views [15]. Finding implied constraints can be used to avoid extensive manipulations in cases where local propagation does not suffice. Stating constraints in a different way with the same meaning can help the system to solve the constraints locally. Otherwise it may have to resort to techniques such as relaxation.

14.3 Quantum labels

In this paper we focus on providing ranges of solutions and discrete alternative solutions in under-constrained situations, by means of a quantum approach. In combination with geometric reasoning and delayed satisfaction, this gives a more powerful problem solving capability than usually provided by constraint systems. The quantum approach is first introduced in [18]. We first give an informal presentation, but the main ideas are formalized in the subsequent sections.

Constraints work on constraint variables, which are geometric primitives such as a point, a line (segment), or a circle.

Our approach to geometric constraint satisfaction is based on reduction of both problem complexity and solution domain complexity by geometric simplification. Problem

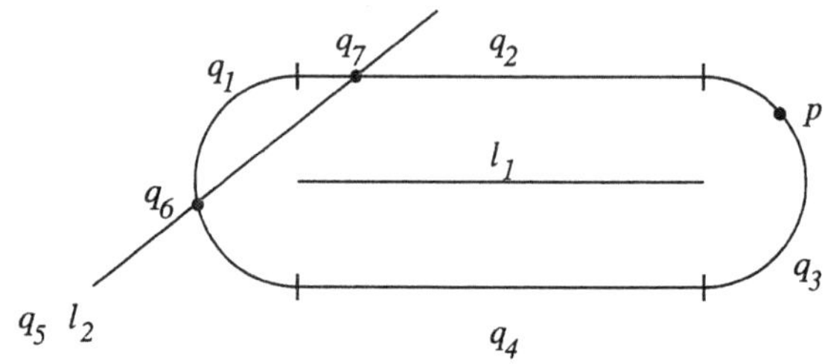

FIGURE 14.1. Quanta resulting from the constraints distance(p, l_1, d) and on(p, l_2).

complexity reduction is achieved by expressing geometric primitives in terms of characteristic parts. For example, the position of a circle is reduced to the position of a point: the circle center. Constraint solving amounts to recording the solution of the constrained geometric primitives, in terms of their characteristic parts.

The unit of the solution domain is a 'quantum', which is itself a geometric primitive:

> A quantum is a geometric primitive describing a part of the solution set with uniform geometrical characteristics.

We use the term quantum because it captures both 'interval' and 'region', and is more descriptive than 'solution set'. The nature of a quantum is that it describes a sharply bounded quantity of some phenomenon, in our case the geometric properties of a solution set. Note that quanta are not indivisible, nor need they be disconnected.

For example, the locus of a point p having a fixed distance to a line segment l_1 consists of the union of two line segments and two half circles, see Figure 14.1. This solution is split into the quanta $q_1, \ldots, q_4$, representing the two line segments and two half circles. These quanta represent alternative sets of solutions. If p is further constrained to lie on a line l_2 (another quantum), the final solution is restricted to the intersection of l_2 with all the alternative solutions so far. Since calculating the intersection of a line and a circle differs from intersecting two lines, we take advantage of the subdivision into quanta $q_1, \ldots, q_4$.

Some simple constraints and the associated quanta are listed below:

TABLE 14.1. Examples of quanta in 2D

primitive	constraint	primitive	quanta
point	distance	point	circle
point	distance	line segment	2 line segments + 2 half circles
point	left/right of	line	half-plane
circle	tangent	line	2 lines
circle	tangent	circle	2 circles

The second variable is fixed.

In general, a solution set may not be representable with a fixed set of geometric primitives, especially in 3-space. However, the quantum approach is useful in 3D as well: a case study in 3D is presented in [17]. Some useful constraints in 3D are listed in Table 14.3.

A set of constraints and geometric primitives can be represented as a network, such that the constraint node is connected to its variables. The network can be disconnected, in which case the variables in one connected component and the variables in other connected

TABLE 14.2. Examples of quanta in 3D

primitive	constraint	primitive	quanta
line	angle	line	cone
line	angle	plane	plane
point	distance	point	sphere
point	distance	line	cylinder
line	distance	line	cylinder
sphere	tangent	line	cylinder
sphere	tangent	sphere	2 spheres

The second variable is fixed.

components are not related by constraints. Constraints are represented in rectangular boxes, geometric primitives in circular nodes. A quantum for a variable v, consistent with constraint $c(\ldots, v, \ldots)$, can be considered as a label on an edge between (v) and [c]. Being a geometric primitive, a quantum is represented as a circle but with the edge that it labels drawn through.

In order to represent unions and intersections of quanta, the special Boolean set operation nodes <∪] and <∩] are used. These nodes are operators whose operands are the quantum labels of edges coming in at the flat sides of the nodes and whose results are the quantum labels of the edges going out at the sharp ends of the nodes. If there are no quantum labels on the edges going out, these quanta are not (yet) computed. Similarly, if there are no quantum labels on the incoming edges, they are not yet generated or computed. In both cases, the logical relation between these forthcoming quanta is already incorporated in the network.

Between a constraint and an associated variable is a sequence of (zero or more) quanta interleaved with Boolean set operation nodes, on a path of edges. In the direction from the constraint to the variable each quantum includes the next one. Figure 14.2 shows a network that corresponds to the example of Figure 14.1. The network shown is an intermediate result in the satisfaction process.

Our approach to geometric constraint satisfaction is thus based on solution set inference, and performs propagation of known states (see the next section).

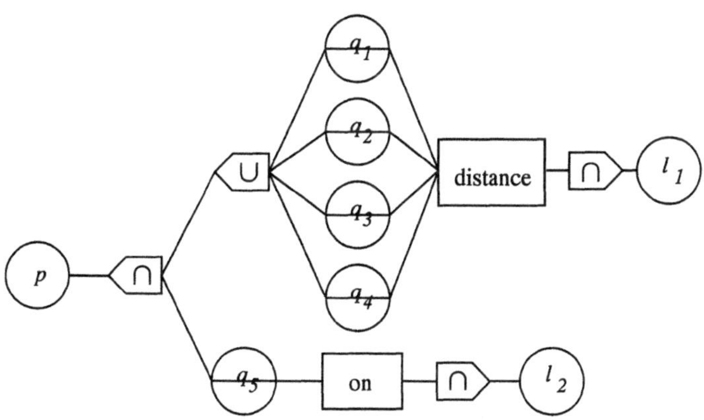

FIGURE 14.2. Intermediate network corresponding to Figure 14.1.

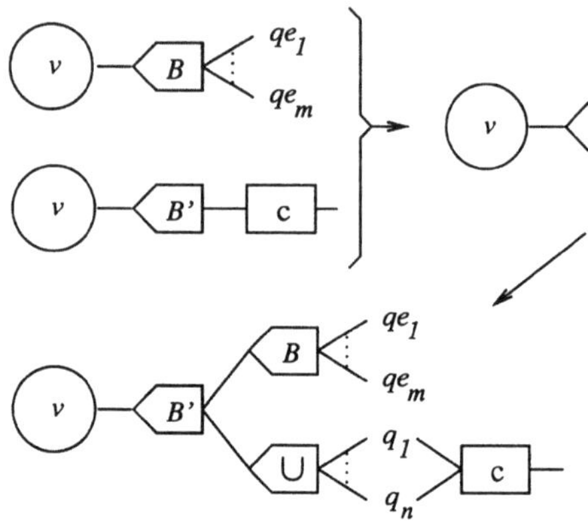

FIGURE 14.3. Addition of $c(v, \ldots)$ to a network, and generation of quanta.

14.4 Formal model

Our model consists of a set of constraints C, a set of constraint variables V, a set of quanta Q, a quantum generating function G, and a function Tolerate(c) that satisfies a constraint c and performs propagation. The constraint variables and quanta are geometric primitives.

A constraint is a relation between a number of variables from V. A constraint can involve any number of variables, depending on its type. Constraints are multi-directional in the sense that each of the involved variables constrains the others. In contrast to many other systems, many variables may be undetermined. Constraints are specified incrementally, so that every variable in V is involved in an ordered set of constraints. Each constraint can be either *restrictive* or *alternative*. A restrictive constraint further restricts the degrees of freedom of the involved variables. An alternative constraint provides alternatives to the solution of *previously* specified constraints on that variable. Considering a single variable and only constraints on that variable, the constraints thus form an expression of the form

$$((c_1 \text{ AND/OR } c_2) \text{ AND/OR } \ldots) \text{ AND/OR } c_k.$$

A constraint expression like $(c_1 \text{ OR } c_2)$ AND $(c_3 \text{ OR } c_4)$ for example is not possible in this scheme.

The quantum generating function $G : (c, v_i) \mapsto (q_1, \ldots, q_n)$ generates the set of quanta for v_i that are consistent with the constraint $c(v_1, \ldots, v_k)$ and all the variables v_j, $j = 1, \ldots, k$, $i \neq j$, with their current quanta, regardless of the current quanta of v_i. Note that not all combinations of types of constraints and variables are geometrically meaningful; G is defined only for a finite number of combinations. The generation of the quanta involves geometric reasoning, taking into account the nature of the constraint and the current quanta of the variables. Figure 14.3 illustrates the addition of a constraint to a network and the subsequent generation of quanta. Only the part of the network in the neighborhood of variable v is depicted.

When a new constraint is added to the network, the quantum generating function is applied to each constrained variable. For a single variable, the union of the resulting quanta, $\cup(q_1, \ldots, q_n)$, is the solution to that one constraint. If the constraint is restrictive, $q_1, \ldots, q_n$ must be intersected with the variable's current quanta. If it is an alternative constraint, we take the union of the new and the current quanta. In this way we get an expression of quanta, which is the solution set of the constraints so far. In general, this quantum expression is of the form $B(qe_1, \ldots, qe_m)$, where B is one of the Boolean set operators union or intersection, and each qe_i is a quantum expression.

After addition of a new constraint, generating new quanta $q_1, \ldots, q_n$, the solution set becomes $B'(B(qe_1, \ldots, qe_m), \bigcup(q_1, \ldots, q_n))$. For a restrictive constraint B' is $\bigcap$, for an alternative constraint B' is $\bigcup$. This expression is simplified according to the following network transformation rules:

1. $\bigcup(\bigcup(q_1, \ldots, q_n)) = \bigcup(q_1, \ldots, q_n)$
2. $\bigcap(\bigcup(q_1, \ldots, q_n)) = \bigcup(q_1, \ldots, q_n)$
3. $\bigcup(\bigcup(qe_1, \ldots, qe_m), \bigcup(q_1, \ldots, q_n)) = \bigcup(qe_1, \ldots, qe_m, q_1, \ldots, q_n)$
4. $\bigcup(\bigcap(qe_1, \ldots, qe_m), \bigcup(q_1, \ldots, q_n)) = \bigcup(\bigcap(qe_1, \ldots, qe_m), q_1, \ldots, q_n)$
5. $\bigcap(\bigcup(qe_1, \ldots, qe_m), \bigcup(q_1, \ldots, q_n)) = \bigcup_{i,j}(q_i \cap qe_j)$
6. $\bigcap(\bigcap(qe_1, \ldots, qe_m), \bigcup(q_1, \ldots, q_n)) = \bigcup_{i=1}^{n} \bigcap(qe_1, \ldots, qe_m, q_i)$

Rules 1 and 2 apply when the constraint is the first one connected to v_i.

Each qe in the other rules is either a single quantum or an intersection of other qe-s, because of the effect of rule 3. Any intersection $q_i \cap qe_j$ on the right-hand sides of equations 5 and 6 that can be computed, is evaluated into a union of quanta: $q_i \cap qe_j = \bigcup(q'_1, \ldots, q'_k)$. It may be difficult to compute the intersection, for example if one of the quanta is a circle that can float around. In that case we leave the intersection unevaluated (see also Section 14.8). This is why a qe_j can be an intersection of quanta or other qe-s: $qe_j = \bigcap(qe'_1, \ldots, qe'_k)$.

The right-hand side of equation 6 may be split into an intersection that can be computed and a part that remains unevaluated.

Figure 14.4 shows the network transformations that correspond to the evaluation rules.

14.5 Propagation

Let $S_i = G(c, v_i)$. We let the procedure Evaluate(v_i, S_i) compute intersections, transform the network accordingly as described in the previous section, and return EMPTY if the resulting solution set of v_i is empty, CHANGED if it has been changed, and UNCHANGED if it is unchanged. We can call $G(c, v_i)$ and apply Evaluate(v_i, S_i) for each variable of c.

If one of the variables gets assigned an empty solution set, the constraint expression is inconsistent — the specification is over-constrained. If the solution set of a variable v_i has been changed and v_i is involved in another constraint c', the other variables of c' need also be updated. These variables must therefore be recorded. The following procedure Revise(c) updates all variables of c, and returns the set of variables whose solution set is changed (adapted from [4]):

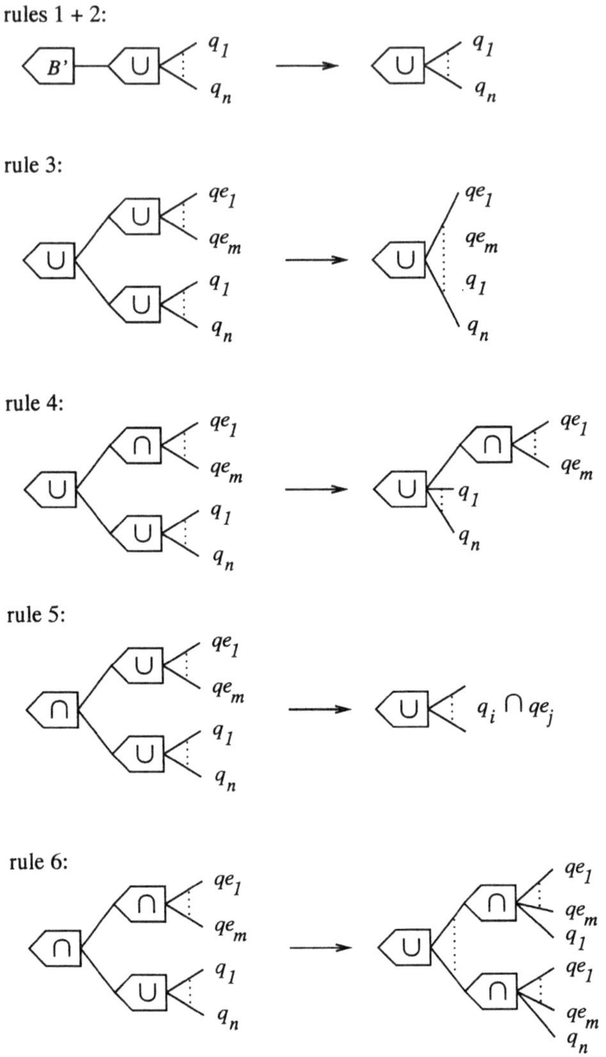

FIGURE 14.4. Network transformations corresponding to the evaluation rules.

```
Revise (c)
{
   Changed = ∅
   for each v_i of c
      S_i := G(c, v_i)
   for each v_i of c
      case Evaluate (v_i, S_i)
      {
         EMPTY: Report (OVER-CONSTRAINED)
         CHANGED: add v_i to Changed
         UNCHANGED:                              // do nothing
      }
   return Changed
}
```

When a new constraint is specified, it must be added to the network, and the resulting changes must be propagated to related constraints. This may cause new changes which in turn must be propagated, and so on. This is performed by the following algorithm:

```
Tolerate(c)
{
   Fifo = [c]                                    // first-in-first-out queue
   while Fifo ≠ ∅
   {
     c' = FirstElementOf (Fifo)
     remove c' from Fifo
     Changed = Revise (c)
     for each v ∈ Changed
        for each c'' ≠ c' having v as parameter
           add c'' to Fifo
   }
}
```

Revise(c) makes the solution set of every variable of c consistent with c and its other variables using their current quanta. That is, the solution is made locally consistent. When Tolerate(c) terminates (see next section), and an over-constrained case has not been detected, the solution sets of all variables are locally consistent (this is equivalent to arc consistency in [10]).

When only restrictive constraints are added to a network, Tolerate(c) amounts to the propagation algorithm of [19], filtering the solution sets to get subsets which are locally consistent.

Because constraints that have variables with changed quanta are placed in a first-in-first-out queue (as opposed to an unordered list), they are guaranteed to be revised again, whether the propagation is finite or infinite. This is called a fair propagation in [6].

14.6 Termination

This section discusses termination of the propagation in Tolerate(c). We treat the cases where c is a restrictive or an alternative constraint separately.

14.6.1 Restrictive constraint

When a restrictive constraint $c(v_1, \ldots, v_k)$ is added to the network, the solution sets of variables do not get larger. More formally, Evaluate($v_i, G(c, v_i)$) is then monotonic in the following sense: let $\{[S_1, \ldots, S_k]\text{Evaluate}(v_i, G(c, v_i))\}$ denote the solution set of v_i after applying Evaluate($v_i, G(c, v_i)$) with S_j the current solution set of v_j, $j = 1, \ldots, k$. Observe that:

1. $\{[S_1, \ldots, S_k]\text{Evaluate}(v_i, G(c, v_i))\} \subseteq S_i$
2. if $S'_j \subseteq S_j$, $j = 1, \ldots, k$, then
   $\{[S'_1, \ldots, S'_k]\text{Evaluate}(v_i, G(c, v_i))\} \subseteq \{[S_1, \ldots, S_k]\text{Evaluate}(v_i, G(c, v_i))\}$.

Because of the monotonicity of Evaluate($v_i, G(c, v_i)$), and the fairness of propagation in Tolerate(c), results from [6] (adapted to our context) apply: if there exists some terminating propagation resulting in a locally consistent solution, then Tolerate(c) will find it. Moreover, the solution is unique.

Still, the propagation in Tolerate(c) need not terminate. A simple example is where two line segments are related by a constraint that requires their lengths to be equal, and a constraint that sets one length to half the length of the other (a pure numeric analogue is the case of two constraints $x = y$ and $x = 2y$). This can lead to an infinite loop halving each line segment in turn.

A loop occurs when a constraint is revised more than once, in one invocation of Tolerate(). When this happens, it is not clear whether this is the prefix of an infinite loop or not, but we must break the loop after a finite number of times. One possibility is to terminate propagation by force when a constraint is addressed a fixed number of times. A more sophisticated approach is to break a loop when there are no more substantial changes made to the solution set. Specifically, when none of the quanta of a variable change type (for example change from a line segment into a point), and no quantum changes more than a certain 'geometric margin', we can break a loop. After a loop has been cut-off, the current quanta of the variables in the loop form a super-set of the real solution. There are several things we can do with such a super-set:

1. present the super-set to the user, which is still better than many graphics systems can provide,
2. compute the solution set numerically, using the super-set to obtain initial values and bounds, if necessary,
3. compute a single solution from the solution set,
4. test whether a solution exists instead of computing a solution (set).

14.6.2 Alternative constraint

If an alternative constraint is added to the network, the solution sets of variables may get larger. This can more easily lead to an infinite loop than when the solution is restricted. When a variable involved in a circular chain of constraints gets assigned alternative quanta, these constraints may produce alternative quanta for their variables ad infinitum. Again, a loop need not be the prefix of an infinite loop, but propagation must be stopped. A loop can be stopped after a constraint is addressed a fixed number of times, or when the increase of the solution set exceeds a certain 'geometric margin'.

A (possibly infinite) increase of the solution may not have been the intention of the user, and the system can ask whether the alternative must be redrawn. In any case, after

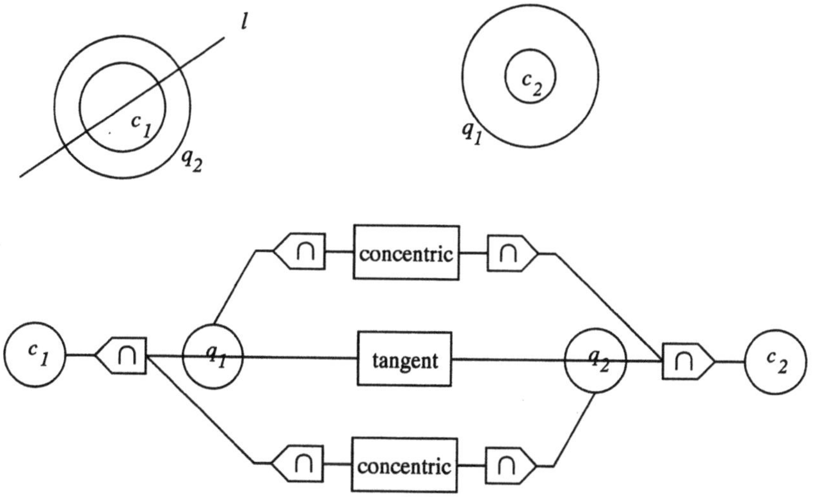

FIGURE 14.5. Implied constraints concentric(c_1, q_2) and concentric(c_2, q_1).

a loop has been cut-off, the current quanta of the variables in the loop form a sub-set of the real solution. We can use this result to:

1. present the sub-set to the user, which is still better than many graphics systems can provide,
2. try to generate a super-set of the solution, and proceed with propagation in order to restrict the solution.

14.7 Implied constraints

Our system uses geometric simplification, for example by shrinking a circle to a point, and regarding the radius as a distance. In this way the constraint that a circle c_1 is tangent to another circle c_2, is regarded as the constraint that the shrinked c_1 has a distance to the shrinked c_2 that is equal to the sum of their radii $r_1 + r_2$. That is, c_1 lies on the circle q_1 with radius $r_1 + r_2$, and concentric with c_2.

In order to simplify geometric reasoning, we exploit implied constraints: constraints that can be derived from other constraints and variables. Implied constraints can operate on variables and quanta. Indeed a quantum is a geometric primitive, and can thus act as a constraint variable, while it still represents a solution to another constraint. The user has no direct access to these implicit constraints.

See Figure 14.5 for a situation where two tangent circles and their corresponding network explicitly represent some useful implied constraints. The quanta q_1 and q_2 are circles concentric with c_2 and c_1, respectively. As c_1 moves around, for example along line l, so does q_2.

An implied constraint is derived from other constraints and variables, and so it does not introduce new facts. Its purpose is to ease the problem solving by an extra simplification. In our example, the coincidence of the centers of c_1 and q_2 causes changes to c_1 to mechanically carry over to q_2.

The type of an implied constraint (restrictive or alternative) must be consistent with the constraint(s) it is derived from, otherwise it would provide incorrect restriction or alternatives. Again, this introduces no more restrictions or more alternatives to the variables of the implicit constraint. For example if parallel(l_1, l_2) AND parallel(l_2, l_3), then parallel(l_1, l_3) is implied, and must also be restrictive.

14.8 Completeness

Completeness of our constraint system depends on the set of geometric primitives, the allowable constraints, the power of geometric reasoning behind the quantum generating function, and the system's competence to intersect quanta. This last aspect is illustrated by the example in Figure 14.5. As the circle c_1 moves along line l, q_2 moves accordingly, thus sweeping out a whole area of solutions. Although the area has a simple shape, its definition in terms of a swept circle makes intersection with subsequent quanta non-trivial. In this example the intersection is clearly computable, but a particular system may not be able to perform this intersection.

When we cannot compute an intersection of two quanta, we leave the intersection unevaluated, and represent it with a $\cap$ node in the network. A following intersection may resolve the problem, as the following extension of our example illustrates. If c_2 is further constrained to lie on line segments m_1 and m_2, and the intersection of m_1 with the swept circle q_2 is left unevaluated, the next intersection $(q_2 \cap m_1) \cap m_2$ is transformed into $(q_2 \cap m_2) \cap (m_1 \cap m_2)$. If $m_1 \cap m_2 = \emptyset$, so is the final solution. If $m_1 \cap m_2$ is a point, it is easier to test if it is included in $(q_2 \cap m_2)$ than to compute $(q_2 \cap m_2)$ itself.

In this way, delayed satisfaction of constraints is handled uniformly, thereby enhancing the satisfaction capabilities of the system.

In our examples, we used only points, lines, and circles as geometric primitives, and tangent(), parallel(), on(), and distance() as constraints. With these collections of primitives and constraints, we can compute the solution sets when the system gets into trouble, by turning the geometric information into equations and solve them algebraically [12] or numerically.

In the case of more general primitives and constraints, the system may resort to some analytical method or simply specify to the user what caused the problem.

14.9 Example

Consider the four-bar linkage of Figure 14.6 (top left), involving points A, B, C, D, and M. The points A and B are fixed and a distance d_1 apart. The constraints on C, D, and M are the following:

constraint 1: the distance between A and C is d_1;
AND constraint 2: the distance between B and D is d_2;
OR constraint 3: the link BD is vertical;
AND constraint 4: the distance between C and D is d_3;
AND constraint 5: M is the midpoint of C and D;
AND constraint 6: the line through A and C has a fixed slope.

These constraints are specified in the given order, and we are interested in the position of C, D, and M.

This problem can be solved algebraically by computing the locus of C, D, and M each time a constraint is specified. As an illustration, set A to $(0,0)$ and B to $(0, d_1)$, without loss of generality, and ignore constraint 3 for the moment. Then the locus of $M = (x, y)$

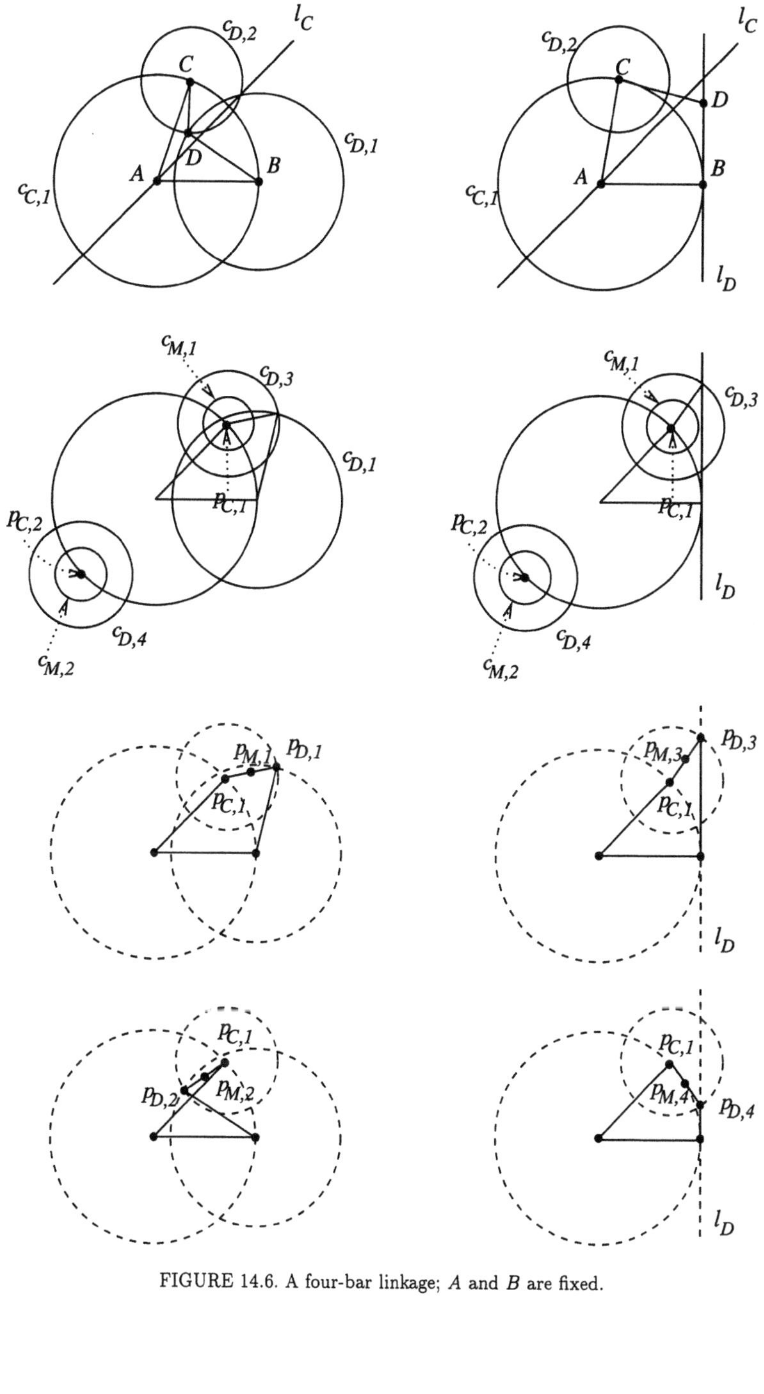

FIGURE 14.6. A four-bar linkage; A and B are fixed.

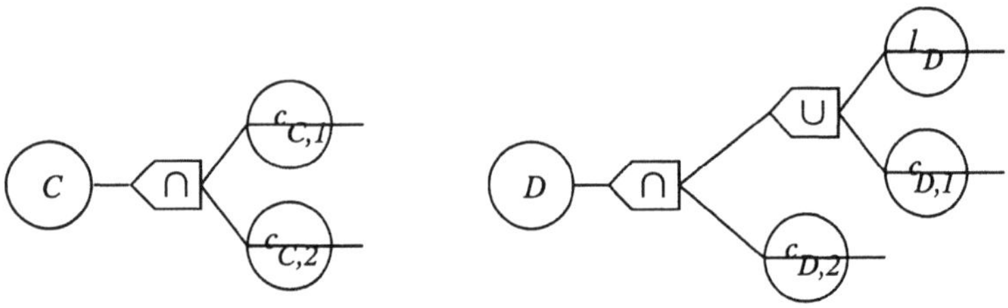

FIGURE 14.7. Partial network in the neighboods of C and D.

just after specifying constraint 5 is given by a polynomial of degree six. If $d_2 = d_3$ this polynomial reduces to:

$$16(x^4+y^4)+32x^2y^2-16d_1(x^3+xy^2)-(12d_1^2+4d_2^2)(x^2+y^2)+(8d_1^3-4d_1d_2^2)x = d_1^2d_2^2-4d_1^4$$

(see [3]).

This equation can be derived by an algebraic manipulation system, but that is of little use here because it does not interact in a simple way with the subsequent constraint.

Alternatively, our constraint management system comes up with the following quanta:

by constraint 1: C is restricted to a quantum that is the circle $c_{C,1}$ centered at A and having radius d_1;
by constraint 2: D is restricted to the circle $c_{D,1}$ centered at B with radius d_2;
by constraint 3: D is restricted to the vertical line l_D through D;
by constraint 4: for C: a circle $c_{C,2}$ with radius d_3 which has no fixed position, but is centered at D; similarly for D: a circle $c_{D,2}$ concentric to C, having radius d_3;
by constraint 5: because M is the midpoint of two points whose positions are unknown, no quantum is generated yet for M;
by constraint 6: since the line through A and C must have a fixed slope, another quantum is associated to C: line l_C.

The quanta $c_{C,1}$, $c_{D,1}$, $c_{D,2}$, l_C, and l_D are shown at the top row of Figure 14.6, where the two possible cases induced by the alternative constraint are depicted in separate columns.

The quanta resulting from constraint 4, $c_{C,2}$ and $c_{D,2}$, are not fixed but can move around. Consequently, it is difficult to intersect these quanta with the other quanta $c_{C,1}$, $c_{D,1}$, and l_D. The intersection remains unevaluated, but is represented in the network, see Figure 14.7.

The next quantum for C, l_C, must be intersected with the current quanta. Application of rule 6 (with $n = 1$) gives $\bigcap(\bigcap(c_{C,1}, c_{C,2}), l_C) = \bigcap(c_{C,1}, c_{C,2}, l_C)$, see Figure 14.8.

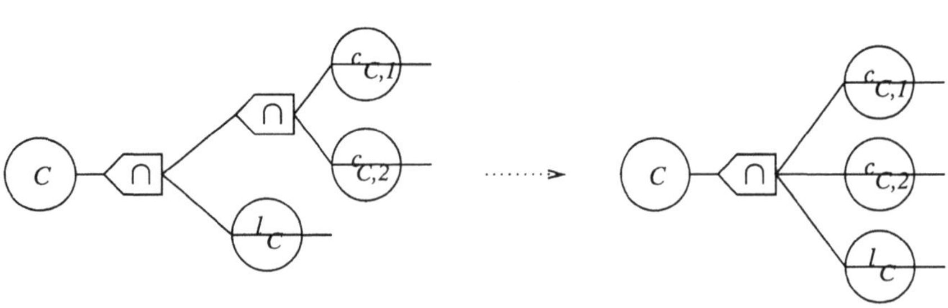

FIGURE 14.8. Addition of quantum l_C.

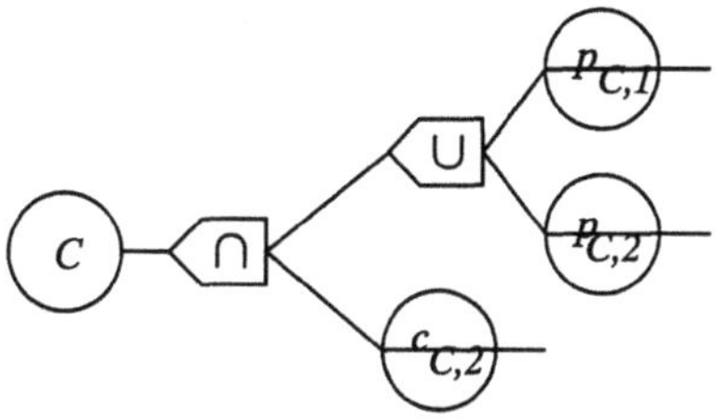

FIGURE 14.9. The intersection with $c_{C,2}$ is delayed.

The intersection $l_C \cap c_{C,1}$ can be computed and gives two points $p_{C,1}$ and $p_{C,2}$, the intersection with $c_{C,2}$ is delayed, see Figure 14.9.

This intermediate result must be propagated to objects related to C. Because C is now restricted to at most $p_{C,1}$ and $p_{C,2}$, the movable quantum $c_{D,2}$ of D is replaced by two circles $c_{D,3}$ and $c_{D,4}$ centered at $p_{C,1}$ and $p_{C,2}$ respectively. These circles can be seen as two specific instances of $c_{D,2}$. Also, since M is the midpoint of C and D, it is now restricted to two circles $c_{M,1}$ and $c_{M,2}$, concentric with $c_{D,3}$ and $c_{D,4}$ but with half their radius. The new quanta are depicted in the second row Figure 14.6.

Quanta $c_{D,3}$ and $c_{D,4}$ must be intersected with $c_{D,1}$ and l_D, see Figure 14.10. Application of rule 5 gives $\bigcap(\bigcup(c_{D,1}, l_D), \bigcup(c_{D,3}, c_{D,4})) = \bigcup(\cap(c_{D,1}, c_{D,3}), \cap(c_{D,1}, c_{D,4}), \cap(l_D, c_{D,3}),$ $\cap(l_D), c_{D,4})$, which can be computed, yielding $p_{D,1}, \ldots, p_{D,4}$.

Again this intermediate result must be propagated. Since D is now restricted to the posotions $p_{D,1}, \ldots, p_{D,4}$, the quanta $p_{C,2}$ and $c_{M,2}$ are deleted, and $c_{M,1}$ is replaced by $p_{M,1}$ and $p_{M,2}$. The four possible solution configurations are shown in the two lower rows of Figure 14.6.

14.10 Further remarks

Operations on a variable, such as translate, scale, mirror, and project, can be handled by performing proper corresponding operations on its quanta. These new quanta must then be propagated through the network.

For interactive CAD purposes, a constraint must also be removable. In our system, deletion from the network, involves removal of the constraint and the sequence of quanta

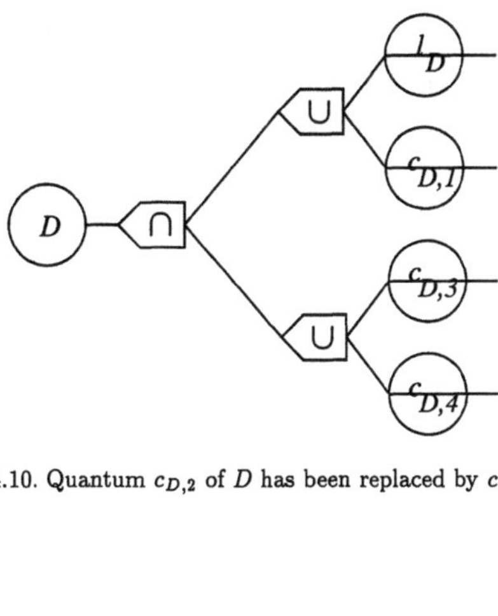

FIGURE 14.10. Quantum $c_{D,2}$ of D has been replaced by $c_{D,3}$ and $c_{D,4}$.

and Boolean set operation nodes towards its involved variables. Sequences from other constraints to the same variables must be 'repaired' by intersecting the proper quanta. The resulting new quanta must be propagated through the new network. Note that the explicitly represented implied constraints must also be removed.

The quantum approach can also be applied to pure numeric constraint problems. Instead of geometric variables and quanta, intervals are concerned. Algebraic reasoning then takes the place of geometric reasoning.

The constraint satisfaction with quantum labels finds whole solution sets. An intermediate result may be a sub-set or a super-set of the true solution set, but the final solution does not depend on the order of satisfaction, so that delayed satisfaction has no side effects. As a result, the declarative semantics of constraints are preserved.

14.11 Conclusions

We presented an incremental approach to geometric constraint satisfaction that is suitable for interactive design and categorizes solutions into quanta. The system keeps the intermediate solutions in the geometric domain, so that new geometric constraints can be interpreted on the same high level of abstraction. This approach has to a number of advantages because our system:

- can handle under-constrained specifications,
- represents both alternative discrete solutions and continuous ranges of solutions,
- can solve simple logical expressions of constraints,
- performs satisfaction locally and incrementally,
- handles delayed satisfaction uniformly,
- supports constraint inference and geometric reasoning,
- preserves the declarative semantics of constraints.

More research is needed to further exploit the capabilities of geometric reasoning. In particular implicit constraints (relations between quanta to which the user has no direct access) can be used by the satisfaction system to solve complex constraints more directly.

Acknowledgements:

This research is part of the IIICAD project, supported by NWO (Dutch Organization for Scientific Research) under Grant NF-51/62-514.

14.12 References

[1] F. Arbab and B. Wang. A Geometric Constraint Management System in Oar. In P.J.W. ten Hagen and P. Veerkamp, editors, *Intelligent CAD Systems III - Practical Experience and Evaluation (Proceedings 3rd Eurographics Workshop on Intelligent CAD Systems, 1989)*, pages 231 - 252. EurographicSeminar Series, Springer-Verlag, 1991.

[2] A. Borning. The Programming Language Aspects of ThingLab, a Constraint-Oriented Simulation Laboratory. *ACM Transactions on Programming Languages and Systems*, 3(4):353 - 387, October 1981.

[3] S.-C. Chou and X.-S. Gao. A Collection of 120 Computer Solved Geometry Problems in Mechanical Formula Derivation. Technical Report TR-89-22, University of Texas, Department of Computer Science, August 1989.

[4] E. Davis. Constraint Propagation with Interval Labels. *Artificial Intelligence*, 32:281 - 331, 1987.

[5] M.J.G.M. van Emmerik. A System for Interactive Graphical Modeling with 3D Constraints. In *Proceedings CGI '90*, 1990.

[6] H.-W. Güsgen and J. Hertzberg. Some Fundamental Properties of Local Constraint Propagation. *Artificial Intelligence*, 36:237 - 247, 1988.

[7] N.C. Heintze, S. Michaylov and P.J. Stuckey. CLP($\mathbb{R}$) and Some Problems in Electrical Engineering. In J.-L. Lassez, editor, *Proceedings of the 4th International Conference on Logic Programming*, MIT Press, 1987.

[8] W. Leler. *Constraint Programming Languages, Their Specification and Generation.* Addison-Wesley, 1988.

[9] V.C. Lin, D.C. Gossard and R.A. Light. Variational Geometry in Computer-Aided Design. *Proceedings SIGGRAPH '81, Computer Graphics*, 15(3):171 - 177, August 1981.

[10] A.K. Mackworth. Consistency in Networks of Relations. *Artificial Intelligence*, pages 99 - 118, 1977.

[11] G. Nelson. Juno, a Constraint-based Graphics System. *Proceedings SIGGRAPH '85, Computer Graphics*, 19(3):235 - 243, 1985.

[12] J.C. Owen. Algebraic Solution for Geometry from Dimensional Constraints. In *Proceedings of the ACM Conference on Solid Modeling*, pages 397 - 407. ACM Press, 1991.

[13] J.R. Rossignac. Constraints in Constructive Solid Geometry. In F. Crow and S.M. Pizer, editors, *Proceedings of the 1986 ACM Workshop on Interactive 3D Graphics*, pages 93 - 110. ACM Press, 1986.

[14] Zs. Ruttkay. A Co-operative Graphical Editor Based on Dynamically Constrained Objects. In C. Laffra and E. Blake, editors, *Advances in Object-Oriented Graphics II (Proceedings of the second Eurographics Workshop on Object-Oriented Graphics, 1991)*, EurographicSeminars Series, Springer-Verlag, to appear in 1992.

[15] G.L. Steele Jr. and G.J. Sussman. CONSTRAINTS. *APL Quote Quad*, 9(4–Part 1):208 – 225, June 1979.

[16] I. Sutherland. Sketchpad: A Man-Machine Graphical Communication System. In *Proceedings of the Spring Joint Computer Conference (IFIPS)*, pages 329 – 345, 1963.

[17] R.C. Veltkamp. Geometric Constraint Management with Quanta. In D.C. Brown, M. Waldron and H. Yoshikawa, editors, *Proceedings of the IFIP TC5/WG5.2 Working Conference on Intelligent Computer Aided Design, 1991*, North-Holland, to appear in 1992.

[18] R.C. Veltkamp. A Quantum Approach to Geometric Constraint Satisfaction. In C. Laffra and E. Blake, editors, *Advances in Object Oriented Graphics II (Proceedings of the 2nd Eurographics Workshop on Object Oriented Graphics, 1991)*, EurographicSeminar Series, Springer-Verlag, to appear in 1992.

[19] D.L. Waltz. *Generating Semantic Descriptions from Drawings of Scenes with Shadows*. PhD thesis, MIT, 1972.

15 Computer Vision, Descriptive Geometry, and Classical Mechanics

Christoph M. Hoffmann

ABSTRACT
The medial-axis transform, also called skeleton, is a shape abstraction proposed by computer vision. The concept is closely related to cyclographic maps, a tool developed by descriptive geometry to investigate distance functions, and to the solution of the eikonal equation. We discuss these connections and their implications on techniques for computing the skeleton.

15.1 Introduction

The medial-axis transform, also called the skeleton of a shape, was proposed in computer vision by Blum as a tool for shape recognition and abstraction [2]. Cyclographic maps were developed in classical descriptive geometry as a means to solve the problem of Apollonius, and, more generally, as a device for studying distance maps, surface curvature, and other basic geometric properties [19]. In the Hamilton-Jacobi theory, the Jacobi S function is related to both of these concepts via the eikonal equation that describes wave propagation [31]. This similarity of concepts is not surprising in view of the fact that all three notions fundamentally relate to measuring Euclidean distance, from a given geometric structure, and Blum [3] was aware of these connections. Nevertheless, how this interrelationship impacts devising algorithms for computing the skeleton, and how one might take it into account when attacking its applications seems to go unreported in the literature.

Let B be a compact set in $\mathbf{R}^2$ or $\mathbf{R}^3$, such as the boundary of a solid, that is smooth almost everywhere. Define the distance from B as follows:

$$\mathrm{dist}(p, B) = \inf\{d(p, q) \mid q \in B\}$$

where $d(p, q)$ is the usual Euclidean distance. A point q of B at distance $\mathrm{dist}(p, B)$ is called a *footpoint* of p in B. The *skeleton* with respect to B is the closure of the locus of all points not in B that have at least two distinct footpoints. Intuitively, the skeleton points have two or more geodesic paths to B. We discuss ways of conceptualizing the skeleton, and of determining it algorithmically.

From now on, we fix a domain S in $\mathbf{R}^2$ with a compact, smooth boundary ∂S. Taking $B = \partial S$, the skeleton with respect to B is just the medial axis of Blum [3], and can be defined alternatively as the closure of the locus of the centers of maximal circles inscribed in S. An inscribed circle is maximal if no other inscribed circle contains it properly. Moreover, it has been noted that the skeleton is a subset of the Voronoi diagram, [16]. If S is polygonal, then the skeleton is obtained from the Voronoi diagram by deleting the edges incident to reflex vertices.

We discuss algorithms for constructing the skeleton in 2D. Many, but not all of them generalize to the three-dimensional problem, and we will comment throughout whether and how the generalizations can be achieved. A specific difficulty is that the surfaces on which faces and edges of the three-dimensional skeleton lie are not easily described using parametric or implicit algebraics that are commonly used by the geometric modeling community. In general, the surfaces can be described exactly and practically only using the dimensionality paradigm [12], as a system of nonlinear equations.

Note that the skeleton is an informationally-complete representation of the domain S, in the sense of [24], provided we associate with each skeleton point p its distance $\text{dist}(p, \partial S)$ to the boundary. It will be convenient to associate the distance as an extra coordinate, to each skeleton point. Thus, the skeleton in 2D is a three-dimensional structure, and the skeleton in 3D is a four-dimensional structure. Note that an association of the distance with every interior point defines the distance surface of Blum [3]. This distance surface also arises in cyclographic maps and in the integration of the eikonal equation.

15.2 Geometric Approaches to the Skeleton

The skeleton can be determined by exact or by approximate methods. The exact algorithms we discuss depend on the geometry of the set B, and so they are sensitive to the permitted shape primitives. Furthermore, extending these algorithms to the 3D problem requires careful consideration of the underlying assumptions and is by no means automatic. Often, only structural elements of the algorithms generalize. Approximate algorithms discretize the set B in various ways and are, thereafter, independent of its geometry. Roughly speaking, to generalize them reduces to devising a discretization of B in 3-space, since the subsequent processing is usually trivial to extend to higher dimensions.

15.2.1 The Two-Dimensional Problem

Perhaps the first efficient algorithm for constructing the skeleton of a polygonal domain is due to Preparata [22]. The underlying organization can be grasped as follows: Consider shrinking the boundary of the domain with uniform speed. In this process, some edges may become smaller and reflex vertices expand as circular arcs. At certain critical distances, however, some offset edges have reached zero length, or different parts of the shrinking boundary have come into contact. These discrete events can be found by proximity computations involving two or three elements of the original boundary. They correspond to branch points of the skeleton. Proper book keeping ensures that not all triples are subjected to this analysis. Preparata's algorithm first locates the branching points of the skeleton, and then constructs the skeleton arcs connecting them. The algorithm determines the skeleton of convex polygons in $O(n^2)$ steps.

Lee [16] gave an algorithm based on the divide and conquer paradigm that computes the Voronoi diagram of a polygonal domain. So, by deleting Voronoi edges incident to reflex vertices, the skeleton is determined. Lee's algorithm takes $O(n \log(n))$ steps. Patrikalakis and Gürsoy [21] generalized Preparata's algorithm to domains bounded by circular arcs and line segments. There is also an algorithm by Srinivasan and Nackman [28], for multiply-connected polygonal domains.

For convex polygons, the skeleton consists of line segments. Reflex vertices introduce parabolic arcs. When circular boundary elements are allowed, the skeleton may also contain segments of the other conic sections.

When the allowed boundary elements are more general, then the skeleton includes curve segments whose geometry has not been studied systematically. The customary response is therefore to approximate either the boundary or the skeleton arcs or both. Exact representations can be given but require the dimensionality paradigm — unless extensive symbolic computation is undertaken; [11].

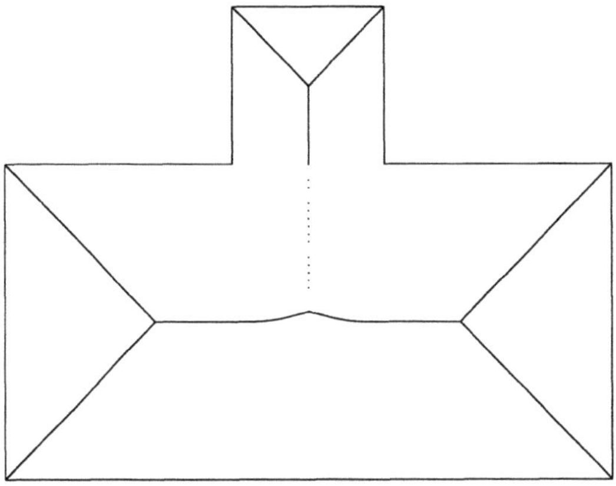

FIGURE 15.1. Dotted skeleton segment cannot be approximated through Delaunay triangulation

Approximations to the skeleton have been based on Delaunay triangulation as follows: Sample the boundary of the domain, obtaining a set of points whose Delaunay triangulation respects the domain boundary. Then the centers of the circumscribed circles of the Delaunay triangles contained in the domain are approximately on the skeleton, owing to the fact that they cannot contain other points of the triangulation and so are approximately maximal inscribed circles. One can now select only those triangles whose circumcenters lie near branch points of the skeleton, [23], or make the Delaunay triangulation sufficiently dense so that all circumcenters connect to a reasonable piecewise linear approximation of the skeleton, [34]. Note that certain segments of the skeleton cannot be so approximated, as shown in Figure 15.1. Another method [33] approximates the domain and its interior with boxes, and classifies interior boxes by the boundary element it is provably nearest to. Boxes that cannot be so classified provide an approximation to the Voronoi diagram.

The accuracy of the approximations can be increased using interpolation and iteration, [5]. Moreover, the approximate methods generalize to three dimensions with little change.

15.2.2 The Three-Dimensional Problem

The three-dimensional problem is complicated by the fact that the geometry of the skeleton surfaces is difficult; [8]. Skeleton surfaces of convex polyhedra are simple. However, nonconvex polyhedra already introduce quadric surfaces such as paraboloids and parabolic hyperboloids. More general shape primitives introduce surfaces whose geometry is not well-understood.

We are aware of only one exact algorithm; [13]. It uses many ideas of Preparata's algorithm, and, in particular, organizes the skeleton evaluation by increasing distance from the boundary. Surface construction relies heavily on the dimensionality paradigm and could well be carried out for general boundaries. However, the proximity computations make it advisable to restrict the geometry of the boundary to planes, natural quadrics, and tori. Although proximity calculations can be formulated for general geometries, the resulting systems of nonlinear equations would be difficult to solve and so the skeleton algorithm would become practically unacceptable.

The approximation algorithms all generalize, provided a suitable discretization of the surface can be given. Sapidis and Perucchio [27] describe an algorithm for constructing boundary-conforming triangulations of CSG objects. The algorithm tests whether the triangulation respects the boundary. If not, it determines which additional surface points should be added so as to eliminate tetrahedra that are partially inside and outside the solid. Such a triangulation could be used as the basis for 3D skeleton approximation.

Several papers investigate structural properties of the 3D skeleton; [20, 30]. Such studies are important when devising new skeleton algorithms, and can, in particular, guide evaluations based on marching. Yet it seems that a comprehensive topological characterization of the skeleton in 3-space is not known.

15.2.3 Engineering Applications of the Skeleton

Considerable work focuses on automatic mesh generation using the skeleton. It is argued that the skeleton gives good quantitative criteria for choosing the appropriate mesh density. The approaches to the two-dimensional meshing problem are especially well-developed; e.g., [1, 29]. Armstrong [1] argues that quadrilateral meshes should "flow" along skeleton arcs. Srinivasan [29] gives precise criteria for subdividing polygonal regions, and grades subdomains by the relation between the length of the skeleton arc and its distance from the boundary.

Three-dimensional mesh generation using the skeleton is largely open. However, it appears that the paving approach can utilize skeleton information. This approach generates hexahedral elements layer by layer beginning at the boundary. Each layer is bounded by interior offsets of a certain depth. The skeleton indicates where the paving pattern has to be adjusted because of impending self-intersection of the layer under construction.

In unpublished work, Prinz considers feature recognition and extraction using an approximate skeleton. However, the skeleton is approximated by considering an octree subdivision [26] of the object. The skeleton approach to feature recognition has to address how the local perturbation, introduced by a feature that was added to a basic shape, affects the skeleton. If we should recognize, for example, that the global structure is rotational, will the axis of rotation still be recognizable from the skeleton after adding surface features?

Certain rigid-body transformations of the skeleton are equivalent to forming an offset of the original shape. Thus, the skeleton has potential applications in geometric tolerancing [15, 14]. Stifter [30] uses the (exterior) skeleton for motion planning. By moving on the skeleton, maximum clearance from all obstacles is maintained.

15.3 Image Processing Algorithms for the Skeleton

Skeleton computations in computer vision begin with a digitized picture. In consequence, this problem version considers the distance of points in a regular rectangular grid from a subset of grid points that constitute the contour whose skeleton is to be found. Early algorithms by Rosenfeld [25] and Montanari [17] use a distance function that is only approximately the Euclidean distance. Danielson's algorithm [7], in constrast, computes the Euclidean distance at each grid point with a cumulative error that is only a fraction of the grid spacing. Furthermore, if distance is expressed in multiples of the grid spacing h, the algorithm can be implemented in integer arithmetic.

Danielson's algorithm computes the Euclidean distance transform in two passes over the grid, top-to-bottom and bottom-to-top. In each pass, all rows are scanned twice, in opposite directions. The key to his method is to store at each grid point the distance *amplitudes*, i.e., the distance in the x- and y-directions to the nearest boundary grid point. While the distance from a nearest grid point is not an integer multiple of h, the distance amplitudes always are. Moreover, it is simple to compute the distance of an adjacent grid point using the distance amplitudes. Note that Danielson's algorithm generalizes trivially to 3D grids [4].

If a given set B is not discretized as a set of grid points, interpolatory schemes for computing the distance transform can be devised [5]. For example, the curve B can be approximated by a polygonal arc, where the segments are induced by the grid lines, and the adjacent grid points can be assigned the perpendicular distance to the nearest boundary segment. This improves the accuracy of the distance computation significantly, but the distance amplitudes are no longer multiples of the grid spacing. Further accuracy is obtained using iteration. For a discussion of Danielson's algorithm and such variations see [5].

Once the distance at each grid point is known, there are different methods for extracting the skeleton. Danielson suggests using the skeleton definition as the locus of the centers of maximal disks. His algorithm includes a corrective computation that compensates for errors introduced by the discretization. Other criteria are discussed in Section 6.

15.4 Cyclographic Maps in Descriptive Geometry

Cyclographic maps have been used in descriptive geometry to solve the Apollonius problem. The concept is due to Müller [19]. Let C be an oriented plane curve. For every point p we consider all oriented circles, also called *cycles*, that are tangent to p at C and are oriented at p in the same direction as C is. With each such cycle, we associate a point $q = (x, y, r)$, where (x, y) is the center of the cycle and r its radius. Choosing arbitrarily the counterclockwise orientation as positive, r is signed according to whether the cycle orientation is positive or negative.

When the curve point p is fixed and regular, all oriented cycles tangent at p are associated with points that lie on a line through p, with slope 1 against the (x, y)-plane and are such that the line orthographically projects onto the curve normal at p. See also Figure 15.2. As p varies over C, these lines comprise a ruled surface called the *cyclographic map* $\mathcal{S}(C)$ of C. Since the generators of this surface have constant slope with respect to the plane of C, the surface is developable [32].

We can restrict the surface $\mathcal{S}(C)$ such that it is the graph of a function with the (x, y)-plane its domain. With each point (x, y) we associate the "nearest" point (x, y, r) on $\mathcal{S}(C)$. The surface $\overline{\mathcal{S}}(C)$ so defined is then the distance surface of Blum [3]. It follows, that $\overline{\mathcal{S}}(C)$ can be determined approximately with the Euclidean distance transform. Figure 15.3 shows an example.

When intersecting the surface with a plane $r = d$ parallel to the (x, y)-plane, we obtain an offset of C by d. This is a global offset [5]. Moreover, the points on $\overline{\mathcal{S}}(C)$ that are not continuously differentiable are the skeleton with respect to C. See also [14]. Thus, if the skeleton is known, then the map $\overline{\mathcal{S}}(C)$ can be obtained by adding the ruling from skeleton points to corresponding footpoints on the curve. Curve points can be found directly from the skeleton point and its tangent(s), [19]. Briefly, project the skeleton point $P = (x, y, r)$ orthographically into the (x, y)-plane, obtaining a point $P' = (x, y)$. Find the intersection of the tangent to the skeleton with the (x, y)-plane, obtaining a point Q. Now draw the two tangents from Q to the circle with radius $|r|$, centered at P'. Then the contact points R_1 and R_2 are points of C. See also Figure 15.4.

In principle, cyclographic maps extend to three dimensions [18]. For instance, instead of

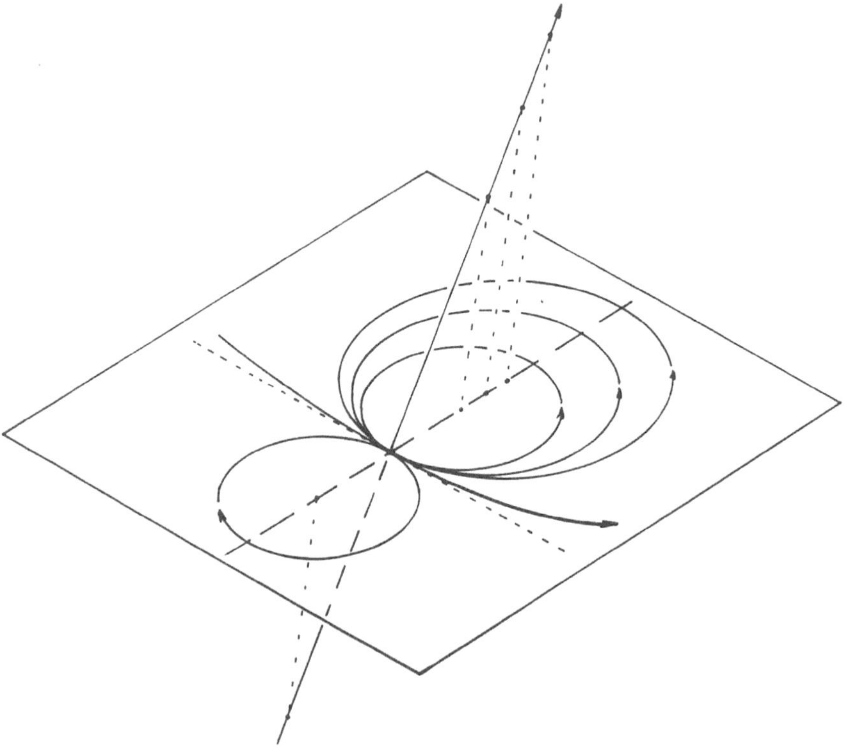

FIGURE 15.2. Cyclographic map at a curve point

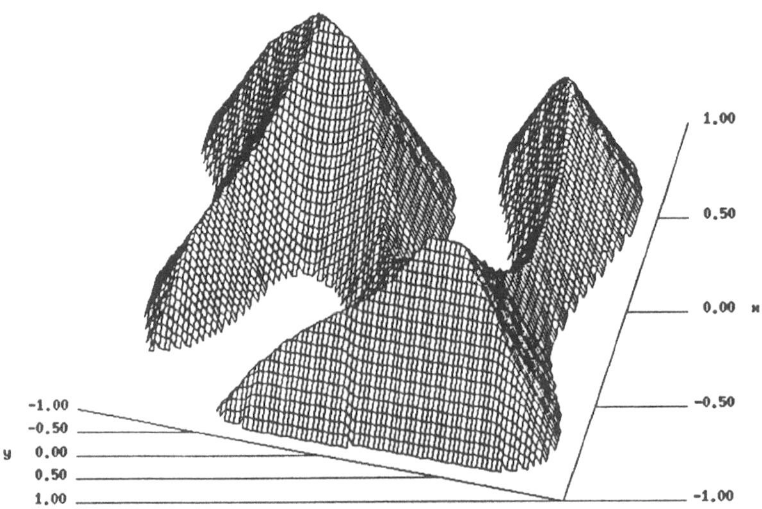

FIGURE 15.3. The surface $\overline{S}$ of a closed curve

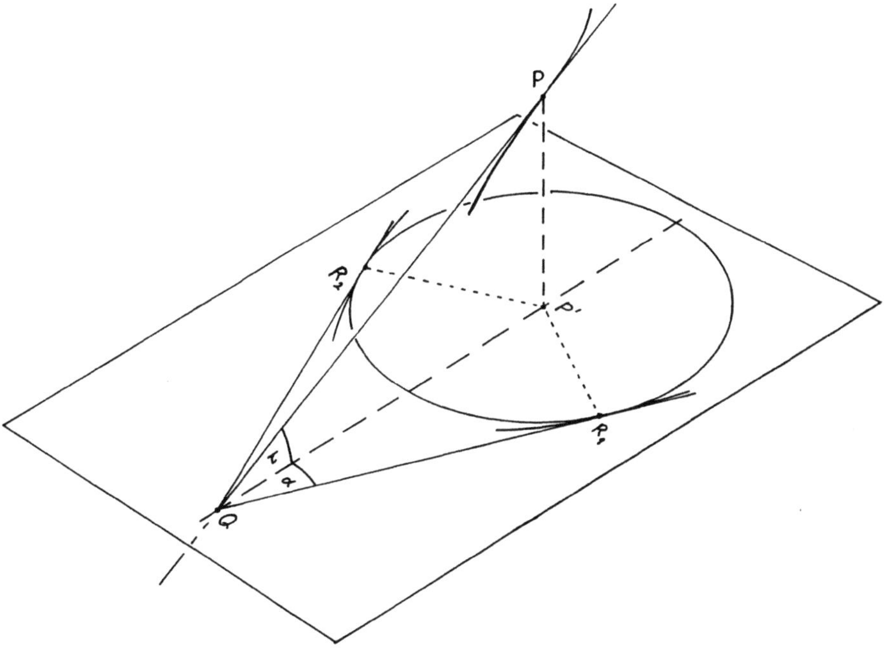

FIGURE 15.4. Recovering C from the Skeleton

oriented circles we now use oriented spheres, adding the radius as additional fourth coordinate, properly signed. The constructions and relationships mentioned before generalize to the three-dimensional case. It is also possible to replace oriented circles with suitably chosen conics. In this case the surface $\mathcal{S}(C)$ corresponds to noneuclidean metrics.

15.5 Classical Mechanics and the Skeleton

Consider a continuum of particles situated at time $t = 0$ on a plane, smooth curve C, moving at constant velocity in a direction locally normal to C. By the conservation of momentum, the kinetic energy of each particle is constant, and will be proportional to the squared velocity components in the principal directions. That is, $v_x^2 + v_y^2 = 1$, where v_x is the velocity in the x-direction, and v_y is the velocity in the y-direction. With $v_x = \partial S/\partial x$ and $v_y = \partial S/\partial y$, we obtain

$$\left(\frac{\partial S}{\partial x}\right)^2 + \left(\frac{\partial S}{\partial y}\right)^2 = 1$$

This is the eikonal equation in geometric optics [9], and $S(x, y)$ is the time when a particle reaches the point (x, y). Since we have assumed constant velocity, $S(x, y)$ is just the distance function assigning to each point in the plane its distance from C.

Let $p = (x, y)$ be a point reached by two or more different particles at the same time. Then p has two or more foot points on C and is therefore a skeleton point. Thus, the derivative discontinuities of the surface S are the skeleton of C. Moreover, the surface S is just the surface $\overline{\mathcal{S}}(C)$ of the cyclographic map. We can thus think of the eikonal equation

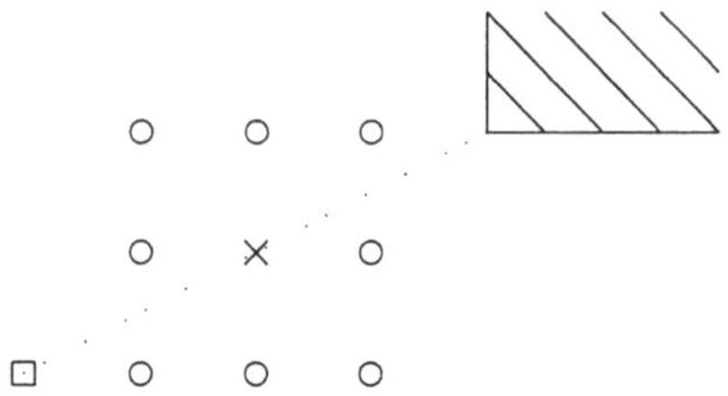

FIGURE 15.5. Disk centered at × is not maximal, yet is not contained in disk centered at the immediate neighbors

as a differential description of the Euclidean distance function; [10]. The three-dimensional form of the equation is

$$\left(\frac{\partial S}{\partial x}\right)^2 + \left(\frac{\partial S}{\partial y}\right)^2 + \left(\frac{\partial S}{\partial z}\right)^2 = 1$$

The eikonal equation suggests computing the distance function using techniques from numerical analysis. Once the distance function has been determined, the skeleton can be extracted either by applying the definition of maximal circles, as suggested in [7], or by finding the tangent discontinuities of the surface, or by processing intersecting characteristics directions during integration.

15.6 Computational Approaches

Once the distance map has been computed, the skeleton can be extracted by applying the maximal-disk criterion. Intuitively, we want to check whether the disk at (i, j) is contained in a disk centered elsewhere.

Given a grid point (i, j) with distance amplitudes (a, b), let $(i + r, j + s)$ be a neighbor grid point with distance amplitudes (a', b'), where $r, s \in \{-1, 0, +1\}$, with h the grid spacing. Let $d_1 = \sqrt{a^2 + b^2}$ and $d_2 = \sqrt{a'^2 + b'^2}$. If $r = 0$ or $s = 0$, and if

$$d_2 - d_1 \geq h$$

then the inscribed disk centered at $(i + r, j + s)$ of radius d_2 contains the inscribed disk centered at (i, j) of radius d_1. Therefore, the point (i, j) cannot be a skeleton point. Similarly, if $r \neq 0$ and $s \neq 0$, and if

$$d_2 - d_1 \geq h\sqrt{2},$$

then the inscribed disk centered at $(i + r, j + s)$ of radius d_2 also contains the inscribed disk centered at (i, j) of radius d_1. Again, the point (i, j) cannot be a skeleton point in this case. If the disk at (i, j) is not contained in the disks centered at each of the eight neighbors, however, then (i, j) is not necessarily a skeleton point. In Figure 15.5, the disk centered at × is not contained in any disk centered at the immediate neighbors. Yet the disk is contained in a disk centered at the grid point indicated by a box.

To decide whether a disk is maximal, we instead attempt to "grow" it. Consider the situation shown in Figure 15.6. We try to enlarge the disk centered at × in the direction of the characteristic through the point. The extrapolated distance d_1 to the boundary along the characteristic is compared with the interpolated distance d_2 obtained from nearby

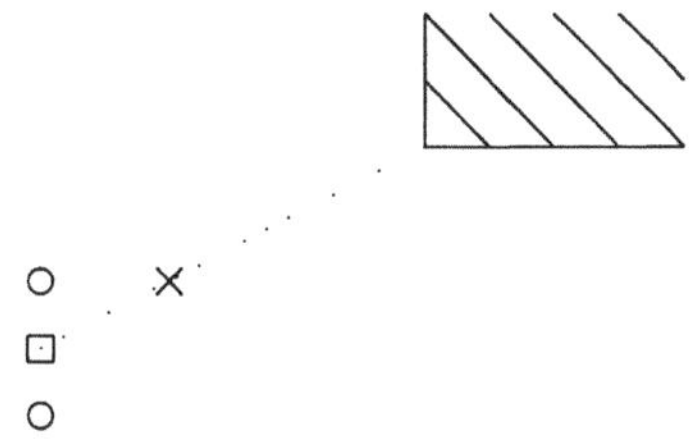

FIGURE 15.6. Disk centered at × is not maximal if it is contained in disk centered at the square.

grid points. If the distances agree, subject to a suitable tolerance, then the disk can be so enlarged, and the grid point × is not a skeleton point.

The computation is as follows: Let d_0 be the distance of the grid point P. Extrapolating from P, we obtain a point Q at distance $d_1 = d_0 + w$ from the boundary, where w is the distance $\overline{P,Q}$; see also Figure 15.7. Note that the slope of the characteristic is 1. The distance of Q can be determined alternatively by interpolating the distances d_a of A and d_b of B, as $d_2 = d_a + u(d_b - d_a)/h$. If the characteristic through P extends to Q on the distance surface, then the two distances d_1 and d_2 will agree fairly closely. If the characteristic does not extend, however, then d_2 should be smaller than d_1 on account of Q being closer to the boundary in a different direction. Let P' and Q' be the points on the distance surface above P and Q respectively, where we assume that Q' is at height d_2. The line $\overline{P',Q'}$ has the slope $\tan(\phi) = (d_2 - d_0)/w$. If $\tan(\phi) \geq t$ for $t = 1.0$, then the characteristic through P extends, and P is not a skeleton point. But if $\tan(\phi) < t$, then the characteristic does not extend. In that case we consider P to be a skeleton point.

When we extract the skeleton from the discrete distance surface, computed by Danielson's algorithm, the discretization of the boundary results in tangent discontinuities of the distance surface that are not skeleton points. Figure 15.8 shows an example in which the tolerated distance deviation is very small, using $t = 1$. If we lower the threshold t, then less points are considered skeleton points. In Figure 15.9, a looser tolerance has been used

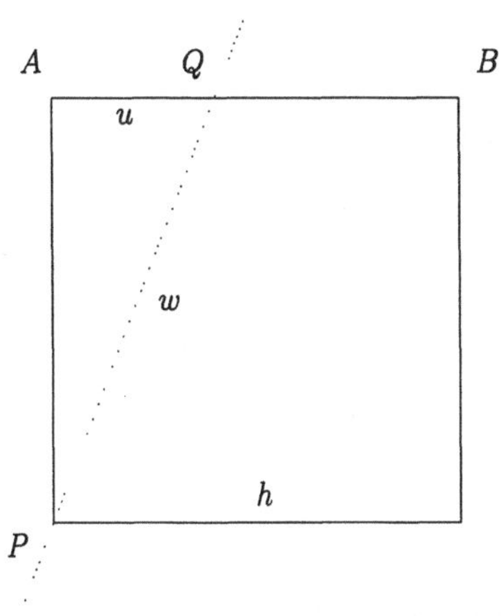

FIGURE 15.7. Extrapolating along the characteristic of P

FIGURE 15.8. Discretization error affecting skeleton construction: threshold $t = 1$.

and fewer nonskeleton points are reported. However, when using too loose a tolerance true skeleton points are also lost, those at which the tangent discontinuity of the distance surface is relatively small.

It is simpler to account directly for the boundary geometry and use iteration to eliminate the discretization errors altogether. We now sketch a preliminary technique for extracting the skeleton, assuming there is a continuous boundary that has been specified. The method includes iteration and interpolation steps, and applies techniques from [6, 12]. First, we give another criterion for finding skeleton points.

Consider a grid square such as the one shown in Figure 15.10. The characteristics of the distance function can be reconstructed from the distance amplitudes, provided that Danielson's algorithm is modified to record signed amplitudes. In the figure, the characteristics are shown as dotted lines. If the characteristics of the corner points intersect in the grid square, then the intersection point is an approximate skeleton point. This intersecting-characteristics criterion works well except near endpoints of the skeleton, because there the characteristics are nearly parallel. Note that the intersection is also

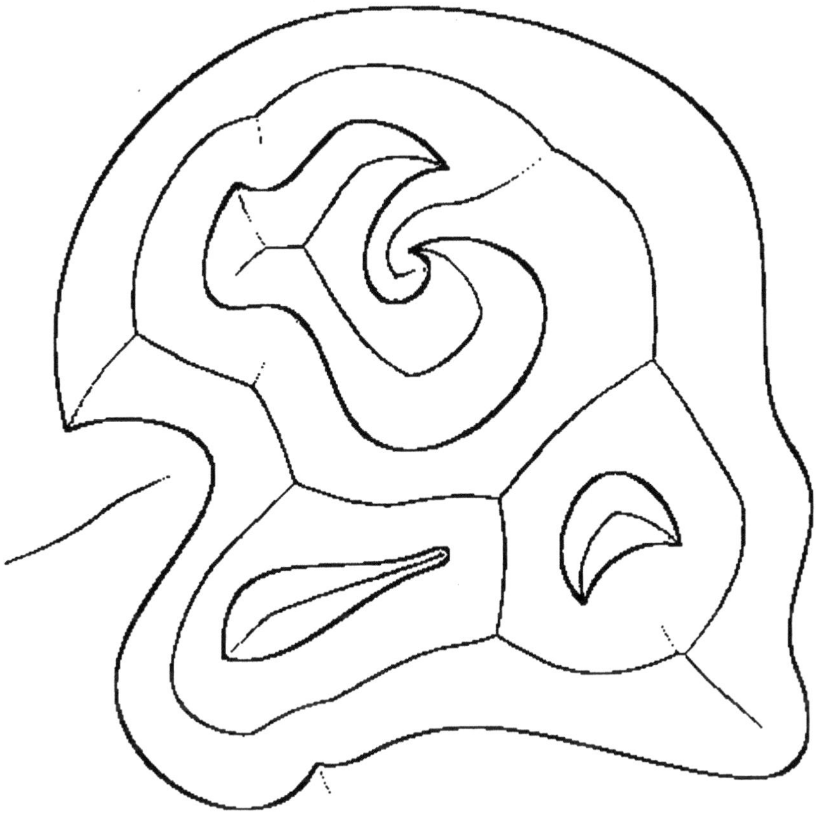

FIGURE 15.9. Effect of looser tolerance; $t = 0.6$

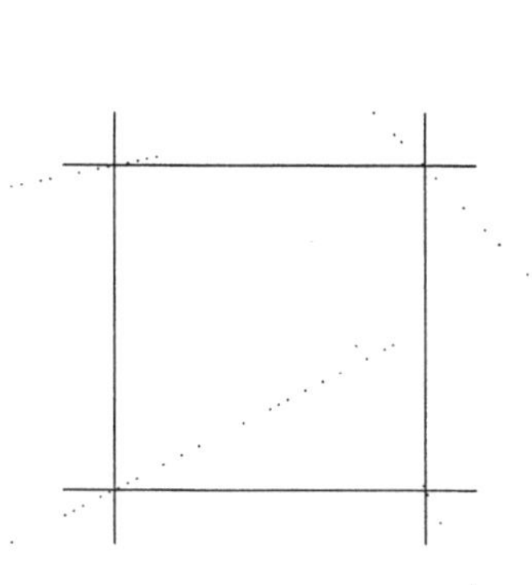

FIGURE 15.10. Approximate skeleton point at intersecting characteristics

difficult to determine for nearly anti-parallel characteristics. However, in that case distance interpolation will locate the intersection quite accurately.

Skeleton points lie on equal-distance curves and surfaces [8]. Since the characteristics localize which parts of the boundary lie nearby, it is not difficult to formulate a system of equations that describes the skeleton locally [13], in terms of the coordinates of the skeleton point, its footpoints, and the distance to the boundary. This system can be solved very effectively with Newton iteration, since the characteristics provide good initial estimates for all quantities.

In three dimensions, matters are more complicated since the characteristics through the vertices of a grid cube will be skew in general. By interpolating the characteristics along grid edges, intersection curves and surfaces can be determined and thereafter refined using iteration in the same way as in the two-dimensional case. Alternatively, the intersections of the skeleton with grid lines can be found using interpolation. Note that the local topological structure of the skeleton can be determined as described in [13]. In essence, our approach negotiates a reduction in the number of grid points against the increased computational work of iteration and interpolation.

Acknowledgements:

This work has been supported in part by by ONR Contract N00014-90-J-1599, by NSF Grant CCR 86-19817, and by NSF Grant ECD 88-03017. Much of the work reported here has been shaped by discussions with C.-S. Chiang, R. Lynch and H. Pottmann.

15.7 References

[1] C. Armstrong, T. Tam, D. Robinson, R. McKeag, and M. Price. Automatic Generation of Finite Element Meshes. In *SERC ACME Directorate Research Conference*, England, 1990.

[2] H. Blum. A Transformation for Extracting New Descriptors of Shape. In W. Whaten-Dunn, editor, *Models for the Perception of Speech and Visual Form*, pages 362–380. MIT Press, Cambridge, MA, 1967.

[3] H. Blum. Biological Shape and Visual Science, Part I. *Journal of Theoretical Biology*, 38:205–287, 1973.

[4] G. Borgefors. Distance Transformations in Arbitrary Dimensions. *Comp. Vision, Graphics Image Processing*, 27:321–345, 1984.

[5] C.-S. Chiang, C. M. Hoffmann, and R. E. Lynch. How to Compute Offsets Without Self-Intersection. Technical Report CAPO-91-35, Purdue University, Computer Science, 1991.

[6] J.-H. Chuang. *Surface Approximations in Geometric Modeling.* PhD thesis, Purdue University, Computer Science, 1990.

[7] P.-E. Danielsson. Euclidean Distance Mapping. *Computer Graphics and Image Processing*, 14:227–248, 1980.

[8] D. Dutta and C. Hoffmann. A Geometric Investigation of the Skeleton of CSG Objects. In *Proc. ASME Conf. Design Automation*, Chicago, 1990.

[9] H. Goldstein. *Classical Mechanics.* Addison-Wesley, Reading, MA, 1986.

[10] A. Goodman. A Partial Differential Equation and Parallel Plane Curves. *American Mathematical Monthly*, 71:257–264, 1964.

[11] C. M. Hoffmann. Algebraic and Numerical Techniques for Offsets and Blends. In S. Micchelli M. Gasca, W. Dahmen, editor, *Computations of Curves and Surfaces*, pages 499–528. Kluwer Academic, 1990.

[12] C. M. Hoffmann. A dimensionality paradigm for surface interrogation. *CAGD*, 7:517–532, 1990.

[13] C. M. Hoffmann. How to Construct the Skeleton of CSG Objects. In A. Bowyer and J. Davenport, editors, *The Mathematics of Surfaces IV.* Oxford University Press, 1990.

[14] C. M. Hoffmann and G. Vaněček. Fundamental Techniques for Geometric and Solid Modeling. In C. T. Leondes, editor, *Advances in Control and Dynamics.* Academic Press, 1991.

[15] C. M. Hoffmann and G. Vaněček. On Alternate Solid Representations and their Uses. In *Proc. DARPA Manufacturing Workshop*, Salt Lake City, Utah, 1991.

[16] D. T. Lee. Medial Axis Transformation of a Planar Shape. *IEEE Trans. Pattern Anal. and Mach. Intelligence*, PAMI-4:363–369, 1982.

[17] U. Montanari. Continuous skeletons from digitized images. *JACM*, 16:534–549, 1969.

[18] E. Müller. Zusammenhang zwischen relativer Flächentheorie und einer Verallgemeinerung der Zyklographie. *Jahresber. Dtsche. Math. Ver.*, pages 155–160, 1923.

[19] E. Müller and J. Krames. *Die Zyklographie.* Franz Deuticke, Leipzig und Wien, 1929.

[20] L. R. Nackman and S. M. Pizer. Three-Dimensional Shape Description Using the Symmetric Axis Transform I. *IEEE Transactions on Pattern Analysis and Machine Intelligence*, PAMI 7:187–205, 1985.

[21] N. Patrikalakis and H. Gürsoy. Shape Interrogation by Medial Axis Transform. Technical Report Memo 90-2, MIT, Ocean Engr. Design Lab, 1990.

[22] F. Preparata. The medial axis of a simple polygon. In *Proc. 6th Symp. Mathematical Foundations of Comp. Sci.*, pages 443–450, 1977.

[23] M. Price, T. Tam, C. Armstrong, and R. McKeag. Computing the Branch Points of the Voronoi Diagram of a Object Using a Point Delaunay Triangulation Algorithm. Draft manuscript, 1991.

[24] A. Requicha. Mathematical Models of Rigid Solids. Technical Report Memo 28, University of Rochester, Production Automation Project, 1977.

[25] A. Rosenfeld and J. Pfaltz. Sequential Operations in Digital Picture Processing. *Journal of the ACM*, 13:471–494, 1966.

[26] H. J. Samet. *Design and analysis of Spatial Data Structures: Quadtrees, Octrees, and other Hierarchical Methods.* Addison–Wesley, Redding, MA, 1989.

[27] N. Sapidis and R. Perucchio. Domain Delaunay Tetrahedrization of Arbitrarily Shaped Curved Polyhedra Defined in a Solid Modeling System. In *Proc. ACM/SIGGRAPH Symp Solid Modeling Found and CAD/CAM Applic*, Austin, Tex, 1991.

[28] V. Srinivasan and L. Nackman. Voronoi diagram of multiply connected polygonal domains. *IBM Journal of Research and Development*, 31:373–381, 1987.

[29] V. Srinivasan, L. Nackman, J. Tang, and S. Meshkat. Automatic Mesh Generation using the Symmetric Axis Transformation of Polygonal Domains. Technical Report RC 16132, IBM Yorktown Heights, 1990.

[30] S. Stifter. An Axiomatic Approach to Voronoi Diagrams in 3D. *Journal of Computer and System Sciences*, to appear, 1991.

[31] G. Strang. *Introduction to Applied Mathematics.* Wellesley-Cambridge Press, Wellesley, MA, 1986.

[32] K. Strubecker. *Differentialgeometrie III.* Sammlung Göschen Bd. 1180/1180a, Walter de Gruyter, Berlin, Germany, 1969.

[33] A. Wallis, D. Lavender, A. Bowyer, and J. Davenport. Computing Voronoi Diagrams of Geometric Models. In *IMPA Workshop on Geometric Modeling*, Rio de Janeiro, 1991.

[34] X. Yu, J. A. Goldak, and L. Dong. Constructing 3D Discrete Medial Axis. In *Proc. ACM Symp. Solid Modeling Found. and CAD/CAM Applic.*, pages 481–492, Austin, 1991.

16 Discrete Surface Models: Constraint-based Generation and Understanding

Bianca Falcidieno
Caterina Pieñovi
Michela Spagnuolo

ABSTRACT A two level description is proposed for natural surface modelling: a geometric model, in terms of low level primitive entities, and a conceptual model, which is a symbolic surface representation based on its prominent shape characteristics. The geometric and conceptual level can exchange information using two operators: the abstraction operator and the generation operator.
An abstraction mechanism, based on the analysis of the Gaussian curvature, can be applied to a faceted model of the surface in order to identify subsets of primitive entities linked to particular morphological configurations.
A generation operator produces a geometric model through a tessellation technique that can keep constraints imposed by a high level description of the surface morphological characteristics. The level of approximation and quality of the visualization can be controlled by including an appropriate subset of the data points, and/or by using a stochastic perturbation to generate fractal-like details.

16.1 Introduction

Discrete surface models are generally used to represent terrains or other natural phenomena and are fundamental in geographic information processing. However, today's discrete surface modelling systems are still far from being fully developed despite the wide availability of data acquisition techniques and the efficiency of algorithms and machines that produce realistic pictures. The emphasis in research is on improving computation techniques, and only a few authors have tried to investigate the analytical extraction of properties and specific characteristics from geometric models or the generation of generalized surface models ([1, 8, 9, 11]).

Currently, discrete surface models are represented by large amounts of low level geometric primitives: points, edges, polygons (e.g. grids, contours, triangulations). The problem is: how can we deal with this bunch of primitives? We believe that they are important for the development and enhancement of specific algorithms (visualization, boolean operations,..) but must be further explored to provide higher level of abstractions that can relate directly to certain properties of the surface.

Let us give an example. Suppose we want to describe this kind of landscape: two chains of mountains, a narrow valley between them, a river at the bottom of the valley and a broad meadow at the end of the valley. We will have to transform this concise description into a lot of geometric primitives, worrying about coordinates and edges if we wish to translate this image into a traditional surface model and put the landscape model into computer language.

This bottom up approach makes sense for the development of specific algorithms (e.g. data capture and conversion, visualization) but can create short-sighted and narrow solutions when it comes to develop comprehensive modelling systems. Therefore, this approach should be combined with a top down method to help users and should allow them to forget the low level geometric primitives. The desired surface will be described with specific features and properties, using perhaps natural languages. The computer will generate the surface model by decomposing the described scene into primitives and building the corresponding geometric model.

Natural surface modelling is here considered an iterative refinement process which requires different levels of interrelated models or abstractions. Therefore, an abstract element is needed that can integrate surface properties or constraints (e.g. meaning) with geometry within the surface description and means are required to organize and manage surface constraints. To solve this problem, we have to find the answer to a series of subproblems:

- identify the properties, the features or the constraints necessary to describe a surface in an effective, concise and intuitive way;
- give them a mathematical, rigorous definition so that they can be easily associated to a geometric representation;
- define a conceptual model to describe the surface using these characteristics;
- define two fundamental mechanisms or operators that map the conceptual model into a geometric model and viceversa.

The goal of our research is to find these answers. In particular, we have focused our attention on geometric models defined as triangulations, and we have devised two basic operators called generation (or refinement) and understanding (or abstraction), to allow the interactive communication between the conceptual and the geometric level of surface description.

16.2 Geometric Model and Conceptual Model

Discrete surfaces are usually given as a set of quantitative observations, typically triplets of coordinates, and qualitative information, such as relationships between sample points. These basic data need to be structured to define a representation suitable for surface modelling.

One of the basic requirements of a surface model is to give good interpolation results, a crucial issue which is often influenced by the distribution of the data points and by the accuracy and technique used to sample the surface. An interpolation based on surface patches preserving the values at the data points is widely considered satisfactory for surfaces with well-distributed sample points. Regularly spaced grids (RSG) are commonly used because of the large availability of data measured through regular sampling (e.g. range images) and because of the easy access to the structure. However, the sampling regularity does not necessarily correspond to morphology regularity, thus regular grids may produce either a redundancy of information in uniform regions or a lack of information in rough areas.

The triangulated irregular network (TIN), which is based on triangular facets with vertices at the data points, is a more general and flexible structure for terrain modelling. This structure offers effective advantages because a TIN is independent of the point spatial distribution and can represent the surface at a variable resolution, thus adapting to the local changes and to the roughness of the terrain. Moreover, a TIN can be easily updated locally, thus permitting the inclusion of additional points at strategic locations or constraints.

Since an arbitrary triangulation does not represent, generally, a "good" triangulation several authors have studied the problen of choosing an optimal triangulation among all the feasible ones. In surface approximation problems, an acceptable triangulation is the one in which most of the triangles are as equiangular as possible, in order to avoid long thin triangles. This property is fundamental for numerical interpolation, since it keeps under control the maximum distance of any interior point from a vertex of the triangle which encloses it. This criterion is satisfied by the *Delaunay triangulation*, considered a standard for surface description [12].

Formally, the Delaunay triangulation of a set of points is usually defined in terms of another geometric structure, that is the *Voronoi diagram*, which takes into account the spatial proximity relationship among the points. Let $V = \{P_i \mid P_i = (x_i, y_i),\ i = 1 \dots n\}$ be a set of points in the plane XY, then the Voronoi diagram of V is a subdivision of the plane into polygons, each associated with a point P_i of V and defined as the region which is closer to P_i than to any other point $P_j, j \neq i$, belonging to V. The dual of the Voronoi diagram of V is a planar graph called *Delaunay graph* of V. Formally, the Delaunay graph of V is $T_D(V) = (V_D, E_D)$ where $V_D = V$ and $E_D = \{(P_i, P_j) \mid P_i, P_j \in V, i \neq j,\ P_i, P_j\ are\ Voronoi\ neighbours\}$, so that it explicitly represents the neighbourhood relationships induced by the Voronoi diagram over V. The Delaunay graph is called the *Delaunay triangulation* of V if, and only if, every face in $T_D(V)$ is a triangle. Otherwise, if a non-triangular face f exists in $T_D(V)$, then the vertices of f lie on a common circle, and the Delaunay graph can be completed to Delaunay triangulation by dividing f into non-overlapping triangles.

The triangulation of a surface described by triplets of coordinates can be built on the basis of the Delaunay triangulation of the data point projections onto the XY plane, by considering that two points on the surface are joined by an edge if, and only if, the respective projections are joined in the triangulation of the projection set. Thus, the adopted geometric model consists of vertices, edges and triangles and is stored as a relational graph which makes the adjacency relationships between facets explicit (see figure 16.1). Given a set S of sampling points, the corresponding surface geometric model is formally defined as the labelled graph $G_S = (N, A)$ where N is the set of nodes and A is the set of arcs, such that:

- the node set is $N = \{t_i | t_i\ is\ a\ triangle\ of\ T_D(S)\}$ and for every triangle t a unique node exists in N (corresponding to t), which is labelled t;
- the arc set is $A = \{(t_i, t_j) | t_i\ and\ t_j\ share\ a\ common\ edge\ in\ T_D(S)\}$ where for every edge e shared by triangles t_1 and t_2 a unique arc exists in A joining the two nodes t_1 and t_2, which is labelled e.

While the chosen geometric model has a well defined characterization, a general purpose conceptual model is harder to define precisely. Indeed, the conceptual model can be considered as a framework through which it is possible to point out some properties of interest the surface may have for a specific analysis. In other words, the conceptual model is the synthesis of a large amount of low level geometric entities resulting from some kind of high level filter, so that the symbolic model has only the information needed to describe the surface with its basic features. Although a detailed definition of a conceptual model depends on the user's requirements, it is possible to define some desirable properties a high level surface description should have. First of all, it should adhere to the underlying geometric model, so that the interchanges between conceptual and geometric models are possible without troublesome mappings. Then, the conceptual model should be combinatorially less complex than the original geometric representation because the

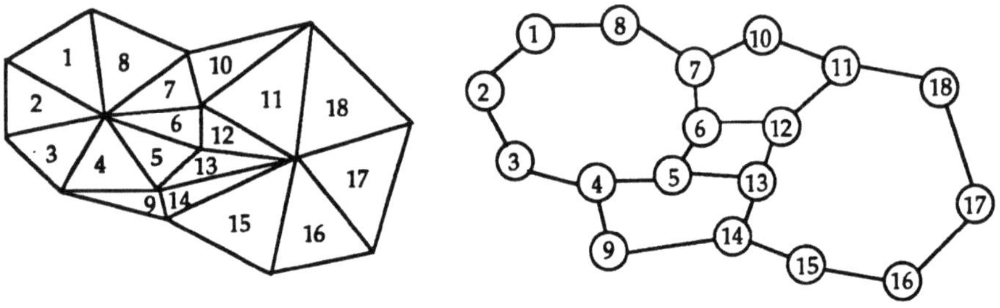

FIGURE 16.1. A triangulation and the corresponding graph model.

conceptual model should be the simplification of the low level description. Moreover, the conceptual model should be easy to use and flexible enough to fit different morphological configurations.

In our approach, the conceptual model is considered as a relational structure that can represent 2-dimensional morphological features (e.g. mountains, valleys, ...), 1-dimensional features (e.g. ridges, ravines, ...) and 0-dimensional features (e.g. peaks, pits, ...). Because the high level description should be based on the geometric model, the conceptual model will be defined by sets of geometric low level entities, i.e. by sets of vertices, edges and triangles. For example, a mountain could be represented by a closed sequence of edges belonging to the mountain border, by an internal point representing its top and by a sequence of edges representing an interior ridge.

16.3 Abstraction or Understanding

By abstraction or understanding we mean the process of analysing the geometric model to recognize some surface properties which can be used to define a higher level surface description based on knowledge about the surface morphology. The first step in this direction is the formal definition of the properties characterizing the surface and a reasonable approach could use analytical tools to "measure" properties of the surface which can be connected with some landscape intuitive descriptors. Gaussian and mean curvatures have long been considered as shape descriptors but their use in the context of discretized surfaces obviously needs an approximation of the underlying surface with a function of class C^2. A piece-wise linear approximation is already given by the triangulation, so our approach exploits the geometric information given by the adopted surface model in order to perform a curvature analysis which can be directly applied to the geometric model. Since each facet of the polyhedral surface approximation is flat, obviously all the curvature is concentrated at the vertices of the triangles and a method to compute it has already been proposed [9].

Instead of considering curvature at the vertices, this paper suggests to use a qualitative analysis of the curvature "along" edges and "around" triangles, to identify the regions of the surface whose shape can be related to the sign of the Gaussian curvature. Thus, *characteristic regions*, i.e. regions with concave, convex, planar or saddle shape, become basic describers of the surface morphology and are identified by considering the geometric relationships between triangles which can be coded as geometric attributes assigned to edges and triangles in the geometric model. The concept of curvature "along" edges is more precisely described in the following definition.

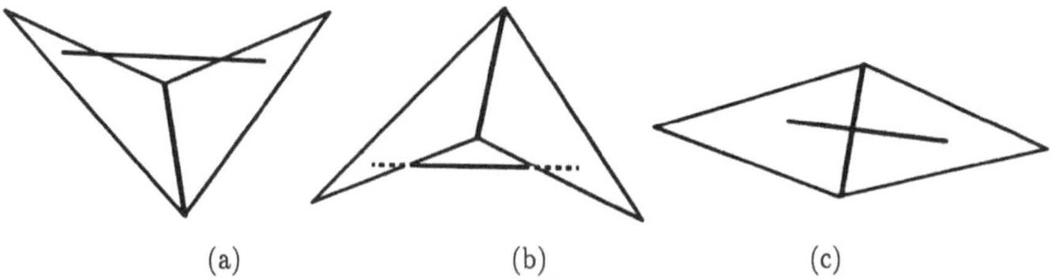

FIGURE 16.2. Example of concave edge (a), convex edge (b) and plane edge (c).

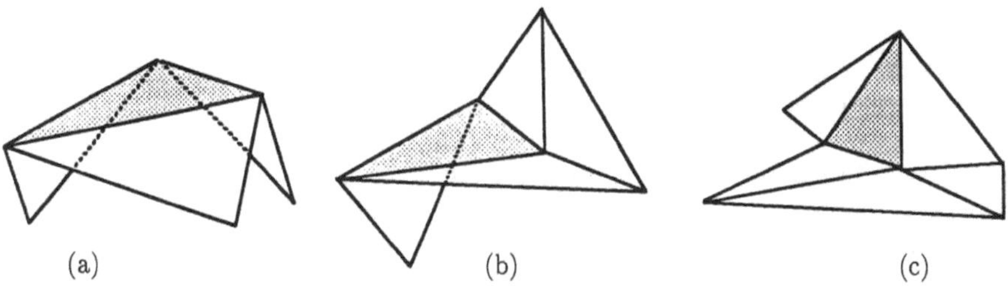

FIGURE 16.3. Example of convex triangle (a), saddle triangle (b) and hybrid triangle (c).

Definition 16.3.1 *The dihedral angle α subtended by two adjacent triangles t_1 and t_2 is said* convex (concave) *if for any two points $p_1 \in t_1$ and $p_2 \in t_2$ the straight line segment p_1p_2 lies completely* below (above) *the surface, otherwise it is* plane *(see figure 16.2). Then, to each arc $e = (t_1, t_2)$ of the surface graph model an attribute is assigned in the set* convex, concave, plane *according to the type of the corresponding angle.*

The graph model representing the surface is updated to the triple $G_S = (N, A, Aa)$, where N and A are the sets previously defined, while Aa defines the set of geometric attributes, according to definition 16.3.1.

Based on these attributed relationships, two sets of characteristic regions are defined, namely *strong* and *weak*, which can be alternatively used in order to perform a characterization which is either closer to a precise analytical description or which aims at the recognition of *global* shape properties. Both classes of regions are formally defined as particular connected components of the graph representing the triangulated surface and are obtained on the basis of a classification which takes into account the curvature around the triangles.

As far as strong characteristic regions are concerned, the classification of triangles follows these rules.

Definition 16.3.2 *A triangle* t *is said to be* convex (concave, plane) *if all its edges are of type convex (concave, plane),* saddle *if all its edges are of different type,* hybrid *in the remaining cases (see figure 16.3).*

Edge attributes have been assigned only to internal edges, so the type of triangles on the surface boundary (border triangles) is determined by the edges that do not belong to the boundary. Besides the triangle attributes, a property is defined for the set of arcs of $G_S = (N, A, Aa)$, which is used to test the *connection* between triangular facets induced by the type of arcs and triangles, as follows.

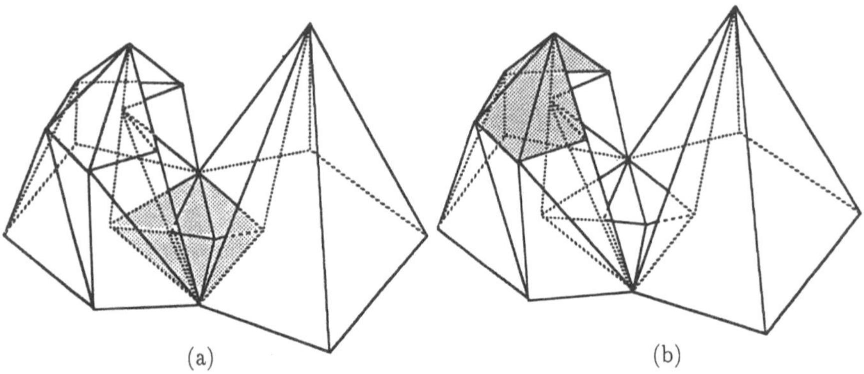

FIGURE 16.4. Example of strong characteristic regions. A concave region is highlighted in (a), while a convex region is highlighted in (b).

Definition 16.3.3 *Let* type *be the function returning the type of a triangle and* $q \in$ {concave, convex, plane, saddle, hybrid}. *Then, an arc* $(t_i, t_j) \in A$ *is said to be* q-connected *if and only if* $\text{type}(t_i) = \text{type}(t_j) = q$.

This connection property is well defined over the set of arcs A, as it is a single-valued property. Moreover, if A_c is the set of arcs complying with the connection property, then $A_c \subseteq A$. Indeed, unless all triangles in $T_D(S)$ are of the same type, at least two adjacent triangles exist such that their types are different, then the edge they share cannot be a q-connected arc. The reason for introducing this property is to give a rule for traversing the graph G_S meaning that it is possible to *walk* from a node t_i to a node t_j only if they are of the same type, that is, if they have similar curvature properties.

Following this classification, strong characteristic regions on a triangulated surface are defined as follows.

Definition 16.3.4 *Let* $q \in$ {concave, convex, plane, saddle, hybrid }, *and* G_S *be the graph surface model of* S. *A* strong characteristic region *of* S *is any connected component of* G_S *obtained removing from* G_S *all arcs not complying with the* q*-connection property.*

In terms of triangulation, characteristic regions identify connected and partially disjoint sets of triangles, whose boundary is defined by one or more closed sequences of edges which do not comply with the q-connection property and correspond to lines where the surface shape changes. An example of characteristic regions is shown in figure 16.4, where both a convex and a concave region is indicated.

Even if non-maximal regions could be of interest for surface description, definition 16.3.4 guarantees that the resulting surface decomposition is *unique*. Indeed, as the q-connection property is well defined, the uniqueness directly results from the definition of connected component of a graph. Obviously, each shape component can be unambiguosly classified according to the type of its interior triangles, and strong characteristic regions of type convex, concave, planar or saddle correspond to areas of uniform Gaussian curvature sign.

Indeed, since all edges belonging to a strongly convex (concave, planar) region are of type convex (concave, planar), it follows that at each vertex in the interior of the region the Gaussian curvature will be non negative [5].

Though the strong characteristic regions have a well defined mathematical meaning, their use in the context of terrain analysis may produce insufficient results. Indeed, in describing a real surface, one would probably use the term "hill" to point to an overall convex area even if it contains small concavities. Thus, if the aim is to find a synthetic surface description, global shapes seem to be more useful than fragmentation of the surface into small regions. Thus, a second set of shape components is introduced to characterize a surface by regions having *weak* curvature properties.

To define these shape components, a new classification scheme for triangles and a new version of the connection property are given.

Definition 16.3.5 *A triangle* t *is said to be* convex (concave, plane) *if at least two of its edges are of type convex (concave, plane), it is termed* saddle *otherwise.*

Definition 16.3.6 *Let* $q \in$ {concave, convex, plane, saddle}, attr *be the function returning the attibute of an arc and* type *be the function returning the type of a triangle according to definition 16.3.5. Then:*

- *if* $q \neq$ *saddle then an arc* $(t_i, t_j) \in A$ *is* weakly-q-connected *if and only if* $\text{type}(t_i) = \text{type}(t_j) = q$ *and* $\text{attr}((t_i, t_j)) = q$*;*
- *if* $q =$ *saddle then an arc* $(t_i, t_j) \in A$ *is* weakly-q-connected *if and only if* $\text{type}(t_i) = \text{type}(t_j) = q$.

Weak characteristic regions are always defined in terms of connected components of the surface graph, as follows.

Definition 16.3.7 *Let* $q \in$ {concave, convex, plane, saddle} *and* G_S *be the surface graph model of* S. *A* weak characteristic region *of* S *is any connected component of* G_S *obtained removing from* G_S *all arcs not complying with the weak-*q*-connection property.*

The difference with strong characteristic regions is that the requirements to classify triangle are weakened, while the rule for walking from a triangle to an adjacent requires triangles compatible with the adjacency type. This implies that in a weak characteristic region of type q it is possible to move from one triangle to any other one through edges of type q. Weak characteristic regions of type concave, convex and plane identify global shapes of the surface, meaning that in the interior of a region type q adjacencies will predominate although chains of non-q edges may exist. Obviously strong characteristic regions of type convex, concave and plane are contained in the set of weak characteristic regions of the corresponding type. Saddle regions extracted using a weak characterization correspond to saddle regions extracted using the strong one, as in both cases they are connected components formed by triangles with three edges of different type. In figure 16.5, an example of weak characteristic regions is shown, applied to the same surface shown in figure 16.4. The concave region includes triangles with at least two edges of type concave and, similarly, the convex regions include triangles with two convex edges. As a result, the characterization covers the complete set of triangles.

Based on these areal features, a conceptual model can be defined taking into account the adjacency between characteristic regions. Therefore, the morphology of a surface can be described by a Characteristic Region Configuration Graph (CRCG) in which characteristic regions correspond to graph nodes, while arcs and hyperarcs correspond to characteristic lines and points which are defined by adjacencies between region boundaries.

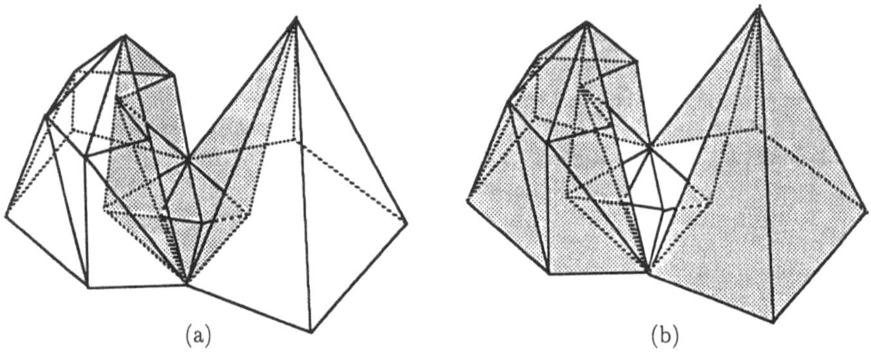

FIGURE 16.5. Example of weak characteristic regions. A concave region is highlighted in (a), while two convex regions are highlighted in (b).

Because of characteristic region definition they do not overlap, thus allowing a unique and unambiguous definition of the node set of the CRCG [6].

Since two regions are partially disjoint they can intersect only in their bounding contours. The edges belonging to the intersections between region boundaries define characteristic lines on the surface, where a variation in the surface shape happens. There are different types of characteristic lines:

- *ridge lines*: lines of local directional maximum on the surface;
- *ravine lines*: lines of local directional minimum;
- *crease lines*: generic creases on the surface.

In this way, the edges shared by two region borders represent the adjacency relationships between two characteristic regions and define the CRCG arc set. Since two regions cannot overlap, it follows that characteristic lines can intersect only at their endpoints. The intersections between characteristic lines define characteristic points on the surface and constitute the CRCG hyperarc set. Characteristic points are classified as minima, maxima or saddle points.

Applying the proposed method to natural terrains create problems since characteristic regions have been defined in nominal geometry and their definition loses meaning if applied to real surfaces. For example, a flat region on a real surface will not, in general, be represented by a set of perfectly coplanar triangles but, instead, by a set of "nearly" coplanar facets having angles neither too convex nor too concave. Therefore, it is necessary to introduce a tolerance parameter in edge and triangle classification criteria in order to ensure a more flexible recognition mechanism. For this reason the definition of a plane angle is weakened through the introduction of a parameter that represents the tolerance within which an angle must be considered plane [5].

16.4 Generation or Refinement

Surface generation and model refinement provide an inverse operator compared to the abstraction process. While the abstraction mechanism gives an interpretation of the geometric model by a subset of primitive entities linked to particular morphological configurations, the generation step is able to produce a geometric model starting from a description of the surface in terms of its high level characteristics. Furthermore, a refinement step can be performed to reach a certain level of approximation or the desired detail in visualization.

Assuming a Delaunay triangulation as the low level geometric model of a surface, the generation technique is a tessellation process able to keep the constraints imposed by the conceptual model, where constraints are here intended as chains of edges which must be preserved in the triangulation. Unfortunately, Delaunay triangulation does not offer such facilities. The problem is that the classical definition of Delaunay triangulation and most of the existing algorithms computing it apply to convex domain only, since the boundary of the domain is implicitly assumed to be the same as the convex hull of the set of data points. Furthermore, the edges of a Delaunay triangulation are defined according to geometric criteria independent of their morphological properties. This is an obstacle to the generation of a model which includes constrained meaningful lines of the surface, such as borders of morphologically uniform regions, sequences of points defining ridges, valleys or other specific points of the surface.

Thus, a definition of *Constrained Delaunay Triangulation* is needed [3]. To this aim, let us call V the set of points in the plane distributed within a multiply-connected planar region R, and π the boundary of R, expressed as a collection of loops:

$$\pi = \{\pi^{(1)}, \pi^{(2)}, \ldots, \pi^{(k)}\}$$

where every $\pi^{(i)}$ is an ordered closed chain of vertices:

$$\pi^{(i)} = [P_1^{(i)}, P_2^{(i)}, \ldots, P_{(qi)}^{(i)}]$$

such that $P_s^{(i)} \in V, s = 1, 2, \ldots, q_i$.

Let γ denote a set of line constraints to be kept in the model, expressed as a collection of chains of vertices:

$$\gamma = \{\gamma^{(1)}, \gamma^{(2)}, \ldots, \gamma^{(n)}\}$$

where:

$$\gamma^{(i)} = [Q_1^{(i)}, Q_2^{(i)}, \ldots, Q_{(ri)}^{(i)}]$$

such that $Q_s^{(i)} \in V, s = 1, 2, \ldots, r_i$.

The boundary π and the set of constraints γ restrict the definition of the Delaunay graph by introducing a new relation between the points of V.

Definition 16.4.1 *Let $P_i \in V$. Any other point $P_j \in V$ is said to be* visible *from P_i with respect to π and γ if and only if segment P_iP_j lies completely inside R or on the edges of π or γ.*

Given a point $P_i \in V$, all the points of V other than P_i may be classified as visible or not-visible from P_i, thus defining a subset W_i of the points of V visible from P_i as:

$$W_i = \{P_k \mid P_k \textit{ is visible from } P_i \textit{ with respect to } \pi \textit{ and } \gamma\}$$

According to the above definition, if π encloses a convex simply-connected region and γ is the empty set, then every point P_j, $j \neq i$ is visible from P_i, and $W_i = V - P_i$.

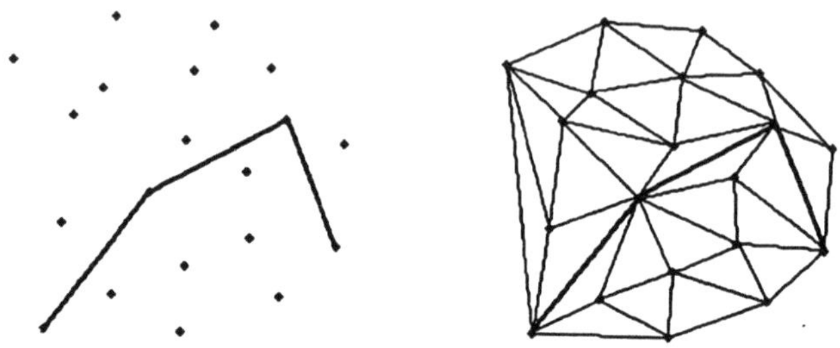

FIGURE 16.6. A constrained Delaunay triangulation including a line constraint.

Visibility defines a binary relation over V, which can be represented as an indirected graph.

Definition 16.4.2 *The* visibility graph *of the set of points V is a pair $G_V = (V_V, E_V)$ such that $V_V = V$ and $E_V = \{(P_i, P_j) \mid P_i, P_j \in V and\ P_i, P_j$ are mutually visible$\}$*

Hence, a *triangulation of V constrained by π and γ*, denoted as $T(V, \pi, \gamma)$, is a graph $T = (V_T, E_T)$ such that $V_T = V$ and E_T is the maximal subset of E_V where no two edges of E_T intersect except at their endpoints. It follows immediately from the definition that T is a planar graph, in which every face is a triangle and every side of π and γ is an edge of T. In order to have a triangulation meeting the Delaunay requirements, the Voronoi neighbourhood condition must be imposed, so that the set of points which can be connected to P_i in a constrained Delaunay triangulation is defined as follows.

Definition 16.4.3 *Let $P_i \in V$ and W_i be the set of points visible from P_i with respect to π and γ. Then:*

$$D_i = \{P_s \in W_i | P_i \text{ and } P_s \text{ are Voronoi neighbours}\}$$

defines the set of points which can be connected to P_i in a Delaunay triangulation constrained by π and γ.

Clearly, D_i is a subset of W_i.

Hence, the edges of a constrained Delaunay triangulation will be line segments joining pairs of points of V which are both mutually visible and Voronoi neighbours. Thus, the graph $G_D(V, \pi, \gamma) = (V_D, E_D)$ such that $V_D = V$ and $E_D = \{(P_i, P_j) | P_i, P_j \in V \text{ and } P_j \in D_i\}$ is called the Delaunay graph of V constrained by π and γ. G_D is uniquely identified by the above definition and, as in the case of the classical definition, it gives a constrained Delaunay triangulation of V if and only if each of its faces is a triangle. Otherwise, it can be completed to a Delaunay triangulation.

In a constrained Delaunay triangulation two kinds of edges can be distinguished, namely *constrained edges* that belong to the boundary π or to the constraint γ, and *Delaunay edges*, which are the remaining edges of the triangulation (see figure 16.6). In surface modelling, the constrained edges usually correspond to parts of surface specific features.

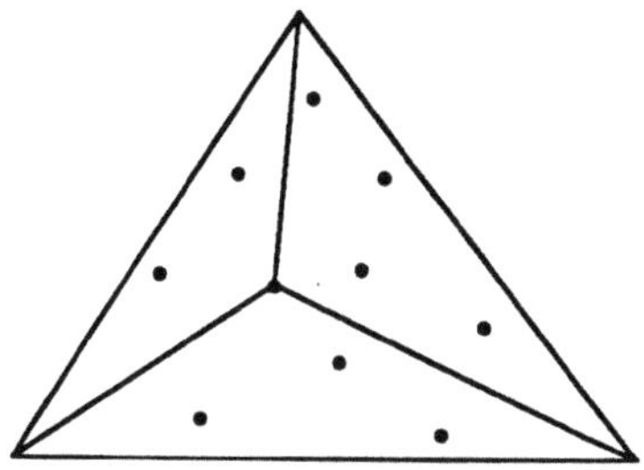

FIGURE 16.7. Example of incremental refinement.

The constrained Delaunay triangulation may be represented as a boundary model in which three types of information are stored, namely topological, geometrical and qualitative information. The geometrical information is given by triplets of coordinates of the data points, while the topology of the surface is represented by a graph in which the nodes are triangular faces of the triangulation, the arcs are edges and the hyperarcs are vertices. The qualitative information is coded by attributes associated with every arc to assign the type of edge (constrained or unconstrained, and the kind of constraint) and with every node to assign the type of region to which the triangle belongs (convex, concave, planar, saddle region).

Refinement of the basic geometric model may be performed in different ways depending on the specific application requirements. As an example, a refinement may be needed to obtain an approximation closer to a sampled surface. This approach may be applied when further information is given, for example when only a subset of the original data points have been included as vertices of the model. Techniques have been defined to include as vertices of the triangulation some additional points which, even if not related to specific properties of the surface, may be relevant for the accuracy of the representation, and their absence would cause an inadmissible deviation from the real surface [2]. Acceptability of a surface representation means the accuracy of the obtained approximation, where accuracy can be seen as a property of a triangulation imposed on the subset of points selected to represent the surface. The vertical deviation of all points not included in the triangulation from the plane of the corresponding triangle is considered to measure the accuracy of the current triangulation.

The criterion adopted to control the accuracy of the surface representation is to incrementally add the internal points whose deviation from the triangulated surface is greater than a given tolerance, which generally depends on the scale. The triangular patches chosen for the insertion of a new point are those in which the vertical distance from the farthest internal point is greater than the required tolerance. Each triangle where this situation occurs is subdivided into three non-coplanar nesting triangles by joining the data point with the maximum error within the triangle to each of its vertices, as shown in figure 16.7

An alternative refinement may be obtained by applying a stochastic perturbation to the basic geometric model when a realistic visualization is needed. This approach permits the inclusion of fractal-like details into the model, without requiring a real approximation to the initial data points. Fractal models of surfaces have been successfully created by

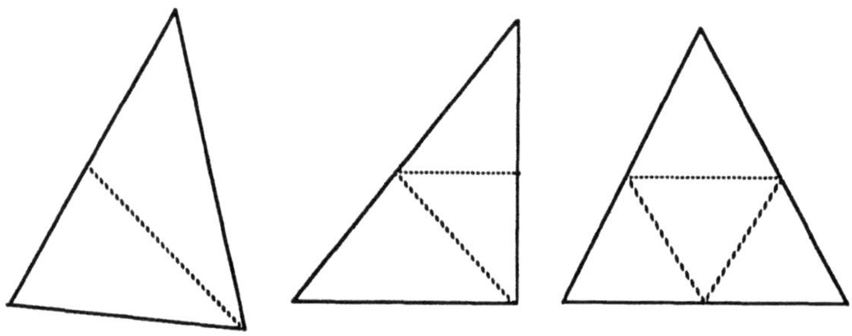

FIGURE 16.8. Possible stochastic subdivisions depending on the length of edges.

methods which use empirically chosen stochastic processes, such as recursive subdivision techniques, where the middle points of the edges are displaced by a stochastic perturbation. Recursive subdivision raises problems of consistency, in particular at the connection of adjacent patches, where a property of continuity is required. This is not an obvious consequence of a recursive subdivision technique, and it is necessary to generate the same point on the two sides of the edge shared by two adjacent triangles. A method to solve this problem is to consider the entire quadrilateral tile formed by two adjacent triangles and to compute the midpoint displacement on the diagonal according to the following formula:

$$z_m = \frac{\sum_{i=1}^{4} z_i^2/d_i^2}{\sum_{i=1}^{4} 1/d_i^2} + RD$$

where z_m is the elevation of the midpoint, z_i are the elevations at the four corners of the tile and d_i their distances from the midpoint.

The stochastic perturbation consists of a recursive triangular subdivision procedure, in which every triangle is subdivided into a suitable set of triangles according to the length of its edges. Given a maximum edge length l, that will in general depend on the scale, every triangle is checked to determine if any, and how many, of its edges are longer than l. These edges are split into two by using their midpoints, and the parent triangle is subdivided in two, three or four child triangles if the number of subdivided edges is one, two or three, respectively (see figure 16.8).

While this kind of approach has been tried and described in the literature [7], [10], its application to a constraint-based model makes it necessary to consider the qualitative information associated with the geometric entities to be perturbed.

In particular, care must be taken in defining the random displacement function:

$$RD(fd,\ et,\ rt)$$

Three parameters are taken into account in defining function RD. First of all, an input factor fd can be assigned with a value in the range $(0., 1.)$. Factor fd is related to the global fractal dimension, because the displayed surface can be made regular and smooth for low values of this parameter, or more irregular and jagged for values close to 1. A second parameter *et* considers the type of edge and permits to keep unchanged the edges classified as constrained. Finally, the third parameter rt takes into account the type of region to which the tile belongs, thus producing different perturbations in concave, convex or plane regions, so that the overall shape of the surface does not change [4].

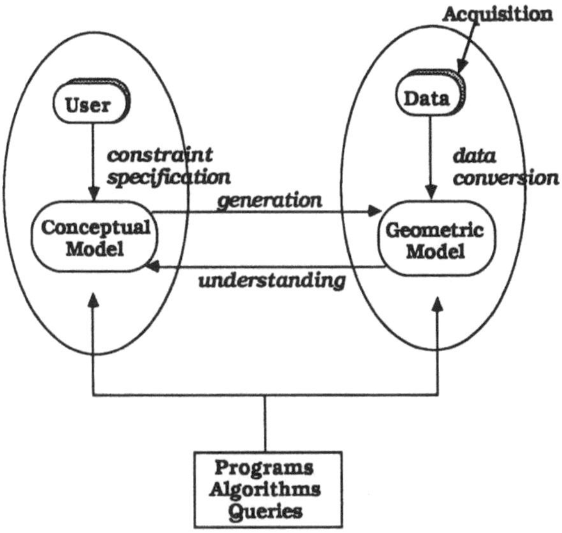

FIGURE 16.9. The architecture proposed for a complete surface modelling system.

16.5 Conclusions

High level models for natural surface definition and analysis have some evident advantages. First of all, they provide a concise description of the surface characteristics thus facilitating the communication between the user and the surface modeler. Secondly, they permit to link non geometric information to sets of geometric primitives. Finally, surface model interpretation like analysis, information extraction or classification may be sped up. On the other hand, a geometric model is needed to organize the information in a lower level structure that can be understood by the computer, thus requiring conversion operators between the two models.

The proposed solution is depicted in figure 16.9 showing the two levels of representation in a system able to deal both with the two kinds of surface models. According to this architecture, users may define the conceptual model of a terrain by specifying its prominent features (or constraints). The system automatically builds the corresponding geometric model by means of a constrained tesselation technique. Alternatively, real data can be structured in a geometric model which can be interpreted by the recognition of its features and converted into a conceptual model. End users or programs can interrogate both models according to the application needs.

In the following figures, examples of the two proposed operators are shown.

In figure 16.10 the basic geometric model of the surface is depicted in the form of a Delaunay triangulation constrained by two mountain tops, a valley and the boundary of the two mountains.

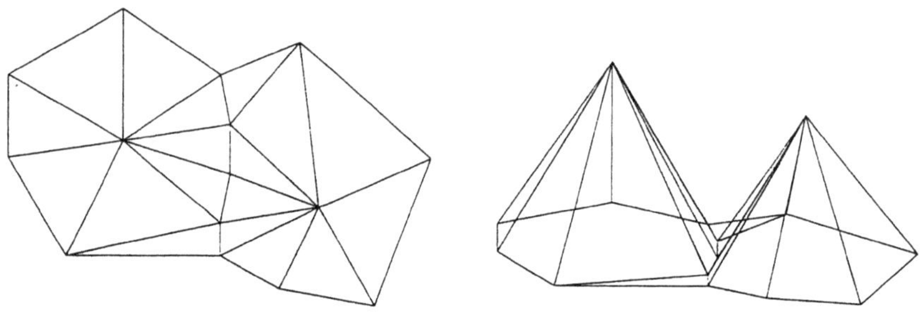

FIGURE 16.10. The basic geometric model.

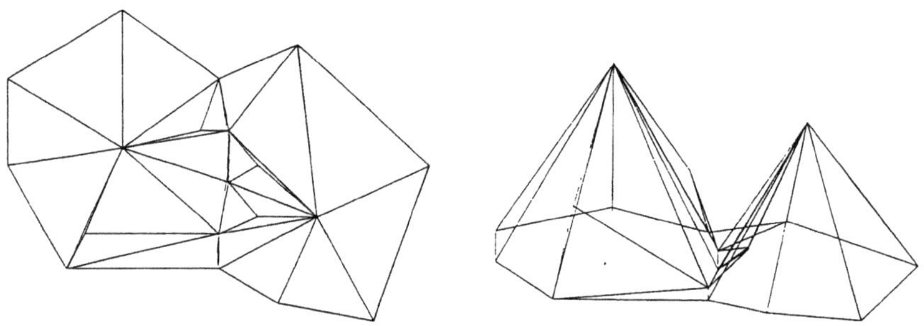

FIGURE 16.11. Refinement by introducing further points.

Figure 16.11 and 16.12 show the refinements obtained by introducing some further sample points and by a stochastic subdivision process respectively.

Figure 16.13 shows the results of the application of the abstraction operator to the surface obtained at a certain level of refiniment. Concave regions are coloured in green (16.13 a) while convex regions are coloured in white (16.13 b).

Acknowledgements:

The authors would like to thank George Nagy and Tapio Takala for their helpful comments and discussions about the topics presented in this paper.

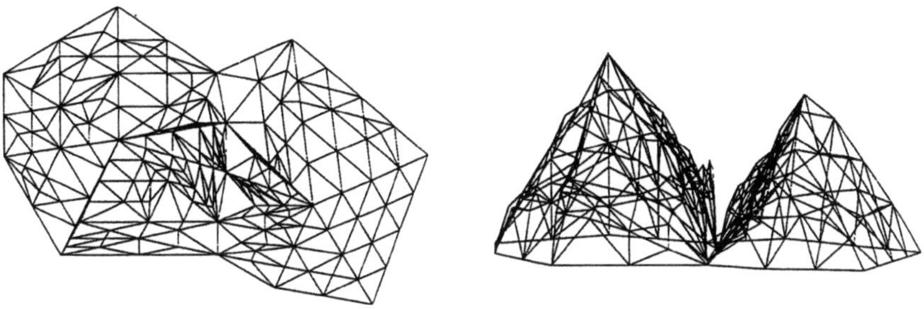

FIGURE 16.12. Refinement obtained by a stochastic perturbation.

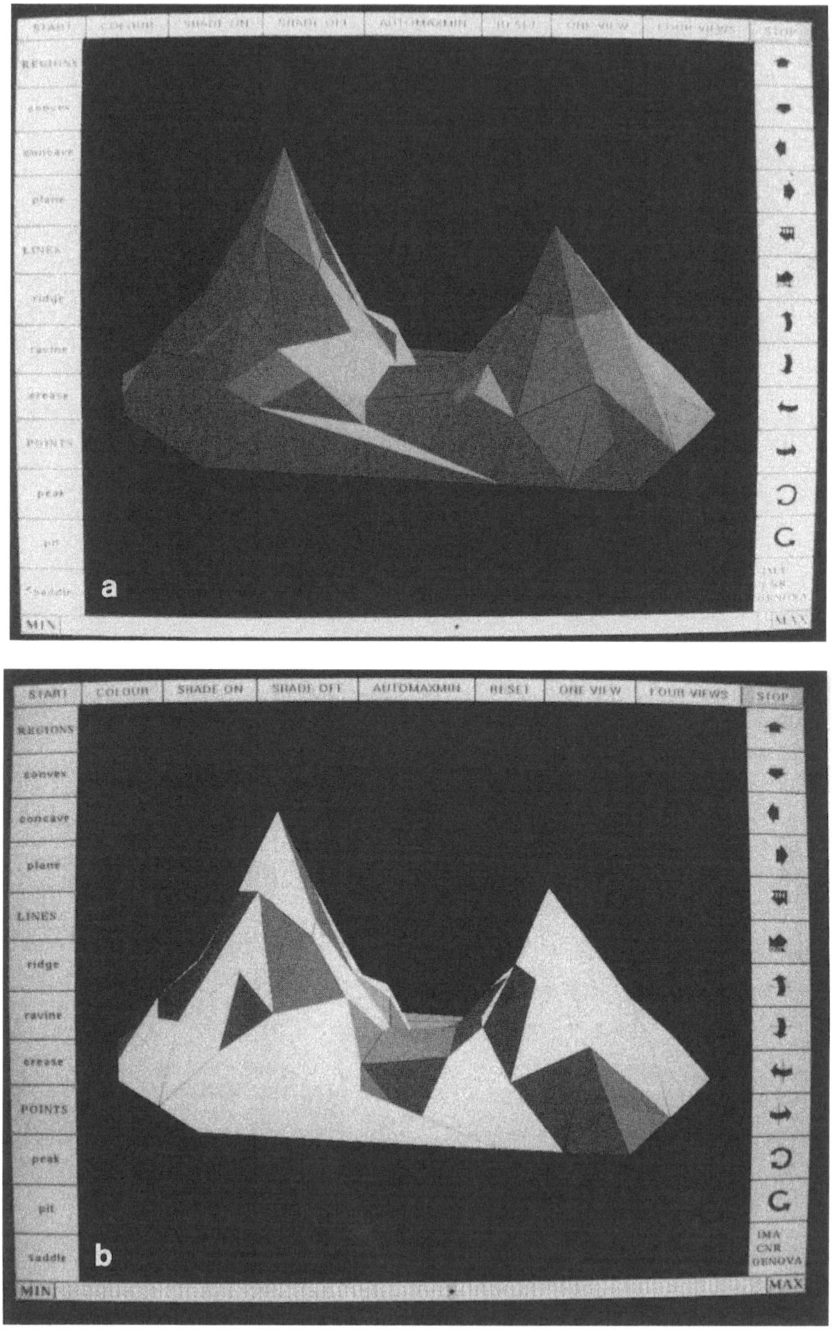

FIGURE 16.13. Example of characterization produced by the abstraction operator. (a) Concave regions green (see cover picture); (b) convex regions white

16.6 References

[1] P.J. Besl and R.C. Jain. Invariant Surface Characteristics for 3D Object Recognition in Range Images. Computer Vision, Graphics and Image Processing *33*, 1986.

[2] L. De Floriani, B. Falcidieno, G. Nagy, C. Pienovi. A Hierarchical Structure for Surface Approximation. Computers and Graphics *8* (2) (1984).

[3] L. De Floriani, B. Falcidieno, C. Pienovi. Delaunay-based Representation of Surfaces Defined over Arbitrarily Shaped Domains. Computer Vision, Graphics and Image Processing *32* (1985).

[4] B. Falcidieno, C. Pienovi. Natural Surface Approximation by Constrained Stochastic Interpolation. Computer-Aided Design *22* (3) (1990).

[5] B. Falcidieno, M. Spagnuolo. A New Method for the Characterization of Topographic Surfaces. International Journal of Geographical Information Systems (1991) (to appear).

[6] B. Falcidieno, M. Spagnuolo. Morphological Decomposition of Natural Surfaces. In *Proc. of SPIE Conference on Curves and Surfaces in Computer Vision and Graphics* Boston, 1991.

[7] A. Fournier, D. Fussel, L. Carpenter. Computer Rendering of Stochastic Models. Communications ACM, *25*, 1982.

[8] R.M. Haralick, T.J. Laffey, L.T. Watson. The Topographic Primal Sketch. Int. J. Robotics Res., *2*, 1 (Spring), 1982.

[9] C. Lin, M.J. Perry. Surface Description Using Surface Triangulation. In *Proc. of the Workshop in Computer Vision: Representation and Control.* Rindge, N.H. August 1982.

[10] B.B. Mandelbrot. Fractal Landscapes without Creases and with Rivers. In *The Science of Fractal Images* H. Peitgen and D. Saupe Eds., Springer Verlag, 1988.

[11] L.R. Nackman. Two Dimensional Critical Point Configuration Graph. IEEE Transactions on Pattern Analysis and Machine Intelligence July, 1984.

[12] F.P. Preparata, M.I. Shamos. Computational Geometry. New York, Springer Verlag, 1985.

17 Visualization of Depth Maps

Jacques Lemordant

ABSTRACT Coordinate systems are presented for fast 3-D transformations of complex surfaces, such as the depth maps produced by a laser camera or by a viewing algorithm. It is shown that these coordinate systems are natural for calculation on depth maps and correspond to a differential geometry approach of the problem. We describe algorithms using these coordinate systems to realize perspective to perspective and parallel to parallel view transformations. These parallel algorithms can be implemented using 1-D transformations only.

17.1 Introduction

3-D transformations of images in scanline order are used for texture mapping onto projections of 3-D surfaces [3, 6]. The texture surface is transformed as a 2-D image until it conforms to a projection of a surface placed arbitrarily in 3-space. The advantage of this approach is that the 2-D transformation can be decomposed into two simpler ones, one in horizontal and the other in vertical scanline order [3, 10, 1, 8]. A depth map is created by computing the depth values of objects from a particular viewpoint [11]. A 3-D transformation on a depth map corresponds to a move of the viewpoint. A depth map is an image-resolution data structure, sometimes without any object-space surface description. For example, laser range-finder systems used mainly in CAD/CAM or computer vision, are one of the easiest means for producing depth maps corresponding to a complex surface. The output of a milling display algorithm is another example of a depth map without any surface description [12].

Once a light source has been specified, a shaded image can be computed from a depth map. The time to compute a shaded image for different viewpoints is critical in some applications, for example to provide realistic visual feedback to users of range-finder systems.

What differentiates depth maps from images is that they do not lie in a plane. This means that if we transform a depth map to a different view, not only its visible surfaces must be recomputed, but moreover the flow on the pixels is dependent upon the depth value. The dependence upon the depth value introduces a foldover problem. To isolate the foldover problem, we have developed two coordinate systems and use them to describe the 3-D transformation on the pixels as a sequence of simpler steps. The first step is a 2-D decomposable flow on the pixels which does not affect and is independent of the depth value, the second and fourth steps affect only the depth value with no spatial displacement of the pixel, while the third step is a 1-D flow which proceeds front to back to treat the foldover problem and eliminate hidden points. This decomposition minimize memory references by accessing data only by rows or columns. In fact, this scheme may be thought of as a parallel algorithm since each row or column could be processed by a different processor without conflict in accessing the data. Similar access methods are used in geographical sciences, specifically in the processing of height fields. A height field is a depth map of a terrain under a parallel projection. In [4, 9], perspective views of these depth maps or 3-D transformations from parallel to perspective view are computed. In this paper, we consider transformations from perspective to perspective view and from parallel to parallel view.

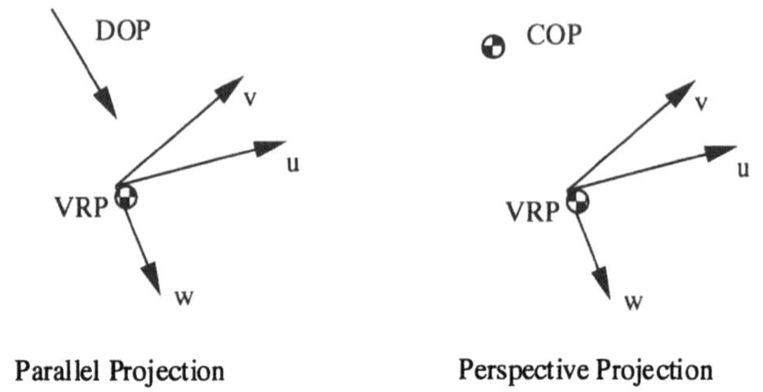

FIGURE 17.1. View System Specification.

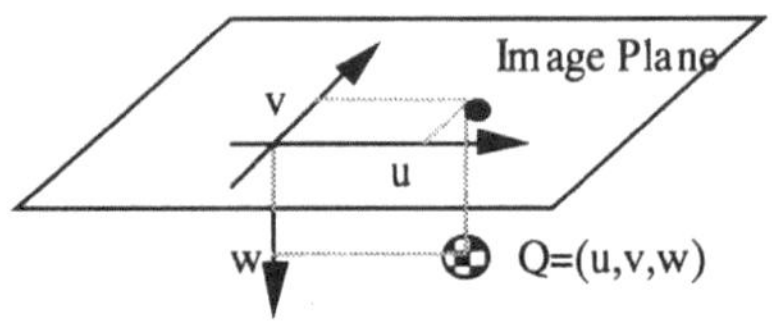

FIGURE 17.2. Cartesian Coordinate System associated to the View System.

17.2 Depth Coordinate Systems of Type I

Throughout this paper, a view system will be specified by a point VRP (for View Reference Point) and three mutually perpendicular directions (u,v,w) that form a left-handed, three-axis coordinate system with the given point as origin. The length of all directions vectors will be supposed to be equal to one. The (u,v)-plane is considered to be the image plane. The geometric projection used to generate the image is specified by a point COP (for Center Of Projection) or a direction DOP (for Direction Of Projection) as shown on figure 17.1. If a perspective projection is used, the object is projected onto the image plane by using lines passing through the point COP. For a parallel projection, lines parallel to the direction DOP are used.

A Depth Map is the representation of a surface in a coordinate system linked to a view system. Figure 17.2 shows the u-v-w cartesian coordinate system associated to the image plane u-v. We propose to use a coordinate system which is not symmetric with respect to the three coordinates and therefore noncartesian. As shown in figure 17.3, it consists of a depth coordinate d and two spatial coordinates u and v in the image plane. We call this system the depth coordinate system of type I. On these figures, P is the point which is the image of Q under the projection on the image plane

At each point P of the image plane is associated a number d which represents the depth of the surface as shown in figure 17.3. The pair (P,d) will be called a *dexel* with "de" for depth and "el" for element. It can be observed that we have chosen Coordinate systems of type I such that the d-coordinate of a point Q is independent of the position of the image plane. Consequently, a move of the image plane with a fixed COP or DOP, induces a geometric flow $g : (P, d) \Rightarrow (P', d)$ on the dexels which is the analogue of the optical flow on the pixels, the depth playing the role of the brightness. This flow is called geometric because only a spatial displacement is involved. Figure 17.4 shows this geometric flow, which is a spatial reorganization of the dexels.

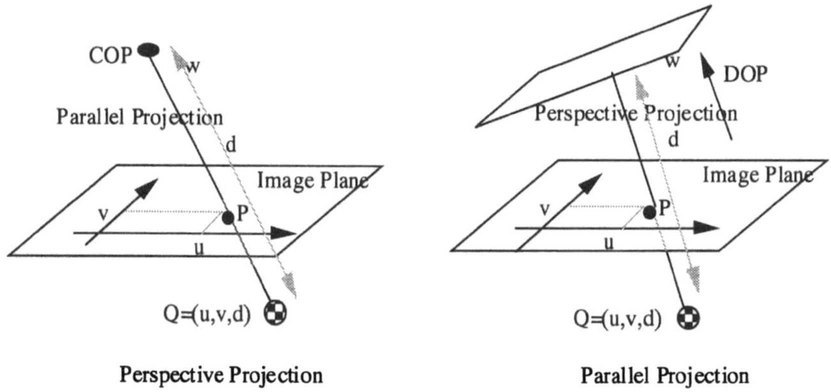

FIGURE 17.3. Depth Coordinate System of Type I.

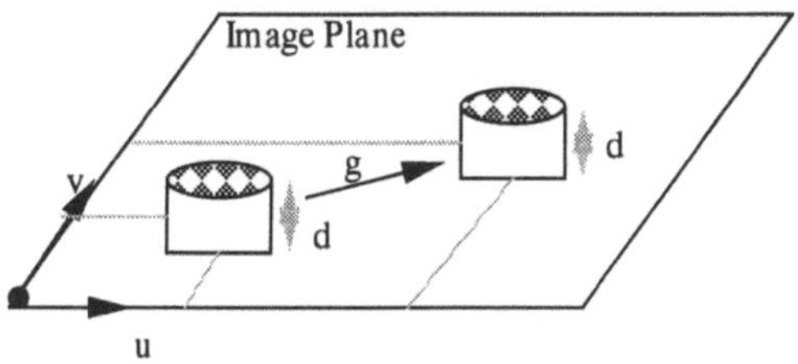

FIGURE 17.4. Geometric Flow on the Dexels.

17.3 The Eye Fiber Bundle

It is apparent that a geometric projection induces a decomposition of the space. We try to find a formalism such that the more suitable decomposition to study 3D transformations on depth maps will be evident. A topological structure, called a *fiber bundle* (see [5]), corresponds to the geometric projectors associated with a view system. In fact, the depth coordinate systems of type I are coordinate systems linked to this structure. A fiber bundle consists at least of a base space, a total space and a map that projects each point in the total space onto a point in the base space. The set of all points in the total space that are mapped onto the same point in the base is called the fiber. The total space can resemble a bundle of fibers. Every fiber in a fiber bundle must have the same topological structure. For example, a sphere defines a two parameter family of rays which can be generated by moving a point on the sphere and tracing out the ray defined by the point and the center of the sphere. Such a family has the structure of a fiber bundle.

17.3.1 Structure of fiber bundles

A fiber bundle is a space that is constructed from two spaces, a *base space* and a *fiber space*, by attaching a copy of the fiber space to each point of the base space. The bundle is the product of the base and the fiber. In figure 17.5, we show a simple example of a fiber bundle. If we pick a particular point on each fiber of the bundle, we define a section of the bundle as shown in figure 17.5.

In the example of figure 17.5, the base and the fiber are both the real number line, a one-dimensional space with a familiar topology. However, in general, the fiber and

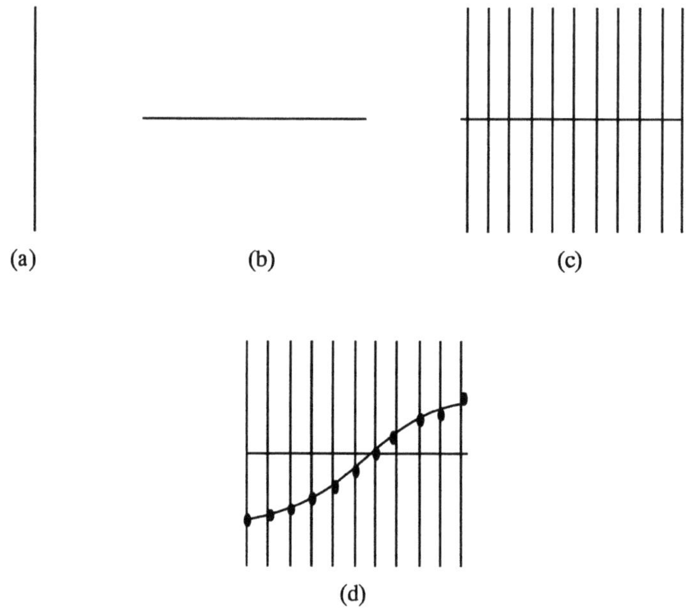

FIGURE 17.5. A trivial fiber bundle. Attaching a copy of the fiber (a) to each point of the base (b) gives the bundle (c). Picking a point in each fiber space defines a section (d).

the base can both be arbitrary, multidimensional spaces. For example, the base space can be a sphere, with the fiber a three-dimensional space. A section of this bundle is a vector field on the sphere. For the simple example given in figure 17.5, familiar geometric ideas such as ordinate, abscissa, and curve are adequate, and the machinery of fiber bundles is unnecessary. However, in the more complicated applications, the familiar ideas become inadequate. The fiber bundle formalism provides a unified, dimension-independent framework that encompasses both simple and complex visualization problems [2].

A fiber bundle may be viewed as a generalization of a function of one or more variables. The base space is analogous to the independent variables of a function; the fiber space is analogous to the dependent variable of a function; a section correspond to the function itself and the fiber bundle is the space in which we graph the function.

For example, consider the function given by

$$v = (v_1, v_2, v_3) = (f_1(u_1, u_2), f_2(u_1, u_2), f_3(u_1, u_2))$$

where $0 < u_1, u_2 < 2\pi$, and f_1, f_2, f_3 are real-valued functions. To interpret this function as a fiber bundle, we emphasize not the individual variables $(u_1, u_2, v_1, v_2, v_3)$ but the spaces in which they take their values. The independent variables u_1, u_2 are coordinates on the surface of a torus. Thus the base space is the surface of the torus T^2. The dependent variables v_1, v_2, v_3 are coordinates of the vector space R^3. The fiber bundle is the product space $T^2 \otimes R^3$. The section is defined by the values of the three functions f_1, f_2 and f_3 at each point of the base.

The kind of bundles that we use in this paper is called trivial bundles. They are locally and globally the Cartesian product of the base and the fiber. Non-trivial bundles, which we will not discuss, contain an additional structure that describes the way in which the

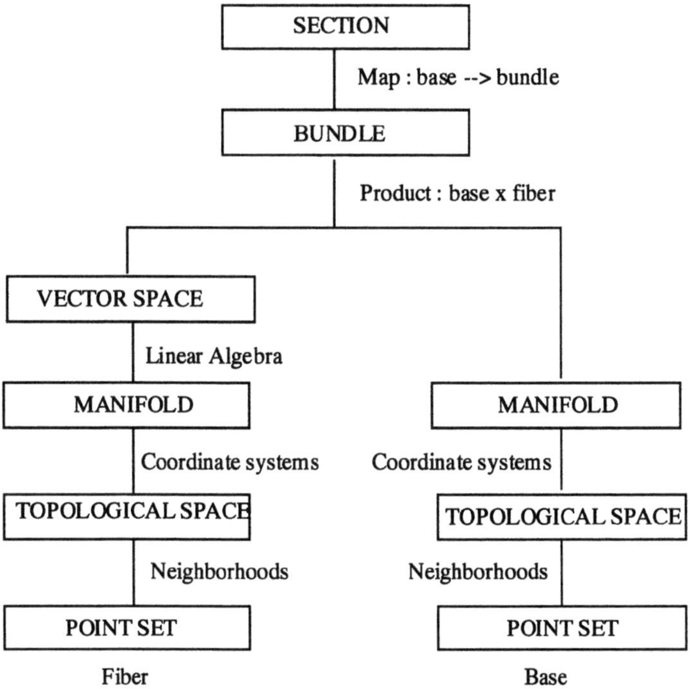

FIGURE 17.6. Layers in the geometry of a vector bundle [2].

bundle is twisted so that it is not a product globally. An example of a non-trivail bundle is the Möbius band.

As we will show, the theory of fiber bundles provides a very general basis to find the decomposition of the space the most suitable for 3-D transformations of depth maps.

17.3.2 Perspective Projection

Definition of the Eye Fiber Bundle. We consider the decomposition of R^3 by the bundle of rays issued from COP as shown on figure 17.7. This decomposition corresponds to a perspective projection.

If we use the points of the unit sphere S^2 with center at COP as an index to these rays, we have:

$$R^3 - COP \equiv \cup_{B \in S^2} \{COP + d(B - COP), d \in R^+\}$$

Let Π be the projection:

$$Q \in R^3 - COP, Q = COP + d(B - COP) \Rightarrow B \in S^2$$

Observe that $\Pi^{-1}(B)$, called the fiber F_B over B, corresponds to a ray from COP. The three-tuple $(R^3 - COP, \Pi, S^2)$ is a fiber bundle with $R^3 - COP$ as total space and S^2 as base space. We call it the Eye Fiber Bundle E. There is a homeomorphism

$$h : E \Rightarrow S^2 \otimes R^+$$

which takes each fiber isomorphically onto R^+. A fiber bundle with such a property is called a trivial bundle. A description of a view system with the product space $S^2 \otimes R^+$

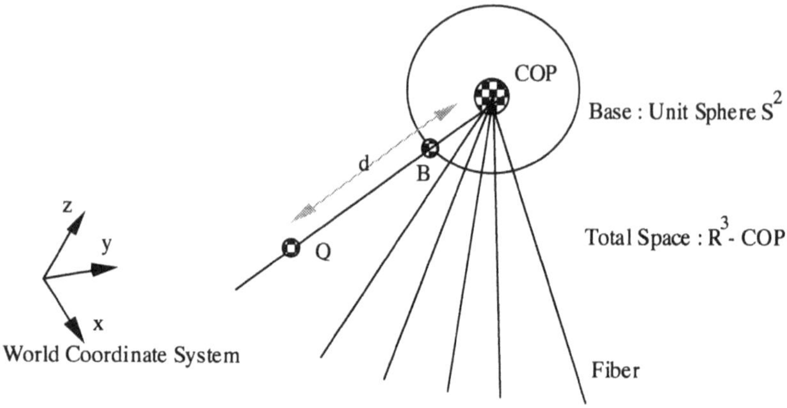

FIGURE 17.7. Eye Fiber Bundle (Perspective Projection).

will be more abstract, e.g., we prefer to speak of the projection on the base that of the projection on the first factor. With the fiber bundle formalism, we have a view model which bring to the fore useful operations. Observe that a depth map can be defined as a section of the eye fiber bundle, a section being a function $s : S^2 \Rightarrow E$ such that $\Pi \circ s =$ identity of S^2

17.3.2.1 The Relation between Type I Coordinate System and the Eye Fiber Bundle.

The Eye fiber bundle is a differentiable manifold: it can be covered by coordinate charts $c_i : E \Rightarrow R^3$ so that on the intersection of two charts c_i and c_j the map $c_j \circ c_i^{-1}$ between subsets of R^3 is differentiable. We have a natural coordinate system f on each fiber Fp:

$$Q \in F_p \Rightarrow f(Q) = d \in R^+$$

Recall that $Q = COP + d(B - COP)$ with $B \in S^2$.

The radius of the sphere being equal to one, d can be interpreted in the world coordinate system as the distance from the point Q to the point COP. There are several means to impose a coordinate system on the base S^2, but we are interested in a coordinate system which corresponds to the depth coordinate system. Therefore, we use the image plane to impose a coordinate system c on S^2. The mapping c is defined on an open hemisphere by:

$$B \in S^2 \Rightarrow c(B) = (u, v) \in R^2$$

u,v being the coordinates of the intersection of the fiber $\Pi^{-1}(B)$ with the image plane as shown in figure 17.8. With these definitions, the coordinate system (f,c) is in fact a type I coordinate system. Coordinate charts (f,c) are charts for the same differential structure on E.

17.3.3 The Change of Coordinate System and the Geometric Flow on the Dexels

We are now in the position to study the geometric flow on the dexels corresponding to a move of the image plane. This geometric flow is given by the application $c_2 \circ c_1^{-1}$ where c_1 is associated to the first image plane and c_2 to the second image plane. The application $c_2 \circ c_1^{-1}$ correspond to a change of coordinates on the base of the view fiber bundle. We show now that the components of $c_2 \circ c_1^{-1}$ are both rational linear polynomials :

$$c_2 \circ c_1^{-1}(u_1, v_1) = ((au_1 + bv_1 + d)/(mu_1 + nv_1 + p), (eu_1 + fv_1 + h)/(mu_1 + nv_1 + p))$$

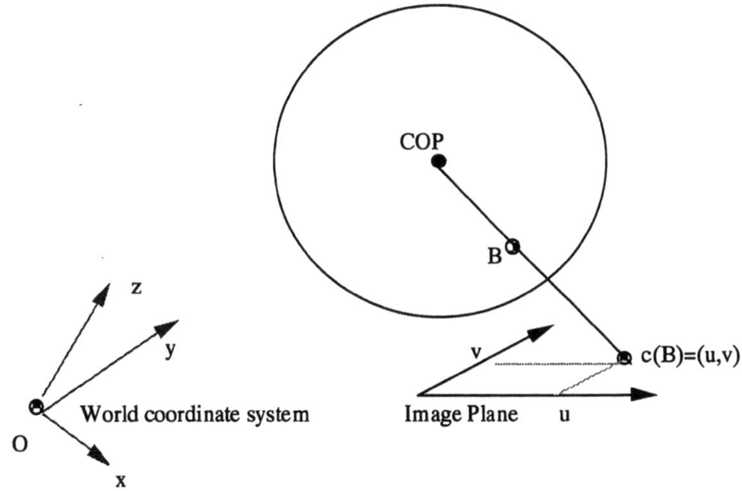

FIGURE 17.8. Coordinate System on the sphere.

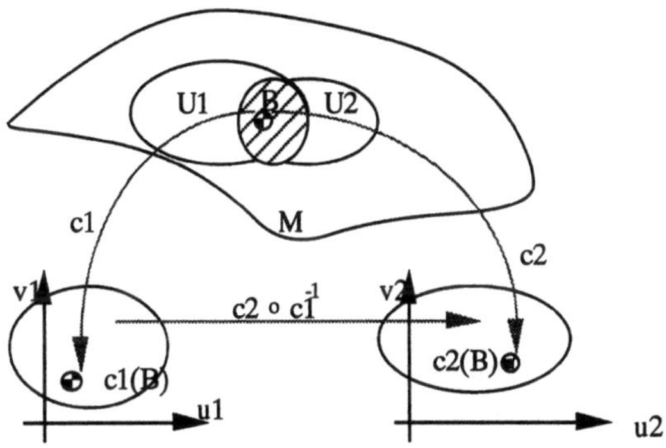

FIGURE 17.9. Change of Coordinate System on a Manifold.

Indeed, it is apparent that the transformation $c_2 \circ c_1^{-1}$ is a projective transform between the two planes (u_2, v_2) and (u_1, v_1). If we consider $c_1(B)$ as a 3D point, we can compute its coordinates in the frame (COP, x_s, y_s, z_s) with $x_s = u_2$ and $y_s = v_2$ as shown in figure 17.10. In homogeneous coordinates, this transformation is expressible as a 4×4 matrix representing the product of the translation, rotation, change of orientation which brings the frame (COP, u_s, v_s, w_s) onto the frame (VRP_1, u_1, v_1, w_1). Then, the point $C_2(B)$ is obtained through a standard perspective projection on the plane (u_2, v_2) parallel to (u_s, v_s).

Thus, the whole transformation is expressible as a 4×4 matrix:

$$\begin{pmatrix} x \\ y \\ z \\ w \end{pmatrix} = \begin{pmatrix} a & b & c & d \\ e & f & g & h \\ i & j & k & l \\ m & n & o & p \end{pmatrix} \begin{pmatrix} u_1 \\ v_1 \\ 0 \\ 1 \end{pmatrix}$$

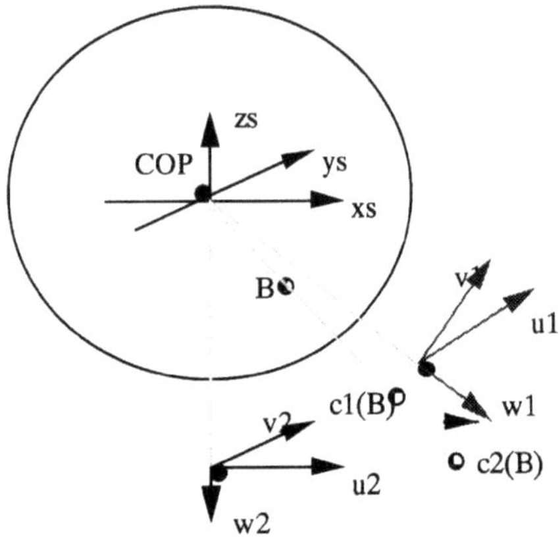

FIGURE 17.10. The Geometric Flow on the Image Plane.

The coordinates (u_2, v_2) of $c_2(p)$ are obtained by dividing through the homogeneous coordinate w and deleting the third coordinate :

$$(u_2, v_2) = (x/w, y/w)$$

Replacing x,y,w, the equation above gives :

$$(u_2, v_2) = ((au_1 + bv_1 + d)/(mu_1 + nv_1 + p), (eu_1 + fv_1 + h)/(mu_1 + nv_1 + p))$$

17.3.4 The Geometric flow as a decomposable Parallel Warping

An application whose components are rational linear polynomials can be decomposed as a product of 1-dimensional transforms. Figure 17.11 below illustrates the decomposition in profile-based mappings of the change of coordinates $c_2 \circ c_1^{-1}$:

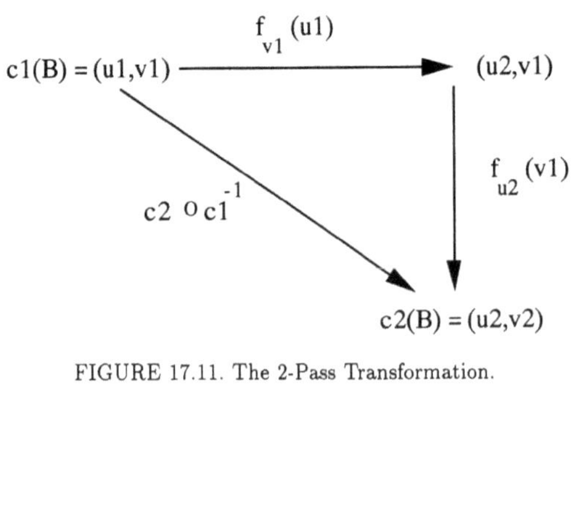

FIGURE 17.11. The 2-Pass Transformation.

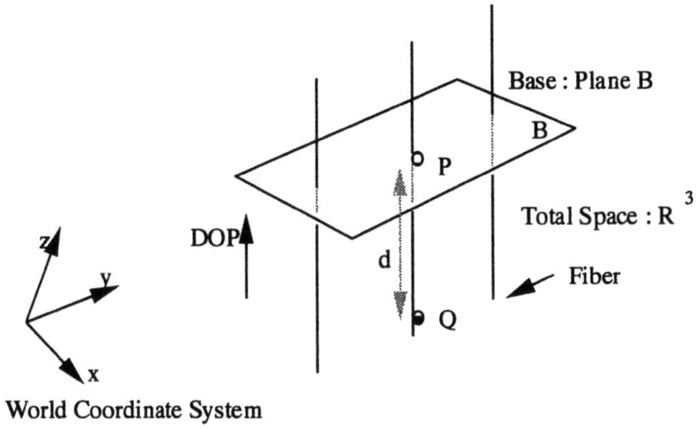

FIGURE 17.12. Eye Fiber Bundle (Parallel Projection).

where (see [3]):

$$f_{v_1}(u_1) = (Au_1 + B)/(Cu_1 + D)$$
$$f_{u_2}(v_1) = (Gv_1 + H)/(Iv_1 + J)$$

with

$$A = a, B = bv_1 + d, C = m, D = nv_1 + p$$
$$E = (b - nu_2)/(mu_2 - a)$$
$$F = (d - pu_2)/(mu_2 - a)$$
$$G = f + eE$$
$$H = h + eF$$
$$I = n + mE$$
$$J = p + mF$$

Thus, we have shown that the computation of the geometric flow on the dexels, which corresponds to a move of the image plane, amounts to computing a change of coordinate system on the base of the fiber bundle defined by COP. This change of coordinate can be implemented as a multi-pass transformation which needs row and line access only.

17.3.5 Parallel Projection

We study now the case of a parallel projection. A plane B defines a two parameter family of lines which can be generated by moving a point on the plane and tracing out the line defined by the point and the vector orthogonal to the plane. The depth coordinate system of type I is a coordinate system on this fiber bundle.

We consider the decomposition of R^3 by the bundle of straight lines parallel to the direction DOP. We will use a plane B orthogonal to DOP as the base plane for these lines.

$$R^3 \equiv \cup_{P \in B}\{P + dDOP, d \in R\}$$

Let Π be the projection :

$$Q \in R^3, Q = P + dDOP \Rightarrow P \in B$$

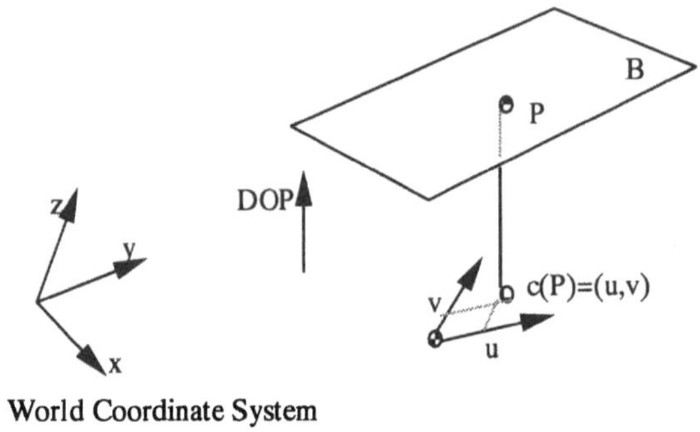

FIGURE 17.13. Coordinate System on the Base.

$\Pi^{-1}(p)$, the fiber over p, is a straight line parallel to DOP. The three-tuple (R^3, Π, B) is a fiber bundle with R^3 as total space and B as base space. A depth map can be defined as a section of this fiber bundle.

17.3.5.1 The Type I Coordinate System and the Eye Fiber Bundle

We have a natural coordinate system f on each fiber F_p :

$$Q \in F_p \Rightarrow f(Q) = d \in R$$

(recall that $Q = P + dDOP$ with $P \in B$). DOP being a vector of length equal to one, d can be interpreted in the world coordinate system as the distance along DOP from the point Q to the plane B. If we use the image plane coordinate system to define a coordinate system c on the base P, we obtain a coordinate system on the fiber bundle which corresponds exactly to the type I coordinate system.

The mapping c is defined by :

$$P \in B \Rightarrow c(P) \in R^2$$

with $c(P) = (u, v)$, u,v being the coordinates in the (u,v) plane of the intersection of the fiber $\Pi^{-1}(p)$ with the (u,v) plane as shown in figure 17.13.

17.3.6 The Geometric Flow and the Change of Coordinate on the Eye Fiber Bundle

With the type I coordinate system, the eye fiber bundle is a differentiable manifold. Indeed, let c_1 and c_2 be two coordinate systems on the base of the fiber bundle as shown in figure 17.14. The transformation $c_2 \circ c_1^{-1}$ is a projective transform between the two planes (u_2, v_2) and (u_1, v_1). We consider $c_1(P)$ as a 3D point with coordinates in the frame (VRP_1, u_1, v_1, w_1). In homogeneous coordinate, the change of coordinates from the frame $(VRP1, u_1, v_1, w_1)$ to the frame $(VRP_2, u2, v2, w2)$ is expressible as a 4×4 matrix representing the product of the translation and rotation which brings the frame (VRP_2, u_2, v_2, w_2) onto the frame (VRP_1, u_1, v_1, w_1).

The point $c_2(P)$ is obtained through a standard parallel projection on the plane (u_2, v_2). It is easy to see that in this case, the components of $c_2 \circ c_1^{-1}$ are polynomials instead of ratios of polynomials:

$$c_2 \circ c_1^{-1}(u_1, v_1) = (au_1 + bv_1 + d, eu_1 + fv_1 + h)$$

The decomposition in profile-based mappings of $c_2 \circ c_1^{-1}$ is realized as explained for the

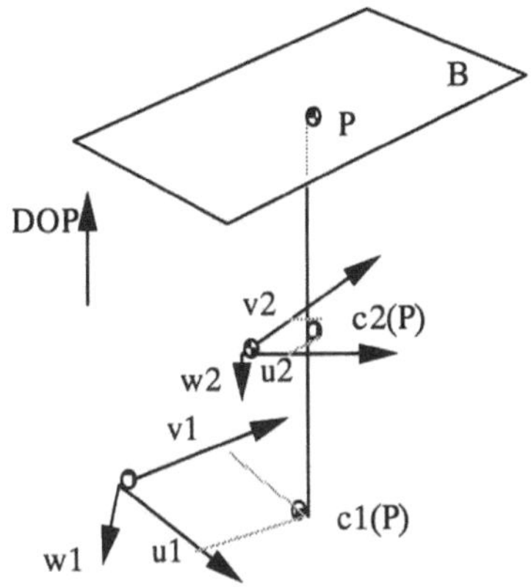

FIGURE 17.14. Geometric Flow on the Image Plane.

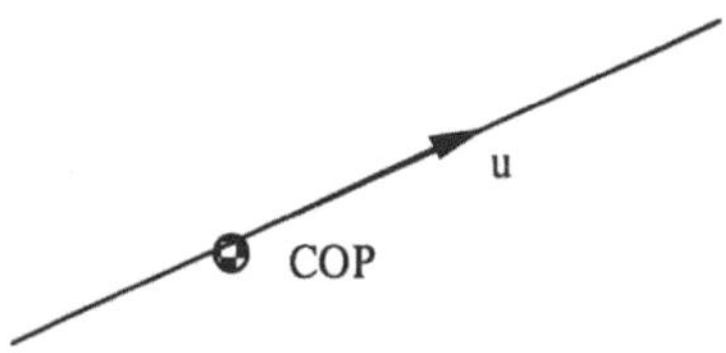

FIGURE 17.15. Moving the Center Of Projection and the Direction Of Projection.

case of ratios of polynomials. Thus, we have shown that the computation of the geometric flow on the dexels, which correspond to a move of the image plane, is in fact a change of coordinate on the Eye Fiber Bundle.

17.4 Depth Coordinate Systems of Type II

Our goal is to see if the transformation of a depth map with respect to a move of the COP or the DOP can also be computed with 1-dimensional transformations. Therefore, we introduce type II depth coordinate systems such that the displacement of COP or DOP induces a geometric flow on the dexels. This is the case because we impose a constraint on the displacement of COP or DOP as shown on figure 17.15. This constraint assures that the flow is geometric and 1-dimensional. The COP must be displaced on a line parallel to the u or v axis and the DOP on a great circle of the Gaussian sphere passing through u or v. With this constraint, the geometric flow, which is a spatial reorganization of the dexels, is parallel to the u or v axis of the image plane. From now, we suppose that u has been chosen.

The type II coordinate systems are defined as shown in figures 17.17 and 17.18, where d is the distance of Q to the straight line (or generator) passing through P with direction u (P is the image of Q under the projection on the image surface).

We use the convention that d is negative for a point Q on the same side of the image surface as COP or the reference plane. In the case of a perspective projection, we consider also the cylindrical image surface with axis parallel to u and centered on COP as shown in figure 17.17.

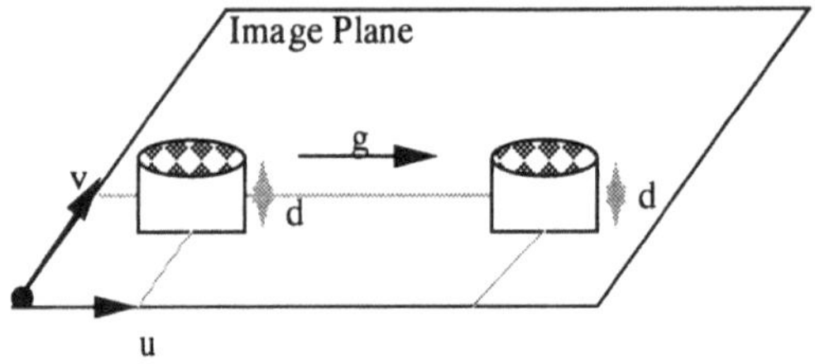

FIGURE 17.16. Geometric Flow on the dexels parallel to the u axis.

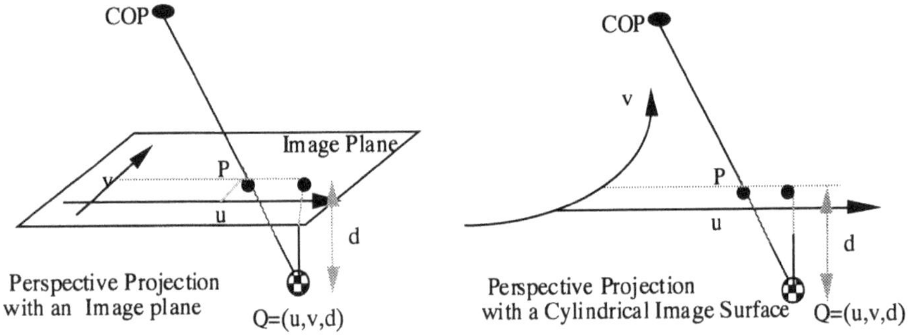

FIGURE 17.17. Depth Coordinate Systems associated to the View System (Type II) (Perspective Projection).

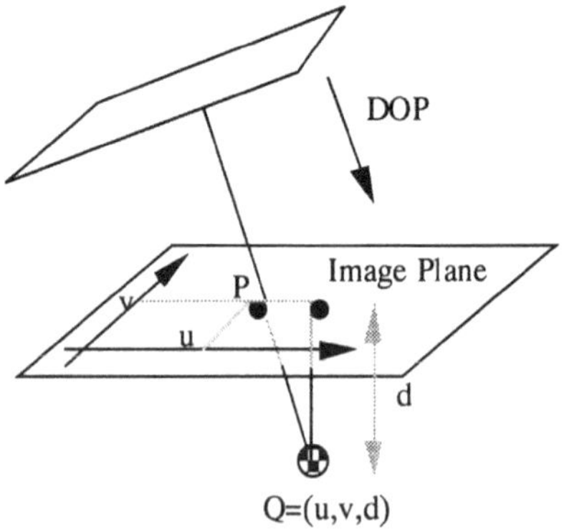

FIGURE 17.18. Depth Coordinate Systems associated to the View System (Type II) (Parallel Projection).

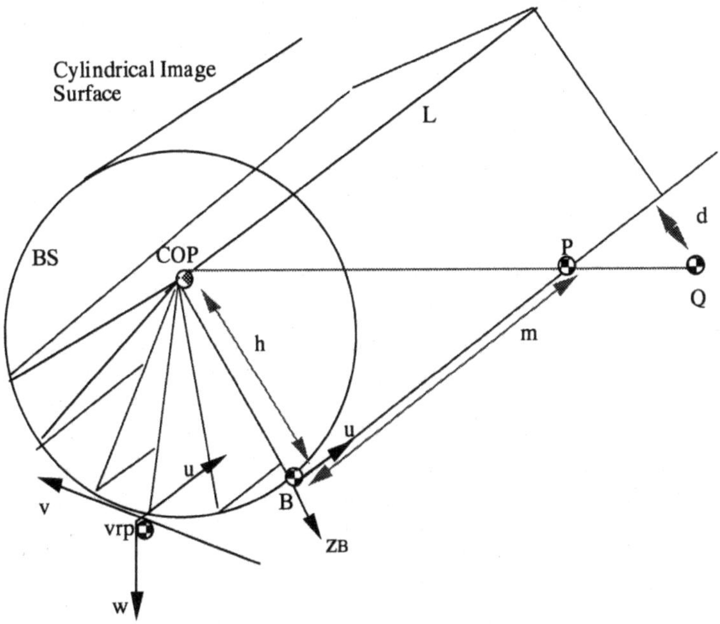

FIGURE 17.19. Image Fiber Bundle.

17.5 The Image Fiber Bundle

The image fiber bundle is, in some sense as we will see later, the dual of the eye fiber bundle. In this section, the key geometric ideas are that, if the center of projection moves parallel to the u axis, the space can be divided into a one parameter family of planes (projecting to a one parameter family of lines parallel to the u-axis) such that the geometric flow in the image surface induced by this move can be studied in each plane separately. Thus the calculation of the depth map in the new view system can be done by a 1-dimensionnal transformation. We call this decomposition of 3-space, the *image fiber bundle* because this decomposition is based on the image surface. Type II depth coordinate systems are the natural coordinates system for this structure.

17.5.1 Perspective Projection

Choose a line L passing through COP with direction u or v, (VRP,u,v,w) being the image coordinate system. By rotating a half plane about an axis lying along the line, we generate a two-dimensional fiber bundle E.

We suppose that the direction of L is u. The circle in the plane orthogonal to L with center at COP and tangent to the image plane (u,v) will be the base BS of the fiber space. Let h be the radius of this circle and B a point on it. The fiber above B is the half plane defined by the axis L and the point B as shown in figure 17.19. We define a cylindrical image surface CIS as the surface of a cylinder with axis L and radius h such that the image plane is tangent to it. The point P of figure 17.19 is on CIS. We prefer to use a cylindrical image surface instead of a plane surface because the conversion from coordinate system of type I to a coordinate system of type II will be easier for a cylindrical image surface uniformly sampled. We can observe that an uniformly sampled image plane will induce an non-uniformly sampled cylindrical image surface. But we can pass from a non-uniformly sampled CIS to an uniformly CIS with an easy 1-dimensional transform along the direction given by the axis of the cylinder. We defined the direction vector z_B by $z_B = (B - COP)/h$. The fiber space E, as a point set , is given by

$$E \equiv \cup_{B \in BS}\{B + dz_B + \mu u, \mu \in R, d \in R^+\}$$

Observe that d is the distance from Q to the straight line passing through B with direction u. Let Π be the projection :

$$Q \in E, Q = B + dz_B + \mu u \Rightarrow B \in BS$$

The three-tuple (E, Π, BS) is a fiber bundle with E as total space and the circle BS as base space. We have a natural coordinate system on each fiber F_B which is defined by the COP:

$$Q \in F_B \Rightarrow (d, \mu) \in R^+ \otimes R$$

where d and μ are defined as shown on figure 17.19. We can observe that d is independent of the choice of COP on the line L. Thus moving the COP on the line L induces a 1-D geometric flow on the dexels along the generator of the cylindrical image surface CIS. This 1-D geometric flow is, in fact, a change of coordinate on the fibers of the eye fiber bundle. Let c_1 and c_2 be the coordinate systems associated to the positions COP1 and COP2 of the COP on the axis L. The 1-D geometric flow is given by the transformation $c_2 \circ c_1^{-1}$ which can be easily calculated :

$$c_2 \circ c_1^{-1}(m, d) = (m + d\delta/h, d)$$

where δ is the scalar product of COP2-COP1 with u .

17.5.2 Parallel Projection

Choose a great circle C on the Gaussian sphere with DOP and u or v on it, (VRP,u,v,w) being the image coordinate system. We suppose that the direction u has been chosen. We consider the family of planes parallel to the plane of the great circle. This family can be generated by translating a plane along the v axis which is used as base of the fiber space. Let B be a point on the v axis. The fiber above B is the plane passing through B and parallel to the plane of C as shown in figure 17.20.

We recall that P is the projection along DOP of the point Q in the image plane (u,v). Let z be the unit vector in the direction orthogonal to u in the plane of C, whose scalar product with DOP is negative. The total space E of the fiber bundle is:

$$E \equiv \cup_{B \in vaxis}\{B + dz + \mu u, \lambda \in R, \mu \in R\}$$

d can be interpreted as the distance from Q to the straight line passing through B with direction u. Let Π be the projection:

$$Q \in E, Q = B + dz + \mu U \Rightarrow B \in v - axis$$

The three-tuple $(E, \Pi, v - axis)$ is a fiber bundle with E as total space and v-axis as base space. We have a natural coordinate system on each fiber F_B :

$$Q \in F_B \Rightarrow (d, \mu) \in R \times R$$

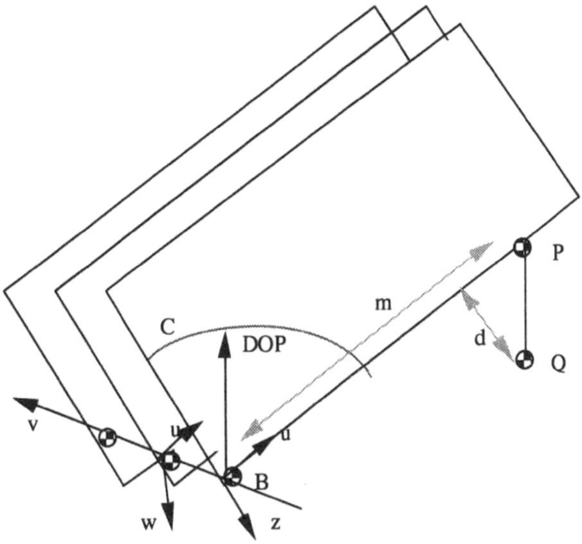

FIGURE 17.20. Image Fiber Bundle.

as shown on figure 17.20. The coordinate d, being independent of DOP, we have shown that, moving the direction of projection DOP on a great circle of the Gaussian sphere, induces a 1-D geometric flow on the dexels of the image plane. This 1-D geometric flow is a change of coordinates on the fibers of the image fiber bundle.

17.5.3 The Duality between the Eye Fiber Bundle and The Image Fiber Bundle

We have some kind of duality between the two kinds of fiber bundle. The Eye Fiber Bundle has the following properties :

Dimension of the base : 2
Dimension of the fiber : 1
The change of coordinate systems takes place on the fiber.

Now, if we exchange the words "base" and "fiber", we obtain the properties of the Image Fiber bundle:

Dimension of the fiber : 2
Dimension of the base : 1
The change of coordinate systems takes place on the base.

17.6 Conversion between Coordinate Systems I and II

17.6.1 Perspective projection

The conversion proceeds along each generator of the Cylindrical Image Surface. Figure 17.21 represents the conversion in the plane determined by COP and the generator j of the CIS. We suppose that the CIS is uniformly sampled, i.e., that the spacing between each dexel is the same for all the generators. Let l_i be the length of the ray issued from

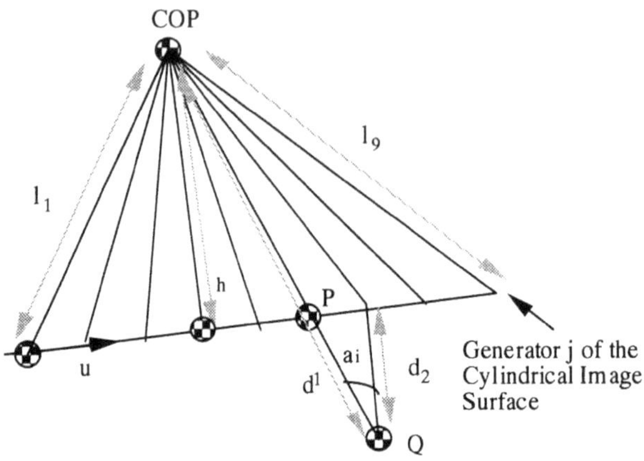

FIGURE 17.21. Conversion Type I - Type II.

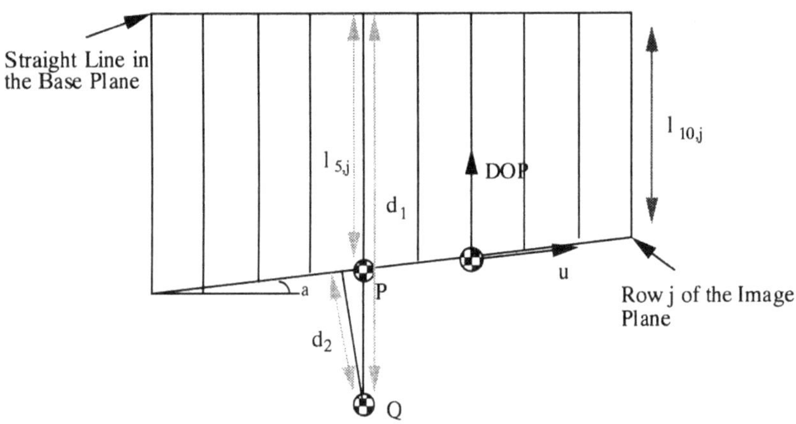

FIGURE 17.22. Conversion Type I - Type II.

COP to the dexel i of the generator j and a_i be the angle between the perpendicular at P to the ray from COP to Q and the generator j of the CIS. We have :

$$\cos a_i = d_2/(d_1 - l_i)$$

Thus, the formulas $d_1 = d_2/\cos a_i + l_i$ and $d_2 = (d_1 - l_i)\cos a_i$ We can observe that the effect of the CIS is to have formulas for the conversion which are independent of the generator that we choose (a_i and l_i are independent of j). RAM Lookup tables can supply the values of the l_i and $\cos a_i$ if we move the image plane in such a way that the distance h from the COP to the image plane (the focal length), is restricted to predetermined values.

17.6.2 Parallel projection

Figure 17.22 shows the conversion in the plane generated by DOP and the row j of the image plane.

Let a_j be the angle between the reference plane and the row j of the image plane. We have :

$$d_1 = d_2 / \cos a_j + l_{ij}$$

and

$$d_2 = (d_1 - l_{ij}) \cos a_j$$

The numbers l_{ij} can be computed from the relation

$$l_{i+1j} = l_{ij} + c_j$$

with $c_j = h \sin a_j$ where h is the distance between two pixels.

17.7 Application: The Implementation of a View System

We will show how to implement the 3D transformations of the depth maps that we have presented to construct a visualization system. Geometrical transformations of (intensity) images are typically specified by a function that maps points from one coordinate system to another coordinate system. In computer graphics, the old coordinate system is referred to as the modeling or local coordinate system, and the new coordinate system as the world or global coordinate system. In image computing, these are often referred to as the source and target coordinate systems, and we will adopt this terminology for depth maps. 3-D transformations of depth maps are different of geometric transformations of images in the sense that the value (or depth) of the transformed depth map at each point in the target coordinate system is not the same as its value at that point's position in the source coordinate system. This transformed depth map's value is unique as long as the geometric transformation is not degenerate (meaning that it maps multiple points to a single point). The key idea behind multi-pass transformations is to replace a single transformation with a sequence of 1-dimensional transformations along orthogonal directions. In a multi-pass transformation only one coordinate at a time is transformed. To perform a multi-pass transformation requires that the original transformation be decomposed into a sequence of transformations that, when composed, have the same effect that the original. Two-pass algorithms have been described for performing (intensity) image transformations. Three-pass algorithms for applying geometric transformations on uniformly sampled volume arrays are presented in [7]. We propose to use multi-pass transformations to transform depth maps.

The problem is very different from the one that we have with volume arrays. A depth map cannot be considered as a section of a uniformly sampled volume array. If the domain, where defined, the depth map is generally uniformly sampled, its depth values are not on a regular grid and must be considered as real values instead of integers. This is the case, for example, with laser range finders where the depth resolution is very large compared to the spatial resolution.

The process of creating a new depth map, corresponding to a 3D transformation, can be conceptualized as a 4-step process as shown in figure 17.23. An example of this process is shown in figure 17.24. It consists of a move of the image plane followed by a change of DOP.

The first step of the 4-step process is a 2-D flow on the pixels which can be decomposed in two 1-dimensional geometric transform of the form x'=Ax+b. This transformation is independent of the depth value which is not changed by the transformation. A is constant, so the relative stretch and compression is the same everywhere. The implementation of these 1- dimensionnal transformations is made with a uniform resampling and is particularly easy.

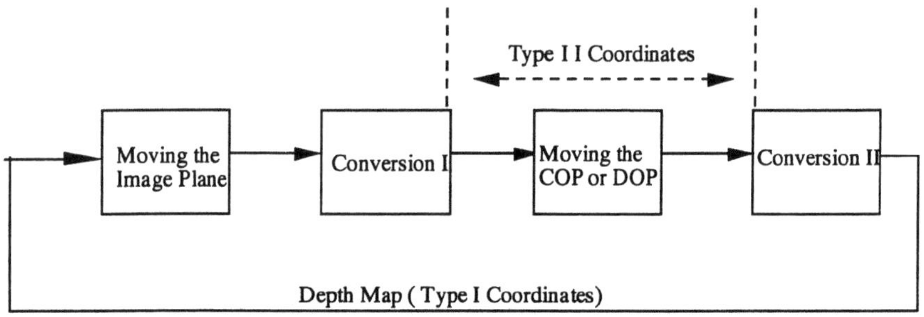

FIGURE 17.23. 4-step process to compute a 3D transformation of a depth map.

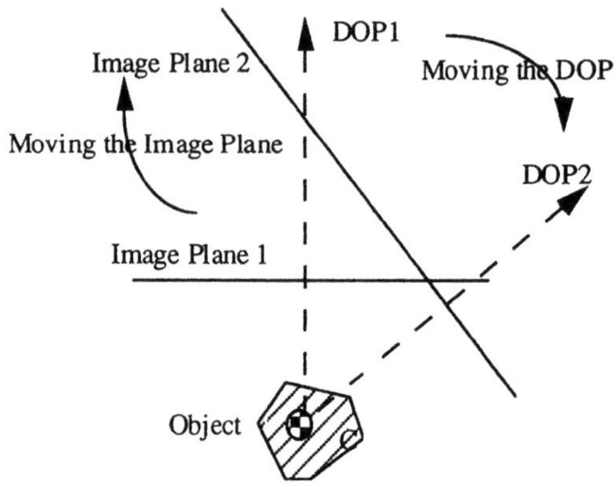

FIGURE 17.24. Decomposition of a 3D transformation.

The second and fourth steps, called conversion, correspond to a change of coordinate system. This step is shown in details in the figure 17.25. No spatial displacement is involved, only the depth value of the array is changed and that can be done by operating on scanlines in the array. These conversions necessite one or two 1-dimensionnal non-geometric transforms depending on the type of the projection.

In some sense, the third step is the most complicated. We have shown that it corresponds to a 1-dimensional geometric transform of the form $x' = A(z(x))x + b$. This transform depends on the depth value, but this value is not changed by the transform (therefore we can call it a geometric transform). A is not constant and is a function of the depth value z at x. The relative stretch and compression will vary on the line. The implementation requires a non-uniform resampling. This transform can present a foldover problem: a move of COP or DOP can produce a set of points which is no longer a section of the eye fiber bundle, two points being on the same fiber, as shown in figure 17.26. To treat this foldover problem and eliminate hidden points, the 1-D flow associated with a move of COP or DOP can be implemented in a front to back method.

We have shown that for a perspective transformation, the 4-step process can be implemented with a seven-pass algorithm, each pass being a 1-dimensional transform, three of them being geometric. The number of step is five for a parallel projection. There are two

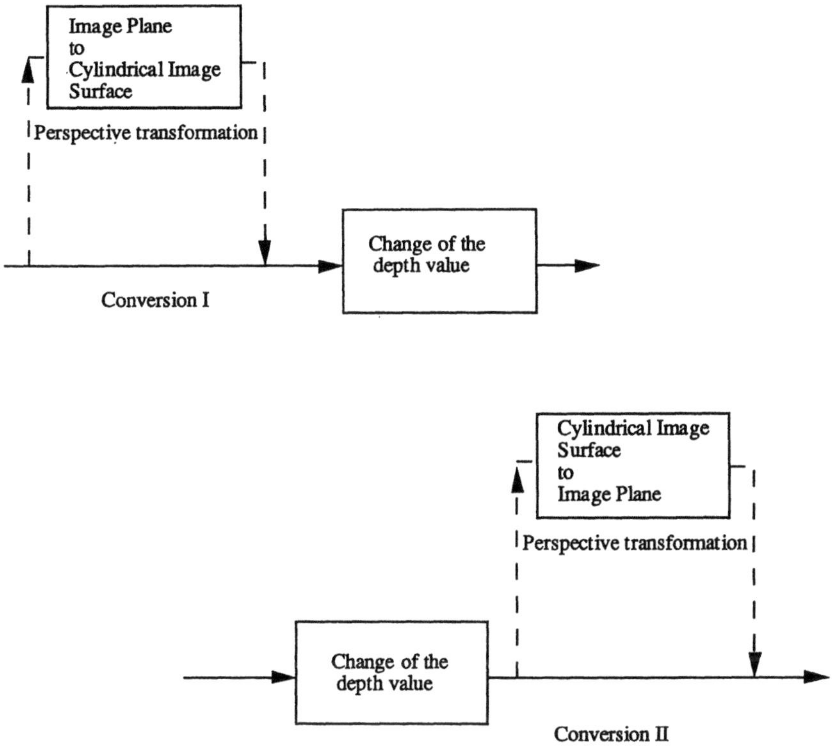

FIGURE 17.25. The two types of conversion.

major advantages to the multi-pass technique. First, the transformation operates on scanlines in the array. This means that read and write access pattern is very simple, involving either consecutive memory locations or by skipping a constant number of intermediate locations. Second, the resampling problem has been reduced from two dimensions to one dimension. This is an advantage because resampling in 1-D is considerably simpler.

17.8 Conclusion

We have presented a new approach to 3-D transformations of depth maps. Our principal result is that 3-D transformations on depth maps can be implemented with parallel 1-dimensionnal transformations. We have tested this four step algorithm on data generated by a laser camera and the results were very satisfactory. The figures 17.27 and 17.28 are examples of a depth map produced by a laser range-finder and transformed by these methods.

Depending on the available hardware, this multi-pass technique may be more or less appropriate. In fact, these multi-pass techniques have efficient parallel implementations. We hope to test them in the future on a parallel scanline processor which is the best machine for multi-pass techniques. We are continuing our work on the conception of a fast multiview information integration system using the same methods with JPEG compression in hardware to store and get the views.

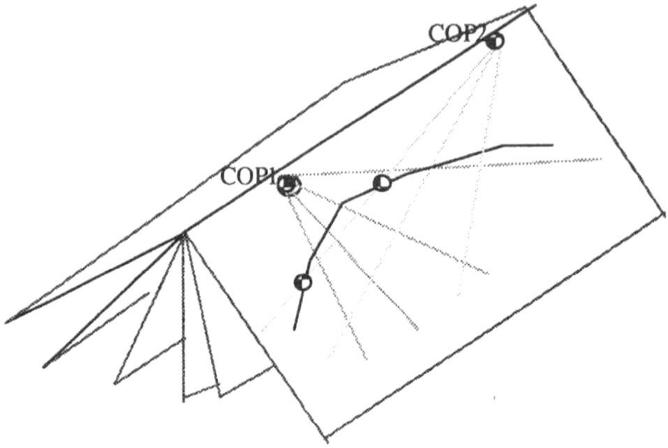

FIGURE 17.26. Moving the COP and the Hidden-surface problem.

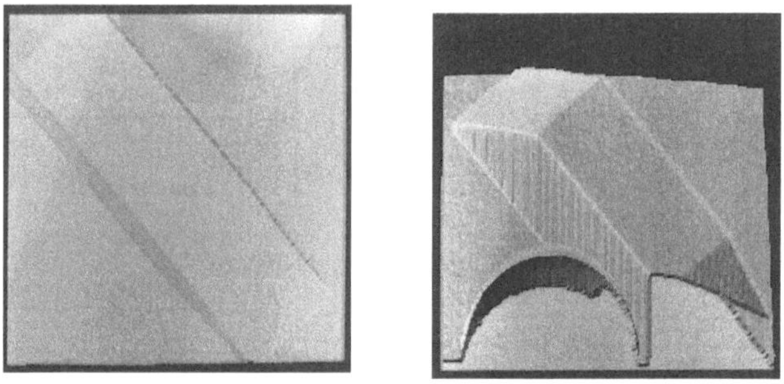

FIGURE 17.27. The original depth map of the block is on the left.

FIGURE 17.28. The original depth map of the Indian mask is the right one in each of these figures.

Acknowledgements:

The depth map of figure 17.28 represents an Indian mask of the Man Museum of Ottawa. This mask was digitalized at the National Research Council of Canada with a laser rangefinder built by the Institute for Information Technology. Particular thanks to the careful reviewers whose suggestions have hopefully clarified the paper.

17.9 References

[1] Paeth A.W. A Fast Algorithm for General Raster Rotation. In *Graphics Interface*, pages 77–81, May 1986.

[2] Butler D. and Pendley M. A visualization model based on the mathematics of fiber bundles. *Computers in Physics*, pages 45–51, sep/oct 89.

[3] Catmull E. and Smith A. R. 3-D Transformations of Images in Scanline Order. *Computer Graphics (SIGGRAPH '80 Proceedings)*, 14,3:279–283, July 1980.

[4] Miller G. The Definition and Rendering of Terrain Maps. *Computer Graphics (SIGGRAPH '86 Proceedings)*, 20,4:39–45, August 1986.

[5] Spivak M. *A Comprehensive Introduction to Differential Geometry.* Publish or Perish, 1979.

[6] Bennett P. and Steven G. *System for Spatially Transforming Images.* United States Patent, September 84.

[7] Hanrahan P. Three-pass Affine Transforms for Volume Rendering. *Computer Graphics (SIGGRAPH '90 Proceedings)*, 24,5:71–78, november 1990.

[8] Heckbert P. Fundamentals of Texture Mapping and Image Warping. Technical Report UCB/CSD 89/516, U. of California at Berkeley, June 1989.

[9] Robertson P. Fast Perspective Views of Images Using One-Dimensional Operations. *IEEE CGA*, pages 47–56, February 1987.

[10] Smith A. R. Planar 2-Pass Texture Mapping and Warping. *Computer Graphics (SIGGRAPH '87 Proceedings)*, 21,4:263–272, July 1987.

[11] Cook R. L. Reeves W. T., Salesin D. H. Rendering Antialiased Shadows with Depth Maps. *Computer Graphics (SIGGRAPH '87 Proceedings)*, 21,4:283–291, July 1987.

[12] Hook T.V. Real-Time Shaded NC Display, Computer Graphics. *Computer Graphics (SIGGRAPH '86 Proceedings)*, 20,4:15–18, August 1986.

18 An Analysis of Digitalization Algorithms for Non-linear Curves

Charles A. Wüthrich

ABSTRACT A brief introduction of the model, based on discrete lattices, used for rasterization onto both the square and the hexagonal lattice is given. Some of the existing rasterization algorithms for non-linear curves, in particular fixed step, forward and adaptive forward differencing, and bisection algorithms are analyzed and ported to perform rasterization onto the hexagonal lattice. A new bisection algorithm, based on the mean value theorem, is proposed. This algorithm naturally adapts the density of the sampling points of the curve to its curvature, increasing the point density where the curvature is greater.

18.1 Introduction

Raster output devices, such as dot matrix printers, laser printers or raster screens, have gained in recent years wide acceptance in the Computer Graphics community. Their display screen technology, which was borrowed from television, represents a relatively cheap alternative to the dedicated hardware employed in vector output devices. The essential difference between raster and vector output devices lies in the display primitive which can be drawn: raster output devices are able to draw only single points, called pels or picture elements, whereas vector output devices are able to draw as basic elements of an image objects of higher complexity such as straight lines or higher order curves. As a raster output device is a collection of single points, the shape of which is quite irrelevant to the mathematics that is going to be used, it can be modelled as a discrete sampling, or alternatively a discrete subtiling of the plane.

While hardware technology has progressed rapidly, the understanding of the mathematical foundations of raster graphics moved at a much slower pace, and only in the latest years research focused on its aspects [11, 12, 6, 19]. The need for such research is justified by the difficulty encountered when performing the operation of rasterization, called also digitalization, i.e., the operation of passing from the real continuous world of images into the discrete space of the raster screen.

Most studies on rasterization have concentrated on traditional output devices, for which the pel disposition is on a square lattice; only lately in [3, 20] attention has been drawn to alternative, namely hexagonal, pel dispositions on the plane. Nowadays rasterization has been defined both for traditional square and for hexagonal lattices, and a reasonable theoretical framework for the developement of algorithms has been developed. This has allowed the migration of Bresenham's algorithm for straight line rasterization to hexagonally based display devices [20].

Unfortunately, digitalization algorithms for higher order curves have not yet been examined exhaustively from the point of view of rasterization. In this paper, we shall address exactly this problem, by analyzing in depth from the point of view of discretization some of the currently used algorithms and by porting them to the hexagonal grid. Forward differencing methods, both using fixed [7] and adaptive [5] step length, as well as recursive subdivision methods, used for example in [1], are examined from the point of view of rasterization. These algorithms are going to be ported to the hexagonal lattice, thus extending the classes of the curves drawable on such devices. Some limitations of existing algorithms are outlined. A new class of algorithms, based on the mean value theorem, which automatically adapt the density of the sampling points to the curvature of the curve, are proposed. The proof of this latter statement is given.

This paper proves that an effective theoretical framework can ease the development of the basic algorithms for drawing on a raster screen. It provides a simple but accurate analysis of the existing rasterization algorithms for higher order curves on raster output devices, and extends existing algorithms to hexagonal lattices, which might represent a practical alternative to today's square-based output devices, since no growth in the algorithmic complexity occurs, and since hexagonally based samplings of the plane are known to be an optimal sampling of the plane [15].

18.2 Digitalization

Only in [11] a first comprehensive definition of rasterization for square-grid output devices was presented. The topological contradictions of traditional square-based output devices have been outlined in [12]. The use of raster output devices having a pel disposition which samples the plane differently from classical square based output devices has been examined quite thoroughly lately.

In [20] and in [19] discrete lattices, that is, the subsets Λ of points $\underline{v}$ of the real plane, the coordinates of which can be expressed as $\underline{v} = n_0\underline{v}_0 + n_1\underline{v}_1$, where n_0, n_1 are integers and $\underline{v}_0, \underline{v}_1$ are fixed vectors of the real plane, were proposed as a model for raster output devices. Lattices have the characteristic of being regular enough to model such devices, and provide a sufficient mathematical framework for actually allowing the development and the porting of algorithms between them. Furthermore, lattices model not only traditionally based "square" output devices by simply setting $\underline{v}_0 = (1,0)$ and $\underline{v}_1 = (0,1)$, but also "hexagonal" devices by setting $\underline{v}_0 = (\frac{\sqrt{3}}{2}, \frac{1}{2})$ and $\underline{v}_1 = (0,1)$.

The vectors $\underline{v}_0$ and $\underline{v}_1$, called *lattice generators*, provide a natural system for introducing coordinates for lattice points of the plane: the lattice point coordinates are represented simply by the integer ordered pair (n_0, n_1). Moreover, between two distinct pairs of lattice generators there is always an affine transformation that transforms one generator pair into the other one, and the point (n_0, n_1) of one lattice into the point of the other lattice bearing the same coordinates. Among the lattice points, a simple neighbourhood relation can be introduced: two points λ and λ' of a lattice Λ are said to be neighbours if they are such that their corresponding Voronoi sets have a non-empty intersection.

The existence of a common underlying model for square and hexagonal pel dispositions on the plane allows the definition of a common digitalization scheme for both lattice dispositions, and eases the developing of rasterization algorithms valid for both lattices. The existence of affine transformations transforming one lattice into the other one, makes the task of porting most theorems valid for one lattice into the other lattice space a relatively easy one.

Currently two types of rasterization schemes are mostly used on the square lattice. In grid intersection digitalization schemes [16] a grid of parallel sheafs of equidistant lines passing through the lattice points and parallel to the lattice generators is laid on the plane; the object to be digitized is intersected with these sheafs. The resulting intersection points are approximated to their nearest lattice neighbours. In cellular digitalizations [9] the object is intersected with the Voronoi sets associated with the lattice. The lattice points corresponding to the Voronoi sets that have a non-empty intersection with the object are chosen as digitalization points.

These rasterization schemes were extended to hexagonal lattices in [20]. Let K be a

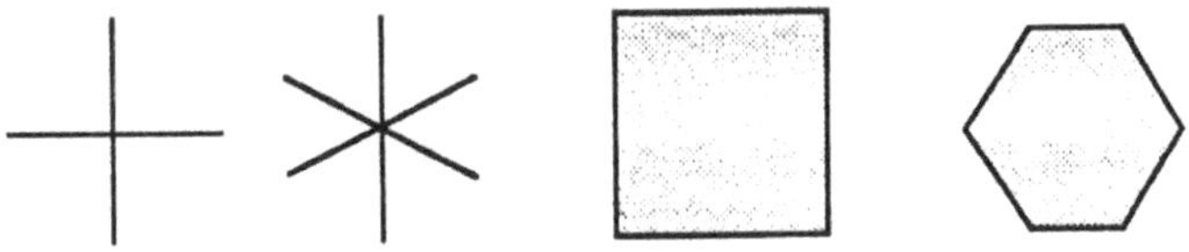

FIGURE 18.1. Basic domains for square and hexagonal digitalization schemes

limited Lebesgue-measurable set of the plane and including the origin $(0,0)$ of the plane. Let K_λ be the translated set of K by the vector $\underline{\lambda} \in \Lambda$.

Definition 18.2.1 *Let $\underline{v} \in \mathbf{R}^2$. The grid point digitalization of $\underline{v}$, $Dig(\underline{v})$, is defined by the following relation*

$$Dig(\underline{v}) = \begin{cases} \{\underline{\lambda}\} & \text{iff } (\underline{\lambda} \in \Lambda) \wedge (\underline{v} \in K_\lambda) \\ \emptyset & \text{otherwise ,} \end{cases} \tag{18.2.2}$$

Definition 18.2.3 *Let $A \subseteq \mathbf{R}^2$. The set*

$$Dig(A) = \bigcup_{\underline{x} \in A} Dig(\underline{x})$$

is called digital image or digitalization of the set A.

Cellular digitalizations, which shall be indicated by Dig_{cell}, are obtained by putting $K = V_\Lambda$, where V_Λ is the Voronoi set of the lattice including the origin. Grid intersection digitalization schemes, which shall be indicated by Dig_{GI}, can be obtained by setting $K = V_\Lambda \cap r_{base}$, where r_{base} is the set constituted by the straight lines passing through the lattice points $(0,0)$ and $(1,0)$, $(0,0)$ and $(0,1)$ in the case of the square grid, while it is the set constituted by the straight lines passing through the lattice points $(0,0)$ and $(1,0)$, $(0,0)$ and $(0,1)$ and $(0,0)$ and $(1,-1)$ in the hexagonal case, and in all the other lattice cases (Figure 18.1).

Grid intersection digitalizations are preferred to discretize curves, as they reduce the problem of rasterization to the intersection of the object with lines of simple form. The major problem with these schemes is that some of the detail may escape the sampling grid, thus generating unwanted effects on the digital output device. Cellular rasterizations are preferred when this happens, especially when a single point or an enumerable non-dense collection of points have to be rasterized. Some newer higher order curve digitalization algorithms combine these two techniques to take advantage of the features of both methods.

18.3 Exact digitalization algorithms

The introduction of grid intersection digitalization schemes also for the hexagonal lattice allowed to port DDA and Bresenham's straight line rasterization algorithms to the hexagonal lattice [20].

Today higher order polynomial and rational curves are getting increasing attention in Computer Graphics [2]. They provide great flexibility when "freehand" surfaces have to be drawn. Special attention has been drawn to the different kinds of splines, which are

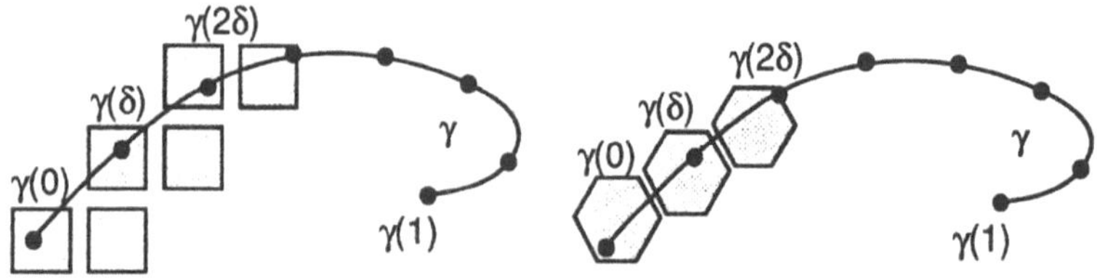

FIGURE 18.2. Fixed step digitalization algorithms

the simplest curves that provide all the flexibility required, but retain a simple analytical expression. But as opposed to straight line rasterization techniques, for which a generally accepted rasterization method is provided, to date there is still no standard method for performing a rasterization of a non-linear curve. Some of the rasterization algorithms used are ad-hoc techniques valid only for a specific class of curves, such as Pitteway's conic sections drawing algorithms [14], or Bresenham's circle drawing algorithm [4], and lack the necessary generality to allow their extension to generic polynomial curves.

The simplest methods for performing curve rasterization are the so-called *exact* methods. Such algorithms compute a discrete set of sample points on the curve; these points are chosen in such a manner that their rasterizations build a discrete curve. According to Rosenfeld [16], a discrete curve is a set of adjacent lattice points $C = \{\lambda_i\}_{i=0}^{k}$. Each of its points has exactly two neighbours in C, with the exception of two points, called endpoints of the curve, that have only one neighbour in C. In general the curve that has to be rasterized is a parametric curve $\gamma(t) = (x(t), y(t))$, with $t \in [0,1]$. Parametric curves are preferred because they are more general than curves expressible as functions of one variable.

The simplest way for performing a rasterization of a parametric curve is by just evaluating the curve points at interval steps of length δ in the parameter space, i.e. by computing the points $\gamma(0)$, $\gamma(\delta)$, $\gamma(2\delta)$ and so on until $n\delta \geq 1$. The resulting points are then discretized using the cellular rasterization scheme (Figure 18.2). Since only cellular digitalizations are required for the actual rasterization of the curve, i.e. for rendering the curve onto the discrete lattice space, and since cellular digitalizations have been defined also for hexagonal lattices, the same digitalization technique can be used to rasterize onto the hexagonal lattice.

Forward differencing (FD) methods [7] are very similar. The basic underlying concept is the same: curve points are evaluated at fixed steps of the parameter and then rasterized with the cellular digitalization scheme. The difference lies in how the curve points are computed. Given a function $g : \mathbf{R} \to \mathbf{R}$, the recursive function

$$\begin{aligned} \Delta_\delta g(t) &= g(t+\delta) - g(t) \\ \Delta_\delta^{i+1} g(t) &= \Delta_\delta^i g(t+\delta) - \Delta_\delta^i g(t) \end{aligned}$$

defines the i-th forward difference of g with step δ. One interesting characteristic of forward differences is the fact that if g is a polynomial function of degree n, then its i-th forward difference is a polynomial of degree $n - i$. For third order polynomials therefore, differences of order higher than three are equal to zero, and if differences up to the third order are used, forward differencing computes exactly points on the curve.

Since both methods presented above use a fixed increment of the parameter for computing the sample points on the curve, they can be named *fixed step* methods. The caveat

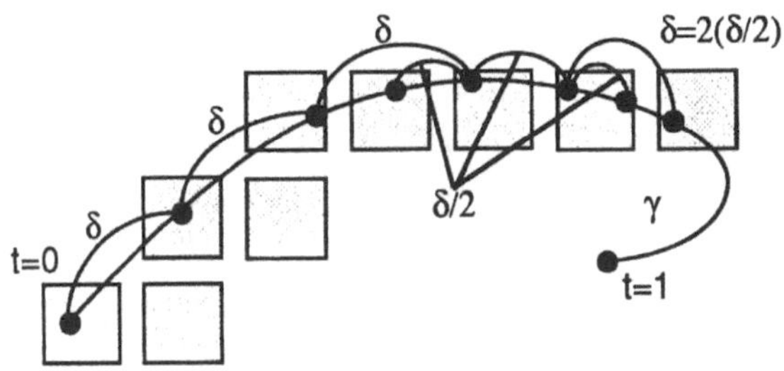

FIGURE 18.3. Adaptive Forward Differencing

by them lies in the fact that the step δ is chosen at the beginning, thus causing subsampling where the curvature is too great, or supersampling where the curvature is small. Generally the curve has to be supersampled to preserve the connectivity of the rasterization, and thus many more sample points than necessary have to be computed. Moreover, since every new point of the digitalization requires either one evaluation of the curve for simple parameter incremental methods, or one evaluation of the forward differences up to a certain order for FD methods, both methods are computationally expensive.

In order to overcome sampling problems, *adaptive forward differencing* (AFD) methods have been introduced [13, 18]. As before, forward differences in the parametric equation of the curve are used and the resulting points are rasterized with the cellular digitalization scheme. But instead of keeping the step fixed, if the new rasterization point obtained coincides with the point obtained in the previous step, the parameter step is doubled, whereas if the new rasterization point is not a neighbour of the point obtained in the previous step, the step is halved. When the new point generated is a neighbour of the previous one, the parameter step is maintained equal (Figure 18.3).

The formulas that rule step doubling in the case of third order polynomials are

$$\begin{aligned}
\Delta_{2\delta} g(t) &= 2\Delta_{\delta} g(t) + \Delta_{\delta}^{(2)} g(t), \\
\Delta_{2\delta}^{(2)} g(t) &= 4(\Delta_{\delta}^{(2)} g(t) + \Delta_{\delta}^{(3)} f(t)), \\
\Delta_{2\delta}^{(3)} g(t) &= 8\Delta_{\delta}^{(3)} g(t).
\end{aligned}$$

Step halving can be easily deduced from the above relations.

Again, AFD methods use cellular rasterization, and since both cellular digitalizations and the notion of neighbour can be introduced easily in the hexagonal lattice [19], there is no problem in porting also AFD algorithms to perform digitalizations onto the hexagonal lattice. At least one forward difference evaluation is required for each rasterization point found: these algorithms are thus expensive, although not quite as expensive as fixed step methods, especially in the case when the derivative of any of the components of the equation of the curve varies strongly in the arc considered [8]. On positive side, recently AFD has been used for designing an algorithm for drawing antialiased cubic curves on the fly [10]. In this algorithm the slope information implicitly contained in the FD equations has been used for antialiasing.

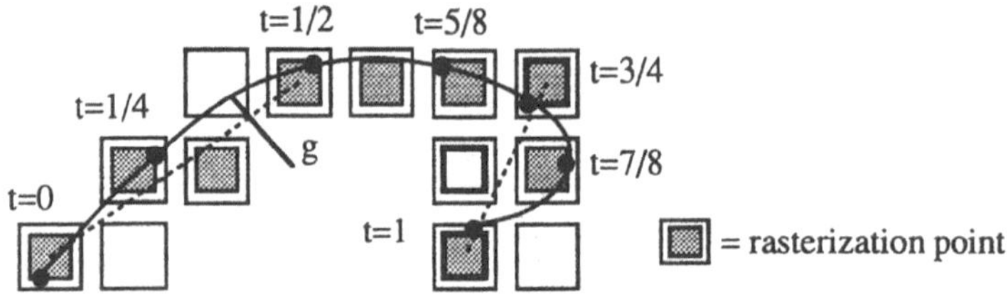

FIGURE 18.4. Recursive bisection

18.4 Recursive subdivision algorithms

A totally different approach to the problem of curve rasterization is used in *recursive subdivision* methods. The parameter interval is divided into $n+1$ parts, and the curve $\gamma(t) = (x(t), y(t))$ $(t \in [0,1])$ is evaluated at the subdivision points. In other words, the points $\gamma(0) = \gamma(t_0), \gamma(t_1), \ldots, \gamma(t_n)$,
$\gamma(t_{n+1}) = \gamma(1)$ are computed ($t_k < t_{k+1}$, $k = 0, 1, \ldots, n-1$). For each subarc $\gamma(t_k), \gamma(t_{k+1})$ $(k = 0, 1, \ldots, n-1)$, a criterion that controls whether the subarc can be approximated by the segment joining the two endpoints is applied. If this happens, then in the subinterval considered the curve is approximated by the segment, otherwise the parameter interval between t_k and t_{k+1} is again subdivided into $n+1$ parts, and the whole process is repeated recursively on its subintervals. The poly-line joining the resulting arc subdivision points is finally used for the rasterization of the curve. The criterion for ending recursive subdivision in the intervals varies according to the algorithm used.

In general, *recursive bisection* of the interval of definition of the parameter is used, as described for instance in the definition of the PostScript language [1]. The curve $(x(t), y(t))$ $(t \in [0,1])$ is divided into two parts by computing the point $(x(\frac{1}{2}), y(\frac{1}{2}))$. The points $(x(0), y(0))$, $(x(1), y(1))$ and $(x(\frac{1}{2}), y(\frac{1}{2}))$ are rasterized with the cellular scheme. If the real straight line joining $(x(0), y(0))$ and $(x(1), y(1))$ is sufficiently close to $(x(\frac{1}{2}), y(\frac{1}{2}))$, i.e. if their distance is smaller than a certain threshold τ, then the straight line is taken as an approximation of the curve in the arc considered, otherwise the subdivision process is repeated recursively by bisecting the subarcs of the curve $t \in [0, \frac{1}{2}]$ and $t \in [\frac{1}{2}, 1]$ until all their subsections can be approximated by straight lines (Figure 18.4). Once this is true, the rasterizations of the lines joining the subdivision points obtained on the curve are computed (18.4.1). This last step is where rasterization is performed, and since only straight line rasterizations are performed, and straight lines can be rasterized onto the hexagonal lattice, for example by using Bresenham's algorithm [20], the same algorithm can provide the means for digitalizing a curve onto hexagonal lattices too. Note that the discretization process described above is equivalent to approximating the real curve with a real poly-line inscribed in it, and then discretizing the resulting poly-line. It shall be therefore be indicated hereafter as the *real bisection* algorithm.

The condition on the closeness of the segment joining the extremes of the subinterval to the new bisection point can be replaced by other closeness criteria. For example, the digitalization of the straight line joining $(x(0), y(0))$ and $(x(1), y(1))$ can be computed: if the rasterization of the new subdivision point $(x(\frac{1}{2}), y(\frac{1}{2}))$ lies on it, then the discretized straight line is taken as part of the rasterization of the curve, otherwise, as before, the subdivision process is repeated recursively (18.4.2). Again, since this last criterion is based on the cellular scheme and on Bresenham's algorithm, it can be ported to perform curve rasterizations onto the hexagonal lattice. This algorithm shall be indicated hereafter as the *discrete bisection* algorithm.

In order to digitalize onto the hexagonal grid, it is always wise to work with the coordinate system defined by the hexagonal lattice generators. The transformation which, given a parametric curve $\gamma(t) = (x(t), y(t))$ in the square lattice, gives its parametric equations in the hexagonal coordinate system, is presented in [20].

Both recursive subdivision methods presented above fail to give correct results when the curve presents an inflection point. This can be overcome by requiring the second derivative of the curve to be non-zero in the interval considered. Moreover, the intervals are just blindly bisected where necessary, whereas the curvature of a curve should play a very important role in the choice of the subdivision process, and should be taken into account for the interval subdivision. In fact, a "natural" algorithm should adapt the sampling density to the curvature of the curve to be discretized, as more samples are needed where the curvature is greater. The next section will present an algorithm which has this characteristic.

18.5 The mean value algorithm

In this section a new class of algorithms, based on the mean value theorem [17], are going to be presented. In brief, the mean value theorem states that given a real-valued continuous function in an interval $[a, b]$ which is differentiable in (a, b), there is a point $\xi \in (a, b)$ at which

$$f(b) - f(a) = (b - a)f'(\xi) .$$

In other words, the point ξ is such that the tangent to the curve in it has the same slope as the straight line joining the points $(a, f(a))$ and $(b, f(b))$.

If the function f does not have any inflection points in $[a, b]$, then the point ξ is also unique. Note that here only real-valued functions of a real variable are considered. The uniqueness of the point where the mean value theorem is satisfied allows the definition of straightforward algorithms for rasterizing higher order curves.

These algorithms, which are recursive subdivision algorithms, differ from the ones presented in the last section only in the choice of the subdivision point, which is here the point satisfying the mean value theorem.

Let $a, b \in \mathbf{R}$ with $a \neq b$, and let f be a real-valued function of class $C^2[a, b]$. Let $f''(x) \neq 0$ in (a, b). The digitalization of f is be computed by the following algorithm, which shall be called the *discrete mean value* algorithm:

Algorithm 18.5.1

1. i = 0; j = 0; $x_{0,0}$ = a; $x_{0,1}$ = b;
   Let $r_{i,j}$ *be the straight line joining* $x_{i,j}$ *and* $x_{i,j+1}$*;*
2. LOOP
   Compute $\bar{r}_{i,j} = Dig_{GI} r_{ij}$*;*
   Using an equation solution method, compute ξ*,*
   zero of the function

$$F(x) = f'(x) - \frac{f(x_{i,j}) - f(x_{i,j+1})}{x_{i,j} - x_{i,j+1}} ;$$

   Compute $y_\xi = f(\xi)$*;*

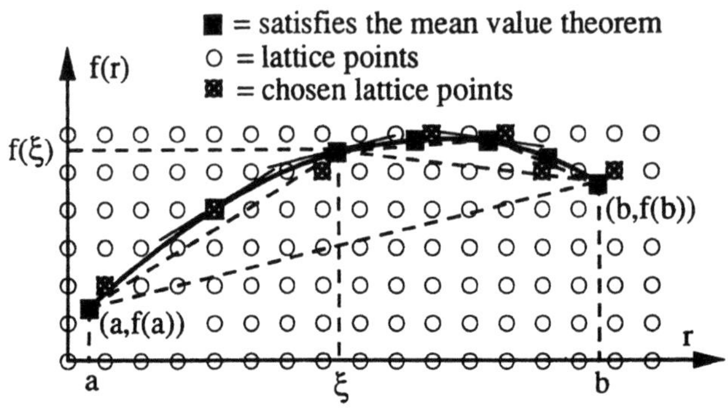

FIGURE 18.5. The mean value algorithm

```
    Compute (X,Y) = Dig_cell(ξ, y_ξ);
    IF((X,Y) ∈ r̄_{i,j}) THEN
      r̄_{i,j} ∈ Dig_GI(f);
      IF ((j + 2) < 2^{i+1}) THEN
        WHILE (j mod 4 = 2)
          i = i - 1;
          j = (j - 2) / 2;
        ENDWHILE
        j = j + 2;
      ELSE
        EXIT LOOP
      ENDIF
    ELSE
      x_{i+1,2j} = x_{i,j}; x_{i+1,2j+1} = x;
      x_{i+1,2j+2} = x; x_{i+1,2j+3} = x_{i,j+1};
      i = i + 1; j = 2j;
    ENDIF
ENDLOOP
STOP
```

Note that the cellular digitalization method has been chosen for rasterizing the point $(\overline{x}, \overline{y})$, as it guarantees the existence of the point (X, Y).

The definition of the mean value equivalent of the real bisection algorithm is very straightforward. Also in this case, the difference lies in the choice of the point satisfying the mean value algorithm as the interval bisection point (18.5.2). This algorithm shall be called *real mean value* algorithm.

Both the algorithms presented above obtain an adaptive sampling width of the interval, depending on the function's curvature. Where the curvature is greater, more sampling points are computed, thus automatically increasing the sampling frequency of the curve where more precision is needed (Figure 18.5). The proof of these statements is the object of the next section.

Also the algorithms based on the mean value theorem can be modified exactly like bisection algorithms to compute the rasterizations onto the hexagonal lattice, since they involve cellular digitalization and Bresenham's straight line algorithm.

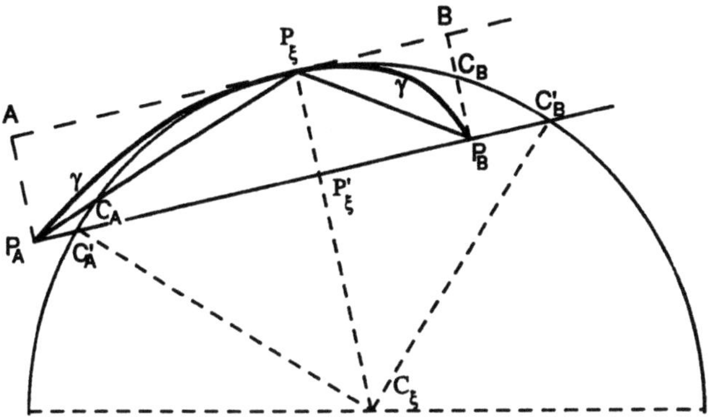

FIGURE 18.6. A curve γ of increasing curvature

18.6 An evaluation of the new algorithm

The main disadvantage of the algorithms based on the mean value theorem is that each subdivision step requires the solution of an equation, as well as an evaluation of the function to be discretized. As analyzed in former sections, AFD methods are not perfect either, as they require at least one forward difference evaluation for each pixel drawn. Mean value algorithms are similar to recursive bisection methods, but unlike the latter ones, which choose blindly the subdivision points, by just halving the interval of definition of the curve's parameter, they select new subdivision points according to the curvature of the curve.

Let γ be a curve such that its equation $y = f(x)$ satisfies in the interval $[a, b]$ the hypotheses of the mean value theorem, and let ξ be the point of $[a, b]$ where the unique mean value theorem is satisfied, i.e. let ξ be the point such that

$$f'(\xi) = \frac{f(b) - f(a)}{b - a}.$$

Let $P_A = (a, f(a))$, $P_B = (b, f(b))$ and $P_\xi = (\xi, f(\xi))$, and let $c(x)$ indicate the curvature function $c(x) = |f''(x)|/(1 + f'^2(x))^{\frac{3}{2}}$ of f. The following theorem holds true:

Theorem 18.6.1 *Suppose that $c(x) \uparrow$ for $x \uparrow$ in $[a, b]$. Then $d(P_B, P_\xi) < d(P_\xi, P_A)$*

Proof: Since P_ξ is the only point satisfying the mean value theorem in $[a, b]$, the straight line t_ξ tangent to the curve γ in P_ξ is parallel to the straight line joining P_A and P_B. Let A and B be the intersections of t_ξ with the perpendiculars to t_ξ passing respectively through P_A and P_B, and let P'_ξ be the foot of the perpendicular to the straight line $P_A P_B$, as in Figure 18.6.

Let c_ξ be the osculating circle for γ in P_ξ, that is, the circle passing through P_ξ, such that its center lies on the normal to γ in P_ξ, i.e. on the positive side of the oriented straight line $P_\xi P'_\xi$, and such that its radius, which is also its curvature radius, is equal to the curvature radius $\rho(\xi) = \frac{1}{c(\xi)}$ of γ in P_ξ. Let C_ξ be the centre of c_ξ.

Consider the intersection C_A of c_ξ with the straight line $P_\xi P_A$. As the curvature of γ becomes greater when x becomes greater, the radius of the osculating circle c_x in $(x, f(x))$ becomes smaller when x becomes greater. The circle c_ξ has a constant curvature, therefore in a left neighbourhood of P_ξ, c_ξ lies between γ and the straight line $P_\xi P_A$. As the curvature radius of γ increases moving on γ from P_ξ to P_A, c_ξ and γ cannot intersect on the arc

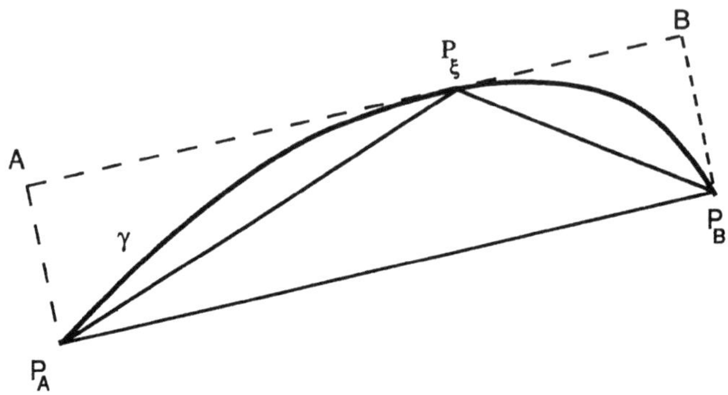

FIGURE 18.7. The mean value point as best choice

$P_\xi P_A$ of γ in a point different from P_ξ. Thus C_A must lie on the straight line segment $P_\xi P_A$. Similarly, the intersection C'_A of c_ξ with the straight line $P_\xi P_A$ lies on the straight line segment $P'_\xi P_A$. Because the curvature of γ increases when x increases, the point C_B, the intersection of c_ξ with the straight line $P_B B$, must belong to the segment $P_B B$, and therefore the point C'_B, intersection of c_ξ with the prolongation on the side of P_B of the oriented straight line $P_B B$, is such that the segment $P_B P'_\xi$ is included in $P'_\xi C'_B$. that, because c_ξ is a circle, and $P_A P_B$ is perpendicular to the radius $C_\xi P_\xi$, so $\overline{P'_\xi C'_A} = \overline{P'_\xi C'_B}$, where $\overline{QR}$ indicates the length of a segment QR. The following inequality is therefore true:

$$\overline{P'_\xi P_A} \geq \overline{P'_\xi C'_A} = \overline{P'_\xi C'_B} \geq \overline{P'_\xi P_B}\,.$$

Note that, since $c(x) \uparrow$ in $[a, b]$, and since, for the mean value theorem, $\xi \in (a, b)$, it must be $\overline{P'_\xi P_A} > \overline{P'_\xi C'_A}$ and $\overline{P'_\xi C'_B} > \overline{P'_\xi P_B}$. Thus $\overline{P'_\xi P_A} > \overline{P'_\xi P_B}$, and therefore $\overline{P_\xi P_A} > \overline{P_\xi P_B}$, which means that $d(P_B, P_\xi) < d(P_\xi, P_A)$. This completes the proof.

Unfortunately theorem 18.6.1 does not tell where exactly the point satisfying the mean value theorem lies, it proves only that the mean value algorithm chooses the bisection points where the curvature is greater.

Although slightly slower than existing algorithms, especially bisection algorithms, mean value algorithms adapt the sampling density naturally, without oversampling as Adaptive Forward Differencing (AFD) algorithms do, and without doing blind bisection as in bisection algorithms. In fact, it is trivial to note that the point satisfying the mean value theorem is the farthest point from the segment joining the subinterval endpoints in consideration, and therefore it is the best possible choice as a bisection point (Figure 18.7). This latter characteristic leads to the fact that the mean value algorithm needs a number of bisection steps at most equal to bisection algorithms to terminate.

Last but not least, the mean value algorithm is independent of the particular parametrization of the curve used. This is important especially when to intervals of equal length in the parameter space correspond arcs of different length on the curve.

18.7 Conclusions

After a brief introduction to rasterization theory and to the main results achieved by it in the latest years, an analysis of currently used algorithms for the rasterization of non-linear curves has been given; in particular fixed step, Forward Differencing and bisection algorithms have been analyzed.

Some of the limitations of existing algorithms, particularly the fact that none of these algorithms takes curvature into account, have been outlined. A new class of algorithms based on the mean value theorem has been proposed for the digitalization of non-linear curves. These algorithms choose denser curve samplings where the curvature is greater. All the algorithms for rasterizing curves have been ported to perform rasterizations onto the hexagonal grid.

Future work shall concentrate on the development of integer algorithms for the direct rasterization of generic curves, since no such algorithm is available at the present time. In order to obtain this, the relations between the derivative of a curve and its Freeman chain codes, which are a discretization of the derivative, have to be studied in depth. Furthermore, studies on the possibility of expressing a discrete curve as a combination of rasterized straight lines have yet to be made, and could give rise to a new way of approaching the problem of curve rasterization.

This work proves that an adequate theoretical framework and a good underlying model provide a solid background where algorithm understanding and development are easier to perform, even on a non-conventional space such as hexagonal output devices. Moreover, it shows that a thorough understanding of digitalization allows the easy individuation of the intrinsic characteristics of the different algorithms used in Computer Graphics, and provides the means for the development of new, exact, and fast digitalization algorithms. Such algorithms are needed by both hardware and software developers in Computer Graphics, since nowadays the complexity of the systems developed is growing at an incredible pace, and perhaps too little has been done to refine existing techniques to perform in the best possible way basic tasks such as discretization. This lack of refinement in fundamental research results in an increase of the global cost of the systems, since it has to be compensated through faster, and therefore more expensive, hardware, and, in the end effect, results in a waste of the system resources.

Acknowledgements:

This work has been partly developed during my stay at the University of Zurich. I wish to thank Prof. P. Stucki who allowed me to start this research at the Multimedia Laboratory of the University of Zurich, and Prof. E. Fiume who allowed me to finish it at the Dynamic Graphics Project of the University of Toronto. Sean Matthews of the University of Edinburgh helped me to give to the listings of the discrete mean value algorithm a decent form. Thanks are also due to my wife Olivia, who has supported me with patience when things did not work as they had to. This work has been partially supported by the Swiss National Research Foundation and by the Information Technology Research Centre of Ontario.

18.8 References

[1] Adobe Systems, Inc. PostScript *Language Reference Manual.* Addison-Wesley, Reading, MA., 1985.

[2] Bartels, R.H., Beatty, J.C., Barski, B.A. *An Introduction to Splines for Use in Computer Graphics and Geometric Modeling.* Morgan Kaufman, Los Altos, CA., 1987.

[3] Bell, S.M.B., Holroyd, F.C., Mason, D.C. A Digital Geometry for Hexagonal Pixels. *Image and Vision Computing,* 7(3):194–204, 1989.

[4] Bresenham, J.E. A Linear Algorithm for Incremental Digital Display of Circular Arcs. *Communications of the ACM,* 20(3):100–106, 1977.

[5] Chang, S.L., Schantz, M., Rocchetti, R. Rendering Cubic Curves and Surfaces with Integer Adadptive Forward Differencing. *ACM Computer Graphics,* 23:157–166, 1989.

[6] Fiume, E.L. *The Mathematical Structure of Raster Graphics.* Academic Press, San Diego, CA., 1989.

[7] Foley, J.D., Van Dam, A. *Fundamentals of Interactive Computer Graphics.* Addison-Wesley, Reading, MA., 1983.

[8] Gonczarowski, J. Fast Generation of Unfilled and Filled Outline Characters. In R.D. Andrè, J. & Hersch, editor, *Raster imaging and digital typography,* pages 14–21, Cambridge, 1989. Cambridge University Press.

[9] Kim, C.E. On the Cellular Convexity of Complexes. *IEEE Transactions on Pattern Analysis and Machine Intelligence,* PAMI-3:617–625, 1981.

[10] Klassen, R.V. Drawing Antialiased Cubic Splines. *ACM Transactions on Graphics,* 10(1):92–108, 1991.

[11] Klette, R. The m-dimensional grid point space. *Computer Vision, Graphics and Image Processing,* 30:1–12, 1985.

[12] Kovalevsky, V.A. Finite Topology as Applied to Image Analysis. *Computer Vision, Graphics and Image Processing,* 46:141–161, 1989.

[13] Lien, S., Schantz, M., Pratt, V. Adaptive Forward Differencing for Rendering Curves and Surfaces. *ACM Computer Graphics,* 21:111–118, 1987.

[14] Pitteway, M. Algorithm for drawing ellypses or Hyperbolae with Digital Plotter. *Computer Journal,* 10(3):282–289, 1967.

[15] Rogers, C.A. *Packing and Covering.* Cambridge University Press, Cambridge, 1964.

[16] Rosenfeld, A. Arcs and Curves in Digital Pictures. *Journal of the ACM,* 20(1):81–87, 1973.

[17] Rudin, W. *Principles of Mathematical Analysis.* McGraw-Hill, London, 1976.

[18] Schantz, M., Chang, S.L. Rendering Trimmed NURBS with Adaptive Forward Differencing. *ACM Computer Graphics,* 22:189–198, 1988.

[19] Wüthrich, C.A. *Discrete Lattices as a Model for Computer Graphics: An Evaluation of Their Dispositions on the Plane.* PhD thesis, University of Zürich, Zürich, 1991.

[20] Wüthrich, C.A., Stucki, P. An Algorithmic Comparison between Square- and Hexagonal-Based Grids. *Computer Vision, Graphics, and Image Processing (CVGIP) Graphical Models and Image Processing*, 53(4):324–339, July 1991.

19 Visualizing Hyperbolic Space

Charlie Gunn

ABSTRACT
Computer graphics opens windows onto previously unseen mathematical worlds. This has been firmly established in the study of chaotic dynamical systems, where significant mathematical discoveries can be directly traced to the advent of visual display of computation. Other realms of science and mathematics stand to derive particular benefit from the powers of computer graphics. For example, non-Euclidean geometry is fundamental to many research areas in mathematics and physics. Heretofore, this has not been amenable to visualization, because standard visualization environments are implicitly Euclidean, or flat. Some of the simplest examples of these curved spaces are non-Euclidean geometries of constant curvature. This paper will describe work in visualizing such geometries, undertaken at the Geometry Supercomputer Project as part of a program to explore 3-dimensional manifolds. In particular, we demonstrate techniques for realistic rendering in three-dimensional hyperbolic space using readily available software tools. We have demonstrated the usefulness of these tools by making a computer graphics movie "Not Knot" which contains several minutes of animation inside hyperbolic space.
The discussion will be structured as follows:

- Previous work
- Mathematical models of hyperbolic geometry.
- Comparison of utility of different models of hyperbolic geometry.
- Rendering in hyperbolic space using custom Renderman [1] shaders.
- Three dimensional topology and hyperbolic geometry.
- A case study: the Borromean rings and the video "Not Knot"
- Conclusion and new directions.
- Appendix 1: Copy of Renderman shader for hyperbolic plastic surface.

19.1. Previous work

It has been long recognized that certain rendering techniques, such as ray tracing, could be used to simulate the behavior of light in a variety of non-ordinary situations. For example, different sorts of camera lenses can be simulated using ray tracing[7]. Recent work by Hsiung [6], demonstrates the use of a custom ray tracer for simulating the visual effects of relativity. Hyperbolic space, like relativistic space, is curved; however, due to the special nature of hyperbolic space it is possible to devise rendering solutions which do not require the use of customized ray tracing software. The rest of this paper will describe what hyperbolic space is, how it can be visualized, and speculate on the significance of this result.

[1]Renderman is a trademark of Pixar, Richmond, CA

19.2. Mathematical models of hyperbolic geometry

Non-Euclidean geometry was discovered simultaneously by three different investigators in the first decades of the nineteenth century: Karl Friedrich Gauss in Germany, János Bólyai in Hungary, and Nicolai Lobachevski in Russia [1]. Each was led to his discovery by attempting to establish to independence of Euclid's Fifth Postulate, the so-called Parallel Postulate, which asserts that given a line and a point in a plane, there is a unique line which goes through the point and doesn't meet the given line. The alternative cases – that there are no such lines, or that there are infinitely many – came to be known as elliptic and hyperbolic geometry, respectively.

The ordinary sphere provides a faithful model of elliptic geometry, but hyperbolic space cannot be isometrically embedded into Euclidean space. Consequently, mathematicians have had to construct different *models* of hyperbolic space which preserve some but not all features of hyperbolic space.

These three geometries exist in all dimensions greater than and including two. The best introduction to the qualitative understanding is the two dimensional case. The situation here is summarized in Figure 1. Readers unfamiliar with hyperbolic geometry are encouraged to take time to study this table before continuing.

Feature	Sphere	Euclidean plane	Hyperbolic plane
Number of lines through point P parallel to (that is, never meeting) L. (P, L)	None	1	∞
S = sum of angles of a triangle	$> \pi$ Area = $(S - \pi)$	$= \pi$ Area not determined by angles. Triangles with the same angles and different areas are called "similar".	$< \pi$ Area = $(\pi - S)$
Circumference of circle of radius r.	$2\pi \sin(r)$	$2\pi r$	$2\pi \sinh(r)$
Divergence of lines (d, t)	Bounded $d < \pi$	Linear $d = kt$	Exponential $d \sim ke^t$
Area of entire surface	4π	Infinite	Infinite
Area of circle radius r	$2\pi (1 - \cos(r))$	πr^2	$2\pi (\cosh(r) - 1)$
Gaussian curvature	+1	0	-1
A portion of a surface in euclidean three-space with the same Gaussian curvature.	"round"	"flat"	"saddle-shaped"
Everyday objects with this curvature	Cabbage leaf, surface of a ball	Paper, surface of cylinder, surface of cone	Kale leaf, cooling tower at power station

FIGURE 19.1. A comparison of 2-dimensional geometries. From [3]

We will limit our discussion to three dimensional hyperbolic space, but what we say here generalizes in a straightforward manner to arbitrary dimension. There are several popular models of hyperbolic space H^3 in Euclidean space R^3. We begin, however, by describing a model which occurs in R^4, the *hyperboloid* model.

19.2.1. The hyperboloid model

Equip R^4 with the indefinite metric: $ds^2 = -dx_0^2 + dx_1{}^2 + dx_2{}^2 + dx_3{}^2$ derived from the quadratic form $Q(\mathbf{x}, \mathbf{y}) = -x_0y_0 + x_1y_1 + x_2y_2 + x_3y_3$. This space is denoted $E^{3,1}$. Q is sometimes known as the Minkowski inner product, and $E^{(3,1)}$, Minkowski space. A point $\mathbf{x}$ is called *spacelike*, *lightlike*, or *timelike* according as $Q(\mathbf{x}, \mathbf{x})$ is positive, zero, or negative. Because all three possibilities occur, Q is an *indefinite* quadratic form. In contrast, the familiar Euclidean inner product is *positive definite* since $Q(\mathbf{x}, \mathbf{x}) = 0$ if and only if $\mathbf{x} = 0$, the null vector.

If we take the connected component H of the timelike hyperboloid $Q(\mathbf{x}, \mathbf{x}) = -1$ such that $x_0 > 0$, we have a model for hyperbolic space. We can think of H as the sphere of radius $\sqrt{(}-1)$ in Minkowski space. The same construction can be done with the standard inner product in Euclidean space. The resulting space is the three-dimensional sphere of radius 1, S^3. All the following constructions can also be made for S^3 and may be more intuitively clear. Consult Figure 1 for the analogous situation in dimension 2.

How do we verify that H is a model for hyperbolic geometry? The first step is to show that the tangent space at a point $\mathbf{x}\varepsilon H$ is the orthogonal complement of $\mathbf{x}$ with respect to Q, that is, $T_\mathbf{x} = \{\mathbf{y} \mid Q(\mathbf{x}, \mathbf{y}) = 0\}$. Next, verify that for $\mathbf{x}\varepsilon H$, $Q(\mathbf{x}, \mathbf{x})$ restricted to $T_\mathbf{x}$ is positive definite. It is a standard result in differential geometry that a space such as H along with a continuous positive definite inner product on the tangent space at each point, gives rise to a Riemannian manifold [9]. This means that there is a well-defined distance between points in the space, and hence that the special curves known as *geodesics* exist. These are shortest paths between two points.

Using standard differential geometric methods to solve for the geodesics shows that the geodesic through $P\varepsilon H$ in the direction S in the tangent space at P is given by the intersection of H with the plane through P, S, and O, the origin. The isometries of H, then, must map planes through the origin to planes through the origin, hence are projective transformations [3]. In addition, they preserve the inner product Q. This defines a subgroup of the matrix group $PSL(4, R)$, denoted $SO(3, 1)$ and called the *Lorentz* group. The analogous construction for S^3 yields the special orthogonal group $SO(4)$.

19.2.2. The projective model

If we now project H onto the $x_0 = 1$ hyperplane [2] we get the Klein, or *projective*, model of three-dimensional hyperbolic space H^3. Figure 2 shows this projection in dimension 2. Under this projection the lightcone $Q(\mathbf{x}, \mathbf{x}) = 0$ maps to the unit sphere S^2 and geodesics map to line segments in the unit ball B^3 with endpoints on S^2. In this model it is easy to see that this space has infinitely many lines through a given point which don't intersect a given line. That this model satisfies the other axioms of hyperbolic geometry can be similarly verified. We can also use differential geometry to compute that the Gaussian curvature is -1 at every point of this surface [9].

Figure 3 shows an example of the projective model of the hyperbolic plane tesselated by right-angled pentagons.

---

[2] a projection well-known in computer graphics practice as *dehomogenization*

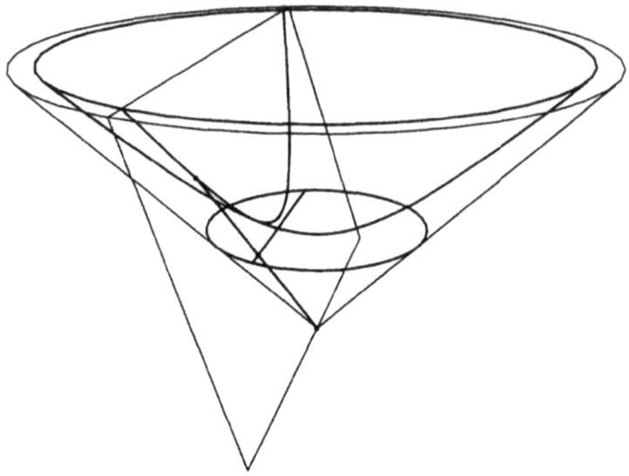

FIGURE 19.2. Relation of the hyperboloid and projective model. The hyperboloid is mapped to the unit disk. Geodesics in H are intersections of planes through the origin with H. They are mapped to straight line segments in the projective model. Figure from [9].

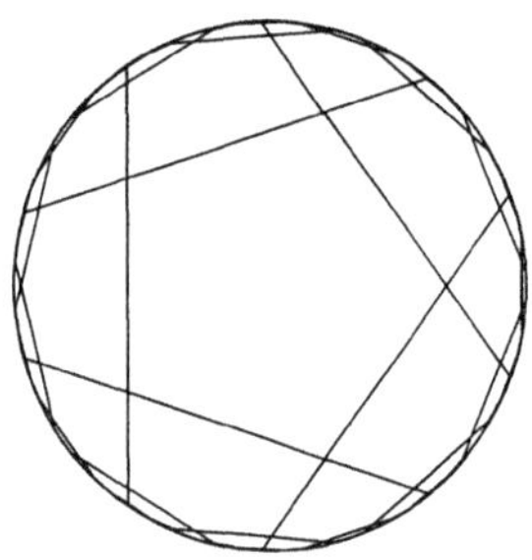

FIGURE 19.3. A tiling of the hyperbolic plane, shown in the Klein model. All the pentagons shown here are congruent and right-angled. The sum of the angles of a hyperbolic polygon is always less than what it would be for a Euclidean polygon of the same number of sides, and the difference is equal to the polygon's hyperbolic area. Figure by Silvio Levy.

19.2.3. Conformal models

However, the best-known model among mathematicians are *conformal* models, those models which accurately represent angles. These include the Poincaré ball model or the upper half space model. This popularity is partly due to the beautiful theory associated to these models in dimensions 2 and 3, based on the matrix groups $SL(2,R)$ and $SL(2,C)$, respectively [2].

The Poincaré ball model is closely related to the projective model. There is a simple transformation from the Klein model to the Poincaré model, given by radial dilation by factor $(1+\sqrt{(}1-r^2)^{-1}$. A geodesic in the Klein model, which is a straight line segment with endpoints on the sphere, is transformed by this map into a circular arc which has the same endpoints but intersects the sphere at right angles. Figure 4 shows the same tesselation of the hyperbolic plane as Figure 3, but using the Poincaré model.

The upper half space model is gotten from the Poincaré model by performing an inversion in a sphere which carries the unit sphere to a plane through the origin. This implies that it is also a conformal model since inversions preserve angles. As a model of hyperbolic space, it is useful in understanding the behavior of the geometry in relation to the points at infinity, which form a Euclidean plane in this model.

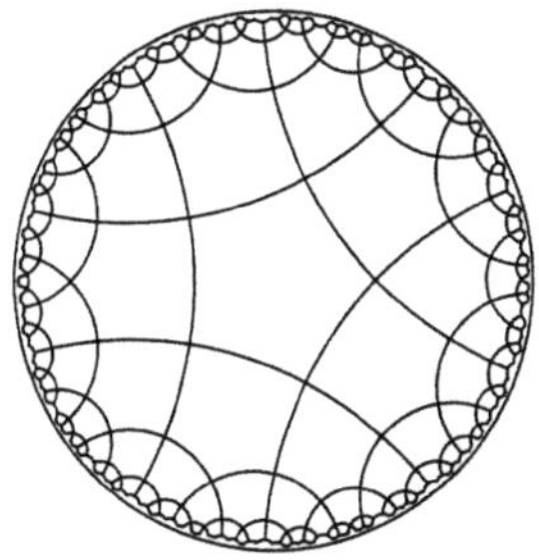

FIGURE 19.4. Tesselation of the hyperbolic plane by regular pentagons shown in the Poincaré disk model. Figure by Silvio Levy.

For a more complete discussion of these models, consult [10].

19.3. Utility of the models for visualization

Despite the prevalence of the conformal model in mathematical literature, our experience indicates that the projective model is the natural choice for use in visualization. Its advantages include:

- Geodesics are Euclidean straight lines.
- Isometries are projective transformations.
- Angle and distance can be easily computed via the Minkowski metric in $E^{3,1}$ rather than the Euclidean metric.

The first property is perhaps the most important for our visualization in 3 dimensions: a light ray that arrives at the camera should appear as a point. A Euclidean straight line is the only curve that has this property. Realistic rendering in the conformal models would require writing a ray tracer to follow the circular path of the geodesics in these models.

The combination of the first two properties implies that hyperbolic *geometry*, considered independently of hyperbolic *shading*, is correctly handled by those rendering systems, which correctly use 4x4 matrices and homogeneous coordinates. Why is this so? The perspective transformation is a standard step in the graphics transformation pipeline. The affine group is also available for standard modeling operations. The perspective transformation, together with the affine group, generate the full projective group $PSL(4, R)$, of which the group of hyperbolic isometries $SO(3,1)$ is a subgroup.

In particular, we have been able to explore hyperbolic 3-space on a Silicon Graphics workstations using a viewing program derived, with only minor modifications, from a Euclidean viewer. For example, a mouse move in the Euclidean case generates a small translation in the direction of the move; in the hyperbolic case the same input generates a hyperbolic translation in the same direction. Figure 5 shows an sample frame from this viewer.

However, as the third property above implies, the shading and lighting provided by these standard systems is not correct. We discuss how to do these correctly in the next section.

There are two main arguments in favor of the conformal model:

- It represents angles correctly, and
- It allows more of the space to be seen at once (compare Figures 3 and 4).

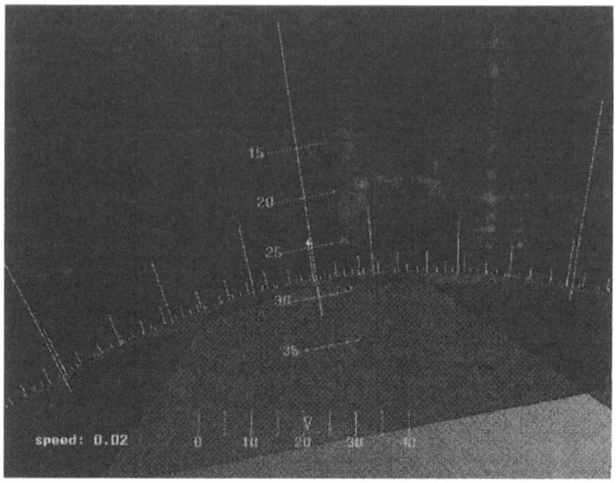

FIGURE 19.5. Hyperbolic geometry in hardware. A frame from an interactive fly-by of the tesselation of the hyperbol c plane by alternately colored right-angled pentagons (central pentagon is colored differently). Program by Linus Upson.

The first argument lacks force, since an observer inside Euclidean space only rarely see angles conformally. For example, if you look at the corners of this page, you will see at most one right angle, and that only if your eye is directly over the corner. You *infer* the correct angle, based on a variety of cues, including motion, stereo vision, and shading. Our experience with using the projective model of hyperbolic space is that viewers learn to do the same there.

The second argument is likewise vitiated by practical considerations. It is now possible, as pointed out above, to implement hyperbolic translations interactively; so the investigator can quickly translate the projective model to center it on any feature that he desires, including any ones which appear visually too small. Furthermore, the phenomenon of rapid decrease in visual angle is an intrinsic feature of hyperbolic space and provides important cues to those within its borders, including mathematical tourists. The conformal models weaken this effect at the expense of realism.

19.4. Hyperbolic shading and lighting using the Renderman shading language

To compute shading and lighting in hyperbolic space requires the ability to redefine the manner in which distance and angle are computed. In particular, the inner product in Euclidean space must be replaced by its counterpart in Minkowski space. This simple change is, unfortunately, not possible in most systems. The one exception which we have found is the rendering interface protocol called RenderMan [11]. This is is a three-dimensional scene description language which can be passed to a rendering server the same way that a Postscript file is passed to a printer. One of the key features of Renderman is its *shading language*, a C-like language which allows users to create arbitrary shading and lighting effects. There are a predefined set of global variables to which the writer of a shader can refer, such as the position of the surface, the normal vector, the color of the surface, partial derivatives, texture coordinates, etc. Other existing graphics standards such as PHIGS constrain the user to adjust a fixed set of parameters within a fixed lighting equation; the shading language of Renderman allows the users to write their own lighting equations.

Using the shading language it is not hard to write shaders to simulate hyperbolic lights and surfaces. [See Appendix 1]. As noted above, the main requirement is replacing one inner product with another. This yields correct distances and angles. These redefined quantities and then substituted into an ordinary Euclidean lighting model, such as plastic, without further change.

There were several difficulties involved in the shading language implementation which deserve mention. First, the data types of the shading language are constrained to be scalar or 3-dimensional vectors. Of course the computations in Minkowski space make heavy use of the homogeneous coordinates. The solution was to create a pair (P, Pw) consisting of a 3-vector and a scalar (the w-coordinate) for each point or tangent vector required. This unnecessarily complicates the shading program. [3] Also, the shading language reflects a Euclidean bias in that it supports the ordinary inverse trigonometric functions Arccos, etc., but not the inverse hyperbolic trigonometric functions, which are required for computing distances in hyperbolic space. These functions can be computed via the *log* and *sqrt* functions, but the underlying Euclidean bias should be corrected.

19.5. Hyperbolic geometry and 3-dimensional topology

One of the outstanding conjectures in mathematics is the Poincaré Conjecture, which asserts that a closed, connected, simply-connected n-dimensional manifold is homeomorphic to the n-dimensional sphere S^n [8]. It has been solved in the affirmative for all n except $n = 3$. Recent work by Thurston and others has established that in many cases it is possible to give a 3-manifold an essentially unique geometric structure, and that by far the most common type is a hyperbolic structure [10]. Hence the study of hyperbolic geometry is central to the understanding of 3-dimensional topology.

What exactly is a hyperbolic structure for a 3-manifold M? One of the most important topological invariants of a space is its *fundamental group*, which is the set of all the closed paths in the space, where 2 paths are considered equivalent if they can be deformed continuously into each other. A hyperbolic structure is an isomorphism of the fundamental group Γ of M into the group $O(3,1)$ of hyperbolic isometries such that M is homeomorphic to the quotient H^3/Γ. A choice of the image of this homeomorphism in H^3 is called a fundamental region for the group. We can choose this to be a polyhedron such that Γ is generated by isometries which pair faces of the fundamental region.

In two dimensions, the theory of compact surfaces tells us that except for the sphere, the Euclidean plane, the projective plane, the Klein bottle, and the torus, all such surfaces have a hyperbolic structure. For example, there is a hyperbolic octagon with 45 degree corners which folds up to make the two-holed torus.

One further concept which is used in the sequel is that of an *orbifold*. This is like a manifold, except it contains some singular points. These singular points, instead of having neighborhoods that look like R^n, as in a manifold, have neighborhoods which look like the quotient of R^n by a finite group. For example, a cone point singularity locally looks like R^n/C_m, a cyclic group of finite order.

A good introduction to the theory of topological manifolds is [12].

---

[3]The version included in the appendix is written as if 4-vectors were available within the shading language.

19.6. "Not Knot" and the complement of the Borromean rings

One rich source of 3-manifolds is as knot or link complements. Given a link L in S^3, consider the set $S^3 \setminus L$ as a 3-manifold and ask if we can give it a geometric structure. Thurston was able to show that in almost all cases, it is possible to give the complement a hyperbolic structure [10]. Based on his work, we have created a computer graphics animation to visualize how to construct the hyperbolic structure on the complement of a link. We chose the particular example of the Borromean rings to serve as the focus. See Figure 6.

FIGURE 19.6. The Borromean rings. This set of three interlocked circles has the property that if you remove any one of them the other two are not linked. Figure by David Broman and Charlie Gunn.

The result is a 16-minute computer animation "Not Knot" [5]. It approaches the manifold of the link complement via a sequence of orbifolds which are *branched* of finite order over the Borromean rings. Figure 7 shows the view from inside the orbifold of order 2.

FIGURE 19.7. Order 2 orbifold over the Borromean rings. The view from inside this orbifold has a Euclidean structure. The fundamental region is a cube. The images of the man are rotated by 180 degrees with respect to the coordinate axes.

The rest of the series have hyperbolic structures with hyperbolic dodecahedra of varying shape as fundamental regions. Figure 8 shows the order 4 orbifold. In the orbifolds, the link is a singular locus at a finite distance from any observer. As the order increases, this locus recedes and shrinks until in the limit it is transformed into a finite set of points at

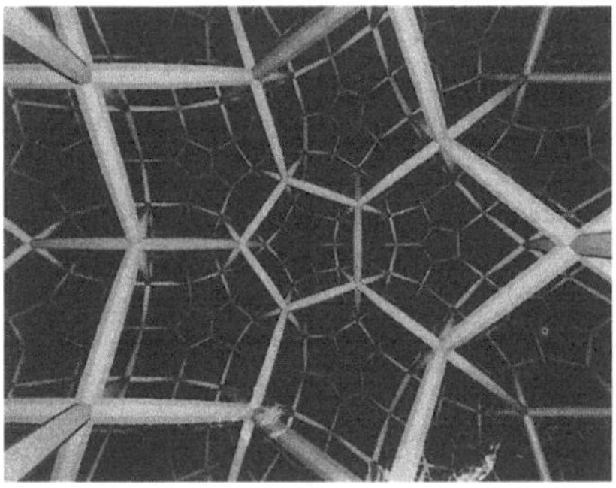

FIGURE 19.8. Order 4 orbifold over the Borromean rings. Insider's view of this orbifold is a tesselation by regular, right-angled dodecahedra.

infinity. Figure 9 shows this final configuration, where the fundamental region is now a rhombic dodecahedron, six of whose vertices lie on the sphere at infinity. The resulting manifold is not compact but does have finite volume.

FIGURE 19.9. Hyperbolic manifold of the complement of the Borromean rings. Insider's view of this manifold is a tesselation by rhombic dodecahedron, one of which has transparent walls in this image. Many copies of this figure meet at every vertex at infinity, for example, the one at left center.

Making the hyperbolic visualizations in "Not Knot" involved several computational tools in addition to RenderMan. Heavy use was made of Mathematica [4] to compute the generators of the groups and the geometry which appears in the animation. The computation of the group elements from the generators was accelerated by the newly developed theory of automatic groups [4].

19.6.1. Further reading

The supplement to "Not Knot" [5] contains a more detailed explanation of the contents of this section.

[4]Mathematica is a trademark of Wolfram Research, Urbana, IL

19.7. Conclusions and new directions

We have demonstrated techniques, using easily available tools, to realistically render scenes in hyperbolic space. This ability arises from the happy coincidence of prevailing computer graphics practice and the projective model of hyperbolic geometry. The movie "Not Knot" represents a first step in exploring the new worlds which are revealed. The movie includes the first realistic fly-throughs of three-dimensional hyperbolic space. Such visual images provide access to mathematical concepts otherwise hidden by formulae. There is much further work to be done in visualizing three-dimensional hyperbolic geometry. There are questions arising from computer graphics, such as:

- How much of the theory of spline surfaces and interpolation extends to hyperbolic space?
- How can we use harmonic analysis on hyperbolic space to derive analogous results to the fractal landscapes and textures so popular in Euclidean space?
- In general, is there a way to isolate the metric in such a way that all software (modeling, animation, and rendering) can be switched between Euclidean and hyperbolic?
- How can these results be extended to other non-Euclidean geometries?

In computer science, hyperbolic geometry may be a natural setting for interacting with certain data structures. Many data structures with exponential growth can be isometrically imbedded into hyperbolic space. For example, n-ary trees can be isometrically embedded in the hyperbolic disk, and then explored using a hyperbolic viewer.

There are also issues emanating from science and mathematics. This ability to render photorealistically in hyperbolic space will aid the intuitive skills of students and researchers in three-dimensional topology. Looking ahead, virtual *hyperbolic* reality promises to provide more intuitive insights into hyperbolic geometry. And we can expect new video games to emerge, ones where it's easier to hide and also proportionately harder to aim – since there's so much more space all around.

In a more general arena, this work represents a first step toward visualization software for general curved spaces, the foundational mathematical object for much of mathematical physics. The challenges here are considerable, since the geodesic structure can be vastly complicated when the curvature varies. But by beginning now to consider these challenges, perhaps we can "put the hooks" into graphics subsystems which will later allow us to design appropriate solutions.

19.8. Appendices

19.8.1. Source code of Renderman shader for hyperbolic plastic surface

```
/*
 * hplastic.sl: a plastic shader for hyperbolic space
 * Author:              Charlie Gunn
 * Description:
 *       The parameters to this shader are the same as the parameters
 *       to the regular plastic shader, but this shader computes
 *       angles and distances using the hyperbolic metric.
 *       This metric is defined on homogeneous coordinates (x,y,z,w)
 *       and is induced by the quadratic form x*x + y*y +z*z - w*w.
 *       This metric is valid on the interior of the unit ball;
 *       the isometries of this metric are projective transformations
 *       that preserve the "sphere" of radius 1  and map the interior to itself.
 *       These are represented by 4x4 matrices hence can be implemented
 *       using the regular geometry/viewing pipeline provided by Renderman
 *       (and other rendering systems).
 * Features:
 *       This version is a hypothetical attempt to use 4-vectors
 *       as if they existed.
 *       A couple of rules are needed to convert
 *       from 3-vectors (regular points) to 4-vectors:
 *       A 3-vector is promoted to a 4-vector by appending a '1' to the 3-vector.
 *       A 4-vector is converted to a 3-vector by "dehomogenizing":
 *       dividing by the 4th coordinate, unless it's zero,
 *       in which case the first 3 elements are unchanged.
 *       Also, define setwcomp(P4, val) to set the new coordinate to a value.
 *       Finally, define an inner product on this new data type:
 *       inpro31(p0,p1) := p0_x*p1_x + p0_y*p1_y + p0_z*p1_z - p0_w*p1_w
 */

surface
hplastic(
        float Ka        = 1,
              Kd        = 1,
              Ks        = 1,
              roughness = .1;
        color       specularcolor = color (1,1,1);
        )
{
        point4      Ph,       /* position vector */
              Nh,       /* normal vector */
              Lh,       /* light vector */
               Ih,       /* eye vector       */
              M;        /* bisector of Lh and Lh */
        float       ss, PdotP;
        point       sP;
        color       total;
        uniform float spec;

        sP = P;
        PdotP = sP.sP;
        /* turn it off if outside unit ball */
        if (PdotP >= 1.0)  {
              Ci = 0;
              Oi = 1.0;
              return;
              }

        spec = 1.0/roughness;
        N = faceforward(normalize(N),I);
        /* make N be "tangent" to P i.e. <P, Nh> = 0*/
        setwcomp(Nh, sP.N);
        Nh = normalize4(Nh);

        /* as a difference vector, I has w cord = 0 */
        /* also, we want the L which points at the eye, not the surface */
        Ih = -I;
        setwcord(Ih, 0);
        Ph = sP;
        Ph = normalize4(Ph);
        Ih =  Ih - inpro31(Ih, Ph) * Ph;
        /* normalize this light vector */
```

```
        Ih = normalize4(Ih);

    total = 0;
    illuminance(sP)
    {
        /* compute specular */
        /* first adjust light vector to be tangent at P */
        /* L is also a difference vector, hence its w-cord = 0 */
        Lh = L;
        setwcomp(Lh, 0);
        Lh = L - inpro31(Lh, Ph) *Ph;
        /* normalize Lh */
        Lh = normalize4(Lh);
        d = inpro31(Lh, Nh);                  /* used in diffuse computation */

        /* now compute bisector of angle between L and I */
        /* important for Lh, Ih to be unit length */
        M = Ih - Lh;
        M = Ih - (inpro31(Ih, M)/inpro31(Lh,M)) * Lh;
        /* detect very small vectors, reject them */
        if (abs(inpro31(M,M)) < .0001)              M = Lh;
        else if (inpro31(M,M) < 0.0)               M = -M;

        /* compute cos(angle between normal and mid-vector H) */
        ss =  inpro31(M,Nh)/hmagnitude(M);

        /* add in ambient and diffuse ... */
        total = total + Os * Cs * Cl * (Ka*ambient() + Kd*d);
        /* ... and the specular too */
        total = total + Os * specularcolor * Cl * Ks*pow(ss,spec);
    }
        Ci = total;
        Oi = Os;
}

float
inpro31(p0, p1)
point4 p0,p1;
{
        return(xcomp(p0)*xcomp(p1) +
                ycomp(p0)*ycomp(p1) +
                zcomp(p0)*zcomp(p1) -
                wcomp(p0)*wcomp(p1));
}

float
hmagnitude(p0)
point4 p0;
{
        return (sqrt (abs (inpro31(p0,p0) ) ) );
}

point4
normalize4(p0)
point4 p0;
{
        float m = hmagnitude(p0);
        point4 p1 = p0;
        if (m) {
                m = 1.0/m;
                setxcomp(p1, m*xcomp(p1));
                setycomp(p1, m*ycomp(p1));
                setzcomp(p1, m*zcomp(p1));
                setwcomp(p1, m*wcomp(p1));
                }
        return(p1);
}
```

19.9. References

[1] C.B. Boyer. *A History of Mathematics.* Princeton University Press, 1968.

[2] E. Cartan. *Leçons sur la géométrie projective complexe.* Gauthier-Villars, 1931.

[3] H.M.S. Coxeter. *Non-Euclidean Geometry.* University of Toronto Press, 1965.

[4] D.B.A. Epstein, J. Cannon, D. Hold, M. Patterson, and W. Thurston. *Automatic Groups.* Jones and Bartlett, 1991.

[5] Ch. Gunn and D. Maxwell. *Not Knot: Movie and Supplement.* Jones and Bartlett, 1991.

[6] P.-K. Hsiung and R.H.P. Dunn. Visualizing relativistic effects in spacetime. In *Supercomputing 89.* IEEE/ACM, Nov, 1989.

[7] N.L. Max. Computer Graphics Distortion for Imax and Omnimax Projection. In *Nicograph '83 Proceedings,* Dec, 1989.

[8] D. Rolfsen. *Knots and Links.* Mathematical Lecture Series No. 7. Publish or Perish, 1976.

[9] M. Spivak. *A Comprehensive Introduction to Differential Geometry,* volume 1. Publish Or Perish, 1979.

[10] W. Thurston. *The Geometry and Topology of Three-Manifolds,* volume 1. Princeton University Press, 1991.

[11] S. Upstill. *The RenderMan Companion.* Addison-Wesley, 1989. chapters 13-16.

[12] J. Weeks. *The Shape of Space.* Marcel Dekker, 1985.

List of Contributors

Maria E. Alonso
Department of Algebra, Faculty of Mathematical Sciences, Complutense University
E-28040 Madrid, Spain

Farhad Arbab
Department of Interactive Systems, Centre for Mathematics and Computer Sciences (CWI)
Kruislaan 413, 1098 SJ Amsterdam, The Netherlands

Paola Cellini
Department of Mathematics, University of Pisa
Via F. Buonarroti 2, I-56100 Pisa, Italy

Marc Coevoet
Computer Science Department, Faculty of Sciences, Free Universiy of Brussels
Pleinlaan 2, B-1050 Brussels, Belgium

Hang L. Du
Department of Computer Science, University of Waterloo
Waterloo, Ontario N2L 3G1, Canada

Bianca Falcidieno
Intitute for Applied Mathematics C.N.R.
Via L.B. Alberti 4, I-16132 Genova, Italy

Patrizia Gianni
Department of Mathematics, University of Pisa
Via F. Buonarroti 2, I-56100 Pisa, Italy

Ronald N. Goldmann
Department of Computer Science, Brown School of Engineering, Rice University
P.O. Box 1892, Houston, Texas 77251, USA

Charlie Gunn
Research Center for Computation and Visualization of Geometric Structures, University of Minnesota
1300 S. 2nd St., Minneapolis, Minnesota 55454, USA

Ivan Herman
Department of Interactive Systems, Centre for Mathematics and Computer Sciences (CWI)
Kruislaan 413, 1098 SJ Amsterdam, The Netherlands

Christoph Hoffmann
Computer Science Department, Purdue University
West Lafayette, Indiana 47907, USA

Nami Kin
Department of Information Science, Faculty of Science, The University of Tokyo
Hongo 7–1–3, Bunkyo–ku, Tokyo 113, Japan

Tosiyasu L. Kunii
Department of Information Science, Faculty of Science, The University of Tokyo
Hongo 7–1–3, Bunkyo–ku, Tokyo 113, Japan

Jacques Lemordant
Artemis Laboratory, IMAG Institute, University of Grenoble
BP 53X, F-38401 Grenoble CEDEX, France

Gábor Lukács
Computer and Automation Institute of the Hungarian Academy of Sciences (MTA SzTAKI)
Kende u. 13–17, Postafiók 63, H-1518 Budapest, Hungary

Yi Luo
Computer and Automation Institute of the Hungarian Academy of Sciences (MTA SzTAKI)
Kende u. 13–17, Postafiók 63, H-1518 Budapest, Hungary

Shashank Mehta
Department of Computer Science, University of Poona
Puna, Maharashtra 411007, India

Teo Mora
Department of Mathematics, University of Genova
Via L.B. Alberti 4, I-16132 Genova, Italy

Maharaj Mukherjee
Department of Electrical, Computer, and Systems Engineering, Rensselaer Polytechnic Institute
Troy, New York 12180–3590, USA

George Nagy
Department of Electrical, Computer, and Systems Engineering, Rensselaer Polytechnic Institute
Troy, New York 12180–3590, USA

Gianfranco Niesi
Department of Mathematics, University of Genova
Via L.B. Alberti 4, I-16132 Genova, Italy

Caterina Pienovi
Intitute for Applied Mathematics C.N.R.
Via L.B. Alberti 4, I-16132 Genova, Italy

Hartmut Prautzsch
Institute for Operating and Dialogue Systems, University of Karlsruhe
Postfach 6980, W-7500 Karlsruhe 1, Germany

Mario Raimondo
Department of Mathematics, University of Genova
Via L.B. Alberti 4, I-16132 Genova, Italy

Daniel Richardson
Department of Mathematics, University of Bath
Claverton Down, Bath BA2 7AY, United Kingdom

Hans–Peter Seidel
Department of Computer Science, University of Waterloo
Waterloo, Ontario N2L 3G1, Canada

Michela Spagnuolo
Intitute for Applied Mathematics C.N.R.
Via L.B. Alberti 4, I-16132 Genova, Italy

Yoshiaki Takai
Division of Information and Graphic Sciences, Faculty of Engineering, Hokkaido University
N13-W8, Kita-ku, Sapporo 060, Japan

Tapio Takala
Department of Computer Science, Helsinki University of Technology
02150 Espoo, Finland

Carlo Traverso
Department of Mathematics, University of Pisa
Via F. Buonarroti 2, I-56100 Pisa, Italy

Remco C. Veltkamp
Department of Interactive Systems, Centre for Mathematics and Computer Sciences (CWI)
Kruislaan 413, 1098 SJ Amsterdam, The Netherlands

Xiaolin Wu
Department of Computer Science, University of Western Ontario
London, Ontario N6A 5B7, Canada

Charles Wüthrich
Computer Systems Research Institute, University of Toronto
6 King's College Road, Toronto, Ontario M5S 1A4, Canada

Xiaolin Zhou
Wilhelm–Schickard Institute for Computer Sciences, University of Tübingen
Auf der Morgenstelle 10, C9, W-7400 Tübingen, Germany

Focus on Computer Graphics

(Formerly EurographicSeminars)

Eurographics Tutorials '83. Edited by P. J. W. ten Hagen.
XI, 425 pages, 164 figs., 1984

User Interface Management Systems. Edited by G. E. Pfaff.
XII, 224 pages, 65 figs., 1985

Methodology of Window Management. Edited by F. R. A. Hopgood, D. A. Duce, E. V. C. Fielding, K. Robinson, A. S. Williams.
XV, 250 pages, 41 figs., 1985

Data Structures for Raster Graphics. Edited by L. R. A. Kessener, F. J. Peters, M. L.P. van Lierop. VII, 201 pages, 80 figs., 1986

Advances in Computer Graphics I. Edited by G. Enderle, M. Grave, F. Lillehagen.
XII, 512 pages, 168 figs., 1986

Advances in Computer Graphics II. Edited by F. R. A. Hopgood, R. J. Hubbold, D. A. Duce. X, 186 pages, 96 figs., 1986

Advances in Computer Graphics Hardware I. Edited by W. Straßer.
X, 147 pages, 76 figs., 1987

GKS Theory and Practice. Edited by P. R. Bono, I. Herman.
X, 316 pages, 92 figs., 1987

Intelligent CAD Systems I. Theoretical and Methodological Aspects.
Edited by P. J. W. ten Hagen, T. Tomiyama. XIV, 360 pages, 119 figs., 1987

Advances in Computer Graphics III. Edited by M. M. de Ruiter.
IX, 323 pages, 247 figs., 1988

Advances in Computer Graphics Hardware II. Edited by A. A. M. Kuijk, W. Straßer.
VIII, 258 pages, 99 figs., 1988

CGM in the Real World. Edited by A. M. Mumford, M. W. Skall.
VIII, 288 pages, 23 figs., 1988

Intelligent CAD Systems II. Implementational Issues. Edited by V. Akman, P. J. W. ten Hagen, P. J. Veerkamp. X, 324 pages, 114 figs., 1989

Advances in Computer Graphics IV. Edited by W. T. Hewitt, M. Grave, M. Roch.
XVI, 248 pages, 138 figs., 1991

Advances in Computer Graphics V. Edited by W. Purgathofer, J. Schönhut. VIII, 223 pages, 101 figs., 1989

User Interface Management and Design. Edited by D. A. Duce, M. R. Gomes, F. R. A. Hopgood, J. R. Lee. VIII, 324 pages, 117 figs., 1991

Advances in Computer Graphics Hardware III. Edited by A. A. M. Kuijk. VIII, 214 pages, 88 figs., 1991

Advances in Object-Oriented Graphics I. Edited by E. H. Blake, P. Wisskirchen. X, 218 pages, 74 figs., 1991

Advances in Computer Graphics Hardware IV. Edited by R. L. Grimsdale, W. Straßer. VIII, 276 pages, 124 figs., 1991

Advances in Computer Graphics VI. Synthesis, Analysis and Interaction. Edited by G. Garcia, I. Herman. IX, 449 pages, 186 figs., 1991

Intelligent CAD Systems III. Practical Experience and Evaluation. Edited by P. J. W. ten Hagen, P. J. Veerkamp. X, 270 pages, 116 figs., 1991

Graphics and Communications. Edited by D. B. Arnold et al. VIII, 274 pages, 84 figs., 1991

Photorealism in Computer Graphics. Edited by K. Bouatouch, C. Bouville. XVI, 230 pages, 118 figs., 1992

Advances in Computer Graphics Hardware V. Rendering, Ray Tracing and Visualization Systems. Edited by R. L. Grimsdale, A. Kaufman. VIII, 174 pages, 97 figs., 1992

Multimedia. Systems, Interaction and Applications. Edited by L. Kjelldahl. VIII, 355 pages, 129 figs., 1992

Advances in Scientific Visualization. Edited by F. H. Post, A. J. S. Hin. X, 212 pages, 141 figs., 1992

Computer Graphics and Mathematics. Edited by B. Falcidieno, I. Herman, C. Pienovi. VII, 318 pages, 159 figs., 1992